Few-Body Systems

Editor in Chief: W. Plessas, Graz

Supplement 15

Nuclear Dynamics: From Quarks to Nuclei

Proceedings of the XXth CFIF Fall Workshop, Lisbon, Portugal, October 31–November 2, 2002

Edited by
M. T. Peña, A. Stadler, A. M. Eiró, and *J. Adam*

Springer-Verlag Wien GmbH

Prof. Dr. M. Teresa Peña
Centro de Física das Interaccoes Fundamentais
Instituto Superior Técnico
Lisboa, Portugal

Prof. Dr. Alfred Stadler
Departamento de Física, Universidade de Évora
Lisboa, Portugal
and
Centro de Física Nuclear, Universidade de Lisboa
Lisboa, Portugal

Prof. Dr. Ana M. Eiró
Centro de Física Nuclear, Universidade de Lisboa
Lisboa, Portugal

Dr. Jiří Adam
Institute of Nuclear Physics
Řež, Czech Republic

Originally published by Springer-Verlag Wein New York in 2003
Softcover reprint of the hardcover 1st edition 2003

Typesetting: Camera-ready by authors

Printed on acid-free and chlorine-free bleached paper
SPIN: 10899687

With 106 Figures

CIP data applied for

ISSN 0177-8811
ISBN 978-3-7091-7301-5 ISBN 978-3-7091-6014-5 (eBook)
DOI 10.1007/978-3-7091-6014-5

Dedicated to Peter Sauer
on the Occasion of His 65th Birthday

Organizing Committee

M. Teresa Peña, CFIF–Instituto Superior Técnico, Lisbon
Alfred Stadler, CFNUL–University of Lisbon and University of Évora
Ana M. Eiró, CFNUL–University of Lisbon
Jiří Adam, Nuclear Physics Institute, Řež–Prague

Invited Speakers

G. Baym (Urbana-Champaign)
S.A. Coon (Department of Energy)
F. Gross (JLab)
A.C. Fonseca (Lisbon)
B. Frois (Min. Recherche, Paris)
H. Garcilazo (Mexico City)
J. Haidenbauer (FZ Jülich)
J.-M. Laget (Saclay)
T.-S.H. Lee (Argonne)
L. Marcucci (Pisa)
G.A. Miller (INT Seattle)
W. Plessas (Graz)
D.O. Riska (Helsinki)
R. Schiavilla (ODU and JLab)
I. Sick (Basel)
R. Timmermans (KVI Groningen)
J.A. Tjon (Utrecht and JLab)
A. Valcarce (Salamanca)

Secretariat

Sandra Oliveira and Dulce Conceição
Department of Physics, Instituto Superior Técnico, Lisbon

Sponsored by

Fundação para a Ciência e a Tecnologia
Fundação Calouste Gulbenkian
Banco Português de Investimento (BPI)
Grupo Teórico para as Altas Energias (GTAE)
Centro de Física das Interações Fundamentais (CFIF)
Centro de Física Nuclear da Universidade de Lisboa (CFNUL)

Foreword

This volume collects the invited and contributed papers presented† at the workshop *Nuclear Dynamics: from quarks to NUCLEI*, which was hosted by Centro de Física das Interacções Fundamentais (CFIF) at Instituto Superior Técnico (IST) in Lisbon, Portugal, from October 31st to November 2nd, 2002. The response to this initiative exceeded the initial expectations of the organizers. Participants arrived to Lisbon, not only from countries within a close vicinity to Portugal, but also from Central and Northern Europe, from Africa, from the United States, from South and Central America, and from Japan.

This meeting was the 20th in a series of schools or workshops organized every fall in Lisbon. Along the years, the series of meetings has covered a wide range of topics in Nuclear and Particle Physics. The 2002 meeting had two unique features:

1) Nuclear Physics at Intermediate Energies

For the first time, the CFIF Fall Meeting focused on nuclear processes at intermediate energies. In physics, an energy range implies a selection of the degrees of freedom which are probed. Experiments and theory at intermediate energies bridge the interesting border between two pictures of reality: the nucleons, and their accompanying cloud of pions, which make up the nuclei forming most of the matter around us, and the underlying quark-gluon structure of the nucleons themselves. The intriguing connection between the two descriptions is tested by experiments using electrons, photons and mesons, or heavy-ion collisions. Throughout the workshop, several talks reported on the latest news from these experimental observations, other talks presented attempts of their theoretical interpretation.

The meeting was organized in the following 10 scientific sessions (in chronological order): *Electroweak Probes (I)* (chair M. T. Peña), *The Nuclear Interaction* (chair A. Arriaga), *Meson Production from Light Nuclei* (chair J. Adam), *Electroweak Probes (II)* (chair George Rupp), *The Nuclear Medium Under Extreme Conditions* (chair J. Dias de Deus), *Constituent Quark Models and Chiral Symmetry* (chair M. Fiolhais), *Relativity and Electromagnetic Observables* (chair F.D. Santos), *Non-Nucleonic Degrees of Freedom* (chair P.U. Sauer), *Relativity, Chiral Symmetry, and Form Factors of Baryons and Mesons* (chair A. Stadler), and *Nuclear Structure and Binding* (chair R. Timmermans).

This scientific program paid special attention to photonuclear, electroweak, and meson production reactions with light nuclei, measured at Jefferson Laboratory (JLab, Virginia), FZ-Jülich (Germany), Mainz (Germany), and MIT-

†With the exception of the talks by S.A. Coon, R. Timmermans and G. Baym who, regrettably, could not provide a written version of their presentations.

Bates (Massachussets), as well to possible future measurements of new neutrino reactions at Fermilab (Illinois) and KEK (Japan), and heavy-ion collisions results from RHIC (New York).

Specifically, the problem of the structure of nuclei and nucleons was considered and recent electron-nucleus scattering data from JLab, taken at momentum transfers in regions up to $Q^2 = 4$ GeV^2/c^2, were presented to the audience. In this context, possibilities and difficulties of disentangling meson-exchange-current contributions from final state interactions were discussed. Additionally, models attempting to describe the baryon spectra and the new baryon form factor data from JLab were shown, and with this respect the role of relativity and the importance of the pion cloud effects were particularly discussed.

Furthermore, the meeting highlighted very recent observations exploring the regime of high momentum transfer in the deep inelastic scattering experiments at HERMES, and the extreme nuclear densities in the heavy-ion collisions at RHIC. The drastic suppression of away-side jets seen in the STAR data will eventually help to filter out theoretical models. Long after the first minutes of the Universe, and far away from exotic stars, this kind of experiments may reveal how the transition between nucleonic and deconfined quark state matter occurs.

2) A Tribute to Peter Sauer

The workshop was also the stage for a tribute to Peter Sauer, and a special session was dedicated to him, on the occasion of his 65th birthday. Peter Sauer finished his Ph.D. in Physics in 1966, at the University of Freiburg/Breisgau, with a thesis in Solid State Physics. After obtaining his Ph.D., he left Germany for research associate positions at Carnegie-Mellon Pittsburg and MIT. He then redirected his research activities to Nuclear Physics. After working on ground-state properties of nuclei in the framework of Hartree-Fock-Bogoliubov and Bruckner theory, and on the nuclear-shell model, Peter decided to approach the complexity of nuclei through the power of exact calculations, which test and control approximations. He focused in particular on few-nucleon dynamics. In 1974, he returned to Germany and settled down at the University of Hannover, determined to place it on the world map of research in Nuclear Physics. Since then, and well before the present era of globalization and networking in science, he established international links and collaborations within Europe, the United States, Japan and Brazil.

Some Impressions on Peter Sauer's Personality

Maybe not too common for a theorist, Peter Sauer always had a strong personal interest in other people, and an eagerness for seizing opportunities and making new life experiences, which is also reflected in his intense international activities. It relates also to his pleasure in traveling and in the arts. For him, art and philosophy are preferred means to aleviate the sometimes heavy burden of the routines of everyday life. We still remember his remarkable speech at the conference banquet of the XVIIth European Few-Body Conference in Évora in 2000, where, inspired by Goethe's ***Faust***, he described episodes in the life of physicists and of the emerging European Union with elegance and philosophical strokes of humour.

With family roots in Eastern Europe, more precisely, born in Breslau (in Silesia, in former East Germany and at present part of Poland), and with a diverse professional activity in the United States, Peter Sauer became clearly a man where two cultures (the west and the east) are interwoven. Although he acts very often with an open and relaxed American style, his eastern origins surface occasionally. For instance, when he got involved in European programs in support of teaching at undergraduate and at graduate levels, he paid particular attention to Eastern Europe. As a co-founder of the European Mobility Scheme for Physics Students, he was the Coordinator for several TEMPUS Programs of Students Exchange, which were specific for students from Central and Eastern Europe.

One of us (M.T.P.) will never forget the day in the fall of 1989 shortly after the profound political changes, when Peter rushed with her, a junior postdoc at the time, across East-Berlin, where the wall was still standing. Peter knew that it was her unique opportunity to witness important changes in Europe right at their epicenter. History was moving very fast. Peter wanted her to still be able to see the best pieces of the historical and artistic graffiti on the Berlin wall. They were to be found in the quarter of Kreuzberg. Naturally, it was far away from the city center, but very close to the watch towers and barbed-wired fences, then still standing like threatening symbols of oppression. It was clear how Peter had longed to go unrestrictedly to this side of the former border.

A Brief Sketch of Peter Sauer's Scientific Work
Peter Sauer was involved, very often side by side with experimentalists, in projects related to microscopic nuclear structure, in particular in experiments that serve as windows to sub-baryonic degrees of freedom (e.g., the Saclay experimental program of the 80's on intermediate energies and few-nucleon targets). One can fairly say that in almost all issues discussed in this meeting, Peter Sauer gave over the years, important and timely contributions. Bernard Frois reminded us of this during his talk in the closing session. Specifically, he emphasized the value of the Hannover group calculations of the electromagnetic structure of ^{3}He and ^{3}H, which revealed clearly the hitherto hidden importance of non-nucleonic degrees of freedom in those nuclei.

Peter Sauer showed often the ability to contribute to his field ahead of time, and not always following the main stream. Illustrations of his leadership are, for instance, his 1975 paper on the role of the neutron charge form factors in electron scattering from the three-nucleon system; his plenary talk, at the 1976 International Conference on Few-Body Problems in Dehli, on the Coulomb problem in few-nucleon systems, with the intention of studying the charge asymmetry and charge dependence properties of the nuclear force; his contributed talk at the same conference where he expanded on trinucleon properties derived from one-boson-exchange potentials, which were still in their infancy. At this point a very special emphasis is due to the comparison between Faddeev and variational three nucleon calculations performed by his group in 1981, in a paper demonstrating the reliability of momentum-space solutions of the Faddeev equations.

As further relevant work of Peter Sauer we mention the study on meson exchange currents and tritium beta decay, published by his group in 1991 and the paper on the radius of the deuteron which, in 1996, established quantitatively the hierarchy of importance of sub-baryonic degrees of freedom (from the meson exchange contributions down to the quark-gluon sub-structure of the nucleon). This Hannover group calculation, at the same time, was able to discriminate between the existing experimental data for the deuteron charge radius, from elastic scattering and atomic physics experiments, favoring the latter one. In 1997, the group also published results on polarized deep inelastic lepton scattering from polarized two and three nucleon bound states. In that work, the Hannover calculations on few-nucleon dynamics, including relativity, provided the appropriate subtraction of nuclear effects, which is necessary to extract the neutron spin structure functions from the measurements. More recently, Peter Sauer succeeded in his long-time goal of investigating the effects of the Δ isobar excitation in few-nucleon scattering. This was achieved by very complex calculations of *nd* elastic scattering, breakup reactions, *nd* radiative capture, and electrodisintegration of the three-nucleon bound state.

Last, but not the least, we refer to Peter Sauer's work on three-nucleon forces and their relation to the role of the Δ-resonance in the nuclear medium, as well as the role of pionic degrees of freedom. He and his group pioneered the investigation of three-nucleon forces of shorter range than $\pi - \pi$ exchange, emphasizing the importance of consistency in the modeling of those forces.

Peter Sauer as a Teacher

Three of the organizers of the workshop belong to the large group of people who, while working with him as postdocs, were exposed to his "Socratic" attitudes, as Dan-Olof Riska at the conference banquet put it. In the final speech at the conference dinner, Franz Gross revealed how impressed he was in the early 80's (before the "powerpoint age") by the scarce number of words in Peter's transparencies for conference talks, and how invariably they were organized in the form of questions. In discussions, Peter Sauer strongly believes in the power of simple and unexpected questions, together with demanding consistency, in probing methods and results. His pedagogical qualities were testified in the meeting, in a marvelous and touching way, by his former students Uwe Oelfke and Ralf Schultze-Riegert. Among other memories, they recalled how thoroughly organized and still appropriate for today's use Peter's notes were, which today exist (unpublished in stacks on Peter's shelves) on a large variety of subjects in Theoretical Physics.

The legacy is out there... To testify it, we included in this volume the three talks presented in the closing session, which was dedicated to tributes to Peter Sauer. As such, they were not restricted to nuclear dynamics at intermediate energies. However, these talks highlighted how a training in Nuclear Physics of high standards, such as the training one could benefit from while working with Peter Sauer, is conducive to very diverse professional choices, within physics and also outside it. Moreover, even though Peter Sauer's work may have affected only a small segment of society, we came out from the closing session with a strengthened certainty: due to the diversified links of Nuclear Physics to Astrophysics, Biomedicine, Industry and Energy Production, Computing and Simulation, it can no longer continue to be inevitable that "the nuclear community tends to be too insular", as commented by the EU Commissioner for Research, Philippe Busquin, in one of his interviews. Still quoting the same Commissioner, "... nuclear scientists have had a tremendous impact on scientific advances in other fields that have benefited society ... ". It is time to make our patrimony more widely known.

Lisbon, February 2003
The Editors

Acknowledgement. The editors want to express their appreciation to the conference secretariat, Sandra Oliveira and Dulce Conceição, for their professionalism and the smooth way they solved many problems. We thank also all the administrative staff of IST and CFNUL, involved in the preparation of the workshop or these proceedings.

Contents

Electroweak Probes (I)

The Nuclear Interaction and Meson Production From Light Nuclei

Electroweak Probes (II)

The Nuclear Medium Under Extreme Conditions

Constituent Quark Models and Chiral Symmetry

Relativity and Electromagnetic Observables

Non-Nucleonic Degrees of Freedom

Relativity, Chiral Symmetry, and Form Factors of Baryons and Mesons

Nuclear Structure and Binding

Tributes to Peter Sauer

Few-Body Systems Suppl. 15, 1–12 (2003)

Few-Body Systems

Meson Exchange Currents in Quasi-Elastic Electron-Nucleus Scattering

Ingo Sick*

Department für Physik und Astronomie, Universität Basel, Basel, Switzerland

Abstract. The excess in the quasi-elastic transverse response function as measured by inclusive electron-nucleus scattering has remained a puzzle for a long time. For light nuclei, where this excess is largest, the corresponding Euclidean response functions are derived from the separated response functions obtained from the world data, and compared to those calculated with Green's function Monte Carlo methods using realistic interactions and currents. Large contributions associated with two-body currents are found for the transverse response, in agreement with data. We find in particular that the large enhancement is only obtained if in the description of the initial (bound) and the final (continuum) state the tensor- and short range $n-p$ correlations are included.

1 Introduction

Much effort has been devoted to understanding quantitatively the roles that two-body components of the nuclear electromagnetic current play. Despite the considerable attention that has been given to this topic, many open questions remain in what concerns quasi-elastic scattering. Complications arise in particular as a consequence of the need to provide an accurate description for both the initial bound- and the final scattering-state wave functions, based on realistic Hamiltonians.

In part, the slow progress is also related to the confusing experimental situation that for some time ofuscated the interpretation of the data. The longitudinal and transverse response functions, obtained during the eighties from a Rosenbluth separation of experimental cross sections, seemed to indicate that there was a gross (up to 40%) lack of longitudinal strength in the main quasi-elastic peak, and a correspondingly too low Coulomb sum rule [1, 2].

The work of J. Jourdan [3] pointed out a number of shortcomings of the previous analyses (which all went into the direction of reducing the Coulomb sum). When using the correct $e-p$ cross section, when including the well

* *E-mail address:* ingo.sick@unibas.ch

known relativistic corrections, when accounting for the fact that the upper cutoff of the integral over the longitudinal response is not ∞ and when doing the Coulomb corrections using the exact calculations of the Ohio group, J. Jourdan finds that there is no violation of the Coulomb sum rule within the experimental error of ~20%. A more precise Coulomb sum can be determined by including, besides the Bates/Saclay data generally employed, the SLAC data [4], which have been taken at more forward angle, and in a more reliable data acquisition mode (event mode, not histogram mode). This allows for a more reliable extraction of the longitudinal contribution. The results show [3] that the Coulomb sum rule is fulfilled within the experimental uncertainty of ~10%, for nuclei between Carbon and Iron.

Much of the theoretical effort of the past two decades was absorbed by this apparent lack of longitudinal strength. The excess of transverse strength — presumably due to two-body currents — observed in the quasi-elastic region turns out to be the genuine problem, see Fig. 1. This has been put in sharp focus by the work on super-scaling [5] which allowed to systematically compare the longitudinal and transverse response functions. This work showed in the most clear way that the transverse strength for nuclei with mass number A=12, ..., 56 exceeds the longitudinal one already in the main quasi-elastic peak ($\psi' \sim 0$) by 20-40%, in addition to the excess of strength occurring in the "dip" between the quasi-elastic and Δ-peaks (related to more complicated processes such as pion production) at large ψ'. This excess of strength in the region of the quasi-elastic peak is discussed in this paper.

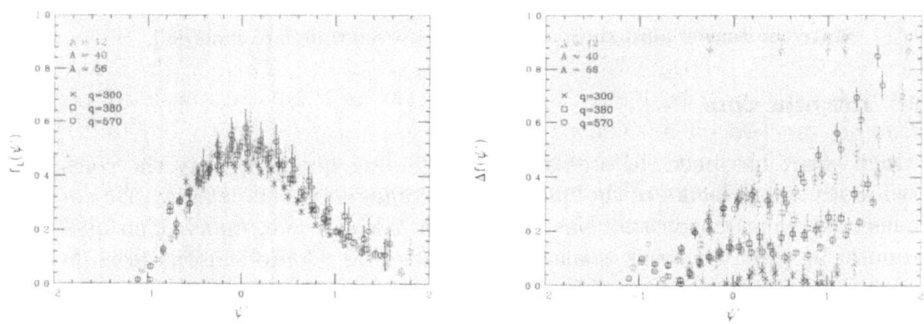

Figure 1. The longitudinal (f_L) and transverse minus longitudinal ($\Delta f = f_T - f_L$) response functions $f(\psi')$ determined for nuclei C, Ca and Fe for various momentum transfers. In the absence of MEC, the Δf response should be zero, and $f_L = f_T$.

Theoretical calculations of two-body contributions in the region of the quasi-elastic peak have been performed by many groups [6]–[20] using different approaches, see Table 1. The overwhelming majority of these calculations find very small contributions due to the dominant two-body terms (pion contact and in-flight, and Δ-excitation diagrams).

These calculations are in general based on an independent-particle initial

Table 1. Various calculations of the MEC contribution (given in %) in the quasi-elastic peak region.

vanOrden+Donnelly	FG	3%
Alberico et al	FG, RPA	5%
Blunden et al	rel. FG, no Δ	6%
Dekker et al	rel. FG	10%
Anguiano et al	IPSM A=4	few %
Fabrocini	NM CBF, 2p2h	15%
Bauer	FG + RPA, no Δ	few %
Van der Sluys et al	IPSM + RPA	20%
Amaro et al	FG	few %

state (shell model (IPSM), Fermi gas model (FG)) and give very small two-body contributions in the quasi-elastic peak as the pion and Δ terms tend to cancel. The calculation of Van der Sluys gives a larger correction, mainly due to the fact that a medium-modified Δ term, fitted to other MEC-dominated properties, has been used [1]. For the calculation of A. Fabrocini [17], see below.

The model study of Leidemann and Orlandini [10], in which the nuclear response was expressed in terms of the response of deuteron-like pairs of nuclear density, first pointed out that it is important to account in the initial state for the tensor correlations between np pairs. Only when these (rather short-range) tensor correlations were included would the two-body terms give appreciable contributions to the quasi-elastic response. This insight was quantitatively confirmed by Fabrocini [17], who calculated the transverse response of infinite nuclear matter using correlated basis function theory including one-particle-one-hole intermediate states.

Carlson and Schiavilla [14] have performed a calculation for ^{4}He using Green's Function Monte Carlo (GFMC) techniques, a realistic (Argonne v_8') N-N interaction and consistently constructed two-body terms. The inelastic response could be accurately calculated in terms of the Euclidean response (an integral over the response function, see below). These"exact"calculations found that the MEC lead to substantial enhancements, in agreement with those observed experimentally. The study of ref. [14] provided a first understanding of the ^{4}He quasi-elastic transverse response, but did not identify those aspects of the calculation responsible for the successful prediction.

2 Response for Light Nuclei

In this paper we study the longitudinal and transverse response functions of light nuclei, ^{3}He and ^{4}He, and compare to GFMC theory. Accurate data for

[1]The calculation of Gadyak *et al.* gives a larger MEC contribution for not-understood reasons, but their IA response appears to be incorrect.

these responses in the region of the quasi-elastic peak have been determined via an analysis of the world data for light nuclei [21]. A simultaneous study of ^{3}He and ^{4}He is particularly interesting as it turns out that the excess in the transverse channel increases very rapidly between A=3 and A=4, a feature which can give us a further handle for the understanding of two-body effects. The study of ^{4}He is especially promising, since the available data (as discussed below) indicate that the relative excess of transverse strength in the quasi-elastic peak is largest for this nucleus. ^{4}He thus is the ideal nucleus for a detailed study of the transverse excess.

The response functions have been extracted for q=300–700 MeV/c in steps of 100 MeV/c for both ^{3}He and ^{4}He [21]. A representative example for the longitudinal and transverse response functions resulting from this analysis of the world data is shown in Fig. 2.

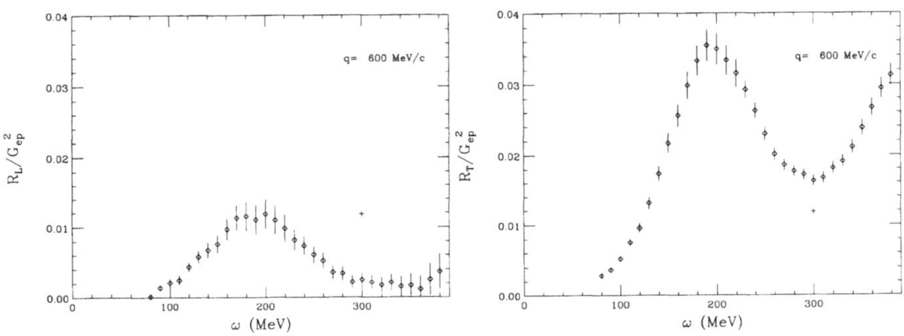

Figure 2. The longitudinal and transverse response functions (divided by the proton charge form factor) for ^{4}He and a momentum transfer of 600MeV/c.

To demonstrate the excess strength of $R_T(q,\omega)$, we have also studied the scaling properties of the response functions. Guided by the approach of [5], we use the variable ψ'. The dimensionless scaling functions $f_{L,T}$ are defined in [5] such that

$$f_{L,T} \equiv k_F \frac{R_{L,T}}{G_{L,T}} \quad , f_L = f_T = f$$

where, in IA, the second equation holds.

We compare in Fig. 3 the scaling functions $f_L(\psi')$ and $f_T(\psi')$ obtained for all response functions extracted from the global analysis of the ^{4}He data. The longitudinal response functions approximately scale to a universal curve over the entire quasi-elastic peak, and do fulfill the Coulomb sum rule. The results $R_T(q,\omega)$ confirm, that the basic problem in quasi-elastic electron-nucleus scattering is the excess strength in the transverse response. f_T exceeds f_L considerably, and the transverse excess is larger for ^{4}He than for ^{3}He (not shown) by almost a factor of two.

The excess of transverse strength occurs at all momentum transfers, and

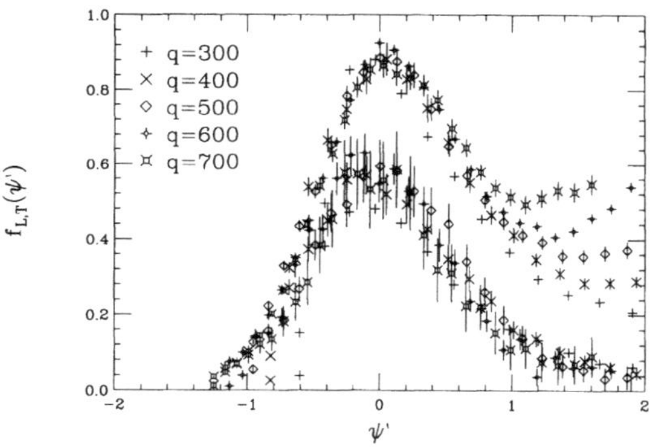

Figure 3. The scaling functions f_L and f_T are shown for all q values for ^{4}He. The upper band of points corresponds to f_T, the lower band to f_L.

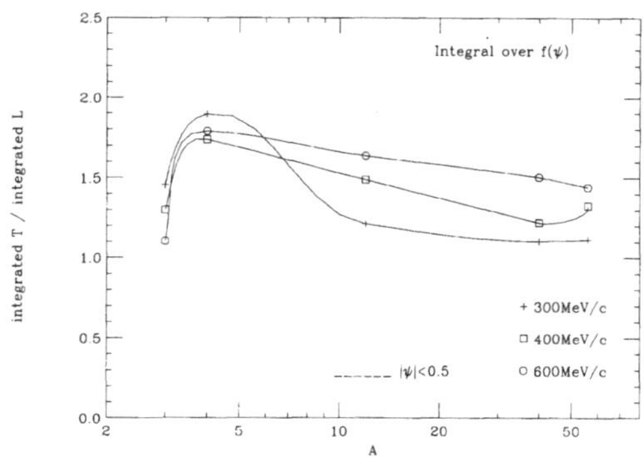

Figure 4. Ratio of transverse to longitudinal integrated strength for ^{3}He, ^{4}He, ^{12}C, ^{40}Ca, and ^{56}Fe: 300 MeV/c: +, 400 MeV/c: □, 600 MeV/c: ∘. Points at the same q are joined by lines.

affects the whole quasi-elastic peak region, extending below the π-production threshold.

The A-dependence of this excess is best seen by looking at the longitudinal and transverse responses integrated over ψ' — those for ^{12}C, ^{40}Ca, and ^{56}Fe have been determined in ref. [5]. We have integrated these responses over the region of ψ' that essentially covers the quasi-elastic peak ($|\psi'| < 0.5$). The resulting ratio of transverse to longitudinal integrated strength is shown in Fig. 4. This figure shows that the excess of transverse strength rises very rapidly between ^{3}He and ^{4}He, and is indeed largest for ^{4}He. It also shows that it is already large at the lowest q, the slow increase at the larger q for the heavier nuclei is mainly due to the fact that the tail of the Δ peak starts to contribute appreciably despite the restricted range of integration in ψ'.

3 Euclidean Response

Here, we are primarily interested in the overall strength of the longitudinal and transverse response. For this, we consider the Euclidean response functions, defined as [14, 22]

$$\tilde{E}_{T,L}(q,\tau) \;=\; \int_{\omega_{\rm th}}^{\infty} \exp[-\omega\tau] \;\; R_{T,L}(q,\omega) \;,$$

where the $R_{T,L}(q,\omega)$ are the standard responses and $\omega_{\rm th}$ is the threshold for the response of the system excluding the elastic contribution. The longitudinal and transverse Euclidean response functions represent weighted sums of the corresponding $R_L(q,\omega)$ and $R_T(q,\omega)$; at τ=0 they correspond to the Coulomb and transverse sum rules, respectively. The derivatives with respect to τ evaluated at τ=0 correspond to the energy-weighted sum rules. Larger values of τ give more weight to progressively lower-energy regions of the response.

For non-relativistic systems, the $\tilde{E}_{T,L}$ can be obtained from:

$$\tilde{E}_L(q,\tau) \;=\; \langle 0|\rho^\dagger(\mathbf{q})\exp[-(H-E_0)\tau]\rho(\mathbf{q})|0\rangle - \exp\left(-\frac{q^2\tau}{2Am}\right)|\langle 0(\mathbf{q})|\rho(\mathbf{q})|0\rangle|^2,$$

where $|0(\mathbf{q})\rangle$ represents the ground state recoiling with momentum $\mathbf{q}$, and sums over spin projections are understood. A similar expression, involving the current operator $\mathbf{j}_T$, holds for $\tilde{E}_T(q,\tau)$.

Below we consider the scaled Euclidean responses

$$E_{L,T}(q,\tau) = \frac{\exp\left[q^2\tau/(2m)\right]}{[G_{E,p}(\tilde{Q}^2)]^2}\,\tilde{E}_{L,T}(q,\tau) \;,$$

where $\tilde{Q}^2$ is the squared four-momentum transfer evaluated at the quasi-elastic peak. This removes the trivial energy dependence obtained from scattering off an isolated (non-relativistic) nucleon, and the q dependence associated with the nucleon form factors. The longitudinal response $E_L(q,\tau)$ is unity for an isolated proton, and the transverse response $E_T(q,\tau)$ is simply the square of its magnetic moment.

The use of the Euclidean response has a great advantage: it can be calculated exactly using Green's function or path integral Monte Carlo techniques. The Euclidian response can be calculated directly from the ground state wave function, hereby including both final state interactions and two-nucleon currents. There is no need to quantitatively describe the final continuum state, which at present is not possible for $A > 3$.

The Euclidian response has the disadvantage that we usually lack a good feeling for this integrated quantity. Model studies [21] show the sensitivity of $E(q,\tau)$ to properties of $R(q,\omega)$, see Fig. 5.

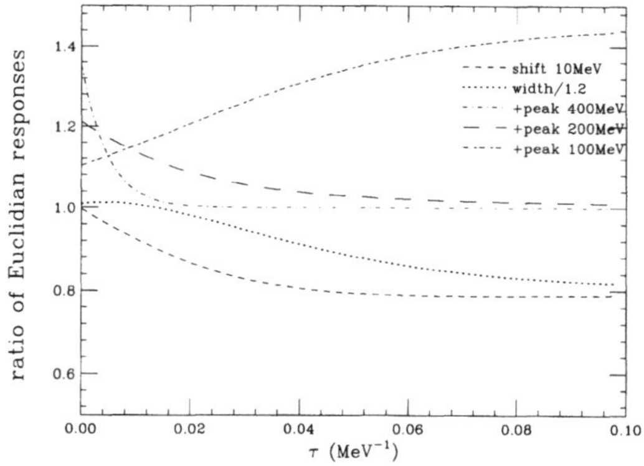

Figure 5. Effect of various modifications of the "normal" response (a peak at energy loss of 150MeV) in the corresponding Euclidian response. Changes include: a 10 MeV shift of the quasi-elastic peak, a change of its width by 20%, an additional peak at an energy loss 300MeV above the quasielastic peak, additional peaks at an energy loss of 100 and 200 MeV.

These studies show that, for the responses that can be extracted from the data, the region $0.01 \leq \tau \leq 0.05$ is the most relevant one for a comparison with theory.

For the nuclei considered here, the ground-state wave functions are obtained with variational Monte Carlo [22]. These wave functions are not exact, but they allow a rather precise characterization of the Euclidean response, as evidenced by comparisons with calculations using the correlated-hyperspherical-harmonics wave functions [23] in A=3.

The Euclidean response can be calculated with Green's function Monte Carlo techniques in an approach similar to the one used for the ground-state wave function. One needs to calculate matrix elements of the type:

$$\tilde{M}(\tau) = \frac{\langle 0|O_2 \exp[-(H-E_0)\tau]\, O_1|0\rangle}{\langle 0| \exp[-(H-E_0)\tau]|0\rangle} \ .$$

For a ground-state calculation of the energy (O_1=1, O_2=H) the matrix element is evaluated by a Monte Carlo sampling of the coordinate-space paths. The denominator is exactly one for an exact ground-state wave function, otherwise there is a correction for finite τ. For a more general matrix element $\tilde{M}$ one keeps another complete set of amplitudes for each operator O_1, each set of amplitudes corresponding to the full operator acting on the ground state. The paths are sampled precisely as in the ground-state calculation [24], and hence unaffected by the operators O_1, O_2. This allows one to calculate the response to a variety of operators (charge, current, different momenta, etc.) simultaneously.

The electromagnetic operator used in the calculation of Carlson and Schiavilla [21] consist of one- and two-body terms for the charge $\rho(\mathbf{q})$ and current $\mathbf{j}(\mathbf{q})$

operators. The one-body operators $\rho_i^{(1)}$ and $\mathbf{j}_i^{(1)}$ have the standard expressions obtained from a relativistic reduction of the covariant single-nucleon current. The two-nucleon operators are the standard ones used successfully in explaining the elastic form factors of light nuclei [25, 26, 27]. They have been, as far as possible, derived in a model-independent way from the N-N interaction.

4 Comparison to Experiment

Euclidean responses have been calculated from the longitudinal (L) and transverse (T) experimental response functions and are shown in part in Fig. 4. The nucleon electromagnetic form factors [28] are already divided out. The integration has been performed up to the energy loss ω where the T-response starts to increase significantly with ω (the corresponding value of ω is indicated in Fig.2 by a +) such as to not include too much of the Δ contribution. For the T-Euclidean response at very small τ the tail of the Δ-peak nevertheless plays a role, so the experimental response in this region is indicated by a dashed line only, and should not be compared to theory.

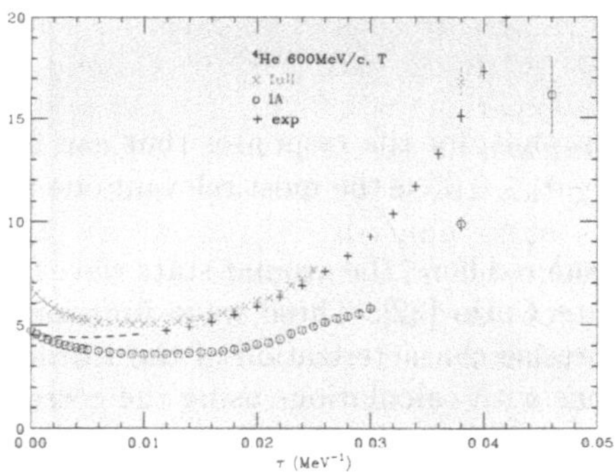

Figure 6. Longitudinal (left half of figure) and transverse Euclidean response of ^{4}He for a momentum transfer of 600 MeV/c.

In Fig. 4 we show the calculated Euclidean responses, obtained both in IA and when including the contributions associated with the two-body charge and current operators. The two-body contributions reduce by a small amount the L-responses, but they increase the T-responses by a very substantial amount at all momentum transfers. The enhancement in the transverse response is found already at quite low ω, as demonstrated by the Euclidean response at large τ. Two-body effects thus are important over the entire quasi-elastic peak, and not only in the "dip-region"on the large-ω side of the quasi-elastic peak which in the past was considered the main place where to see MEC.

At low q the contribution of two-body currents are larger at low ω (large τ) and smaller at large ω (low τ). At large q, this situation is reversed. Theory explains well the rapid increase of two-body contributions between ^{3}He and ^{4}He.

In contrast to most published calculations (for a discussion see Sec. 1), the calculation of Carlson and Schiavilla does give the sizeable two-body contribution required by the data.

We have emphasized above that the transverse excess is largest for ^{4}He, the nucleus which therefore is optimal for a study of MEC. For this nucleus the agreement between theory and experiment is excellent for the L-response, thus implying that an accurate treatment of the nuclear spectrum has been achieved as two-body operators give small corrections in the L-channel. For the ^{4}He T-Euclidean response, the large two-body effects found by experiment are successfully predicted by theory, although the associated contributions are a bit too large in the q-range 400–500 MeV/c. When omitting the $n-p$ correlations, on the other hand, the calculation gives a very small MEC contribution.

5 Bound-nucleon Form Factor

The conjecture of a modification of the form factors of a nucleon embedded in the nuclear medium has received considerable attention during the last decade. This conjecture first arose in connection with the apparent deficit in the Coulomb sum rule (CSR) in nuclei like C, Ca and Fe [1, 2] (see sect. 1). More recently, $(e,e'p)$ experiments [29, 30, 31] yielded separated longitudinal and transverse cross sections, the ratio of which differed from that expected for a free proton. Polarization transfer experiments of the type ($\boldsymbol{e}$,e'$\boldsymbol{p}$) [32, 33] have measured ratios of asymmetries P_x'/P_z' which, in plane-wave-impulse-approximation (PWIA), are proportional to G_{ep}/G_{mp}. After correcting the data for FSI effects — for caveats concerning meson exchange currents (MEC), see below — ratios smaller than those for free protons have been inferred. In particular, this reduction is found to be of the order of ~20% in $(G_{ep}/G_{mp})^2$ for the experiments using ^{4}He as a target. This again has led to speculations about medium-modifications of the proton (charge) form factor.

In this section, we therefore discuss in more detail the contribution of MEC to both the L and T responses at the larger momentum transfers of interest in connection with the above experiments.

At the larger q of interest here, S_L is more difficult to determine. The Coulomb sum S_L^∞ is defined as an integral to $\omega = \infty$. This limit cannot be reached experimentally by (e,e') scattering ($\omega < q$), furthermore the uncertainty of $R_L(q,\omega)$ at large ω becomes large. The strength beyond $\omega_{\max}$ thus has to be calculated and added.

For ^{4}He Morita and Suzuki [34] have calculated the spectral function $S(k,E)$, using the ATMS approach and the Reid soft core v_8 nucleon-nucleon interaction. This spectral function contains components of large momentum k and large removal energy E that lead to strength beyond $\omega_{\max}$. We have used this spectral function to calculate the fraction of strength outside the region the experimental $R_L(q,\omega)$ was integrated over.

In an alternative estimate of the strength at $\omega > \omega_{\max}$, we have used the procedure employed in ref. [35]. It consists in parameterizing the longitudinal response for $\omega > \omega_{\max}$ and in determining the (q-dependent) fall-off at large ω by requiring that the energy-weighted sum rule $W_L(q)$ reproduces the theoretical value, which can be accurately computed.

In Fig. 7 we display the effect of two-body charge operators on the CSR and the integrated transverse response by plotting the quantity

$$\Delta(q) = \left[< 0|\rho^\dagger(\boldsymbol{q})\rho(\boldsymbol{q})|0 > - < 0|\rho^{(1)\dagger}(\boldsymbol{q})\rho^{(1)}(\boldsymbol{q})|0 >\right] / [Z\,\tilde{G}_e^2(\tilde{Q}^2)] \ ,$$

as a function of the three-momentum transfer. This quantity contains interference terms of the type $\rho^{(1)\dagger}(\boldsymbol{q})\rho^{(2)}(\boldsymbol{q}) + \mathrm{h.c.}$ as well as terms involving only the two-body charge operators $\rho^{(2)\dagger}(\boldsymbol{q})\rho^{(2)}(\boldsymbol{q})$.

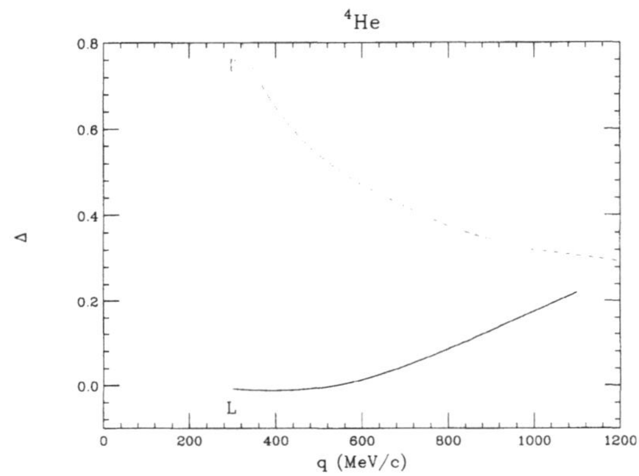

Figure 7. Fractional contribution of two-body currents to the Coulomb sum and the integrated transverse strength.

In Fig. 8 we show the experimental and theoretical results for the CSR. Note that the normalization used here for S_L has here been chosen such that, in the limit in which two–body contributions to the charge operator are neglected, one would find

$$S_L(q; \mathrm{one-body}) \to 1 \qquad q \to \infty \ .$$

Figure 8 shows that both data and calculation increasingly differ from one towards low (large) momentum transfer due to correlation effects (MEC) [36]. Theory and experiment agree within the error bars.

Before addressing the question of the bound-nucleon form factor, we point out some important features of our results [37]:

• Contrary to the usual approximation made when discussing the Coulomb sum — the neglect of two-body charge operators (MEC) — we find that the contribution of MEC to the integral is non-negligible, particularly beyond 600 MeV/c. At large q MEC lead to a significant increase. One should note that the MEC operators are the same ones that are crucial for reproducing the observed

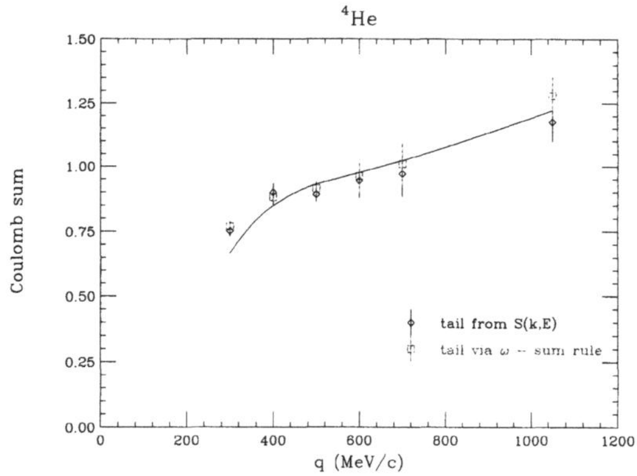

Figure 8. Experimental Coulomb sum compared to theory. The error bars include both the statistical and systematic uncertainties, the calculation includes the contribution of two-body charge operators.

charge form factors of ^{3}He and ^{4}He, for a review see ref. [38].

• In the calculation of the MEC contribution to $R_L(q,\omega)$ and $R_T(q,\omega)$ it is very important to treat the $n-p$ short range and tensor correlations in both the initial and the final state wave functions [21]. When omitting these correlations — as done in most previous calculations of MEC contributions — we find, as these calculations did, a much smaller MEC effect (see also [10, 17]).

• For the transverse response we find large enhancements due to MEC, 65÷30% for q=300÷1050MeV/c, falling only very slowly with increasing q [21].

As evidenced by Fig. 8, data and calculation are in agreement when using the free nucleon charge form factor — no in-medium reduction of G_{ep} is needed to reproduce the measured CSR. We should note that the total integrated strength $S_T(q)$ in the transverse response $R_T(q,\omega)$ is increased by two-body current contributions to a much more significant extent than for $S_L(q)$. The ratio $S_L(q)/S_T(q)$ is less than one, see Fig. 8 of Ref. [21]. One should note that the (e,e') cross section considered here to ~95% [69%] for $R_L(q,\omega)[R_T(q,\omega)]$ is due to the $(e,e'p)$ process that led to the G_{ep}^2/G_{mp}^2 ratios that differ from the free ones. This clearly suggests that the $\simeq$ 20% reduction in the ratio G_{ep}^2/G_{mp}^2 compared to the value for a free proton, which was observed in $(e,e'p)$ and $(e,e'\vec{p})$ experiments [29]–[33], most likely reflects the lack of MEC contributions in the analysis of those experiments.

Acknowledgement. Much of the work described in this paper was done in collaboration with Joe Carlson, Jürg Jourdan and Rocco Schiavilla. This work was supported by the Schweizerische Nationalfonds.

References

1. R. Altemus et al.: Phys. Rev. Lett. **44**, 965 (1980)
2. Z. Meziani et al.: Phys. Rev. Lett. **52**, 2130 (1984)
3. J. Jourdan: Nucl. Phys. **A603**, 117 (1996)
4. D. Day et al.: Phys. Rev. **C48**, 1849 (1993)
5. T. W. Donnelly, I. Sick: Phys. Rev. **C60**, 065502 (1999)
6. T.W. Donnelly et al.: Phys. Lett. **B76**, 393 (1978)
7. J.W. Van Orden, T.W. Donnelly: Ann. Phys. **131**, 451 (1981)
8. M. Kohno, N. Ohtsuka: Phys. Lett. **98B**, 335 (1981)
9. P.G. Blunden, M.N. Butler: Phys. Lett. **B219**, 151 (1989)
10. W. Leidemann, G. Orlandini: Nucl. Phys. **A506**, 447 (1990)
11. M.J. Dekker et al.: Phys. Rev. **C49**, 2650 (1994)
12. J.E. Amaro, A.M. Lallena: Nucl. Phys. **A537**, 585 (1992)
13. W.M. Alberico et al.: Ann. Phys. **154**, 356 (1984)
14. J. Carlson, R. Schiavilla: Phys. Rev. **C49**, R2880 (1994)
15. V. Van der Sluys et al.: Phys. Rev. **C51**, 2664 (1995)
16. M. Anguiano et al.: Phys. Rev. **C53**, 3155 (1996)
17. A. Fabrocini: Phys. Rev. **C55**, 338 (1997)
18. V. Gadiyak, V. Dmitriev: Nucl. Phys. **A639**, 685 (1998)
19. J.E. Amaro et al.: Phys. Rep. **368**, 317 (2002)
20. E. Bauer: Phys. Rev. **C61**, 044307 (2000)
21. J. Carlson et al.: Phys. Rev. **C65**, 024002–1 (2002)
22. J. Carlson, R. Schiavilla: Phys. Rev. Lett. **68**, 3682 (1992)
23. M. Viviani et al.: Few-Body Syst. **18**, 25 (1995)
24. B. S. Pudliner et al.: Phys. Rev. **C56**, 1720 (1997)
25. R. Schiavilla et al.: Phys. Rev. **C45**, 2628 (1992)
26. L.E. Marcucci et al.: Phys. Rev. **C58**, 3069 (1998)
27. I. Sick: Prog. Nucl. Part. Phys. **47**, 245–318 (2001)
28. G. Hoehler et al.: Nucl. Phys. **B114**, 505 (1976)
29. G. van der Steenhoven et al.: Phys. Rev. Lett. **57**, 182 (1986)
30. G. van der Steenhoven et al.: Phys. Rev. Lett. **58**, 1727 (1987)
31. D. Reffay Pikeroen et al.: Phys. Rev. Lett. **60**, 776 (1988)
32. S. Dieterich et al.: Phys. Lett. **B500**, 47 (2001)
33. S. Strauch: World Scientific, in print, (2002)
34. H. Morita, T. Suzuki: Prog.Theo.Phys. **86**, 671 (1991), and priv. com.
35. R. Schiavilla et al.: Phys. Rev. Lett. **70**, 3856 (1993)
36. R. Schiavilla et al.: Nucl. Phys. **A473**, 290 (1987)
37. J. Carlson et al.: Phys. Lett. B, in print.
38. J. Carlson, R. Schiavilla: Rev. Mod. Phys. **70**, 743 (1998)

Few-Body Systems Suppl. 15, 13–23 (2003)

Few-Body Systems

Parity-Violating Effects in Two-Nucleon Systems

R. Schiavilla[1,2], J. Carlson[3], M. Paris[3]

[1] Jefferson Laboratory, Newport News, Virginia 23606

[2] Physics Department, Old Dominion University, Norfolk, VA 23529

[3] Theoretical Division, Los Alamos National Laboratory, Los Alamos, NM 87545

Abstract.
We report recent results obtained for the parity-violating (PV) asymmetries in $\vec{p}p$ elastic scattering, $\vec{n}p$ radiative capture at thermal neutron energies, and deuteron electro-disintegration in quasi-elastic kinematics. The available data on the $\vec{p}p$ longitudinal asymmetry are used to put constraints on some of the PV vector-meson couplings occurring in the DDH model for the PV nucleon-nucleon interaction. In the $\vec{n}p$ capture, we show that a measurement of the associated asymmetry will lead to an essentially model-independent determination of the PV pion coupling to the nucleon. Finally, in quasi-elastic electron scattering we demonstrate that the asymmetry induced by hadronic weak interactions (and associated currents) is two orders of magnitude smaller than that originating from interference of the γ- and Z-exchange amplitudes.

1 Introduction

Two-nucleon experiments are the clearest probes of hadronic parity violation. A recent experiment has measured the longitudinal asymmetry in $\vec{p}p$ elastic scattering [1], while another one is underway to measure the photon asymmetry in low-energy $\vec{n}p$ radiative capture [2]. This experiment has been designed to provide a definitive measurement of the weak πNN coupling constant, which determines the longest-range part of the parity-violating (PV) nucleon-nucleon interaction.

This interaction–and associated electromagnetic currents–also induce an asymmetry in deuteron disintegration at quasi-elastic kinematics with polarized electrons, in addition to the asymmetry originating from the interference of the γ- and Z-exchange amplitudes. Indeed, it has been speculated that these hadronic weak interaction effects could cloud the interpretation, in terms of

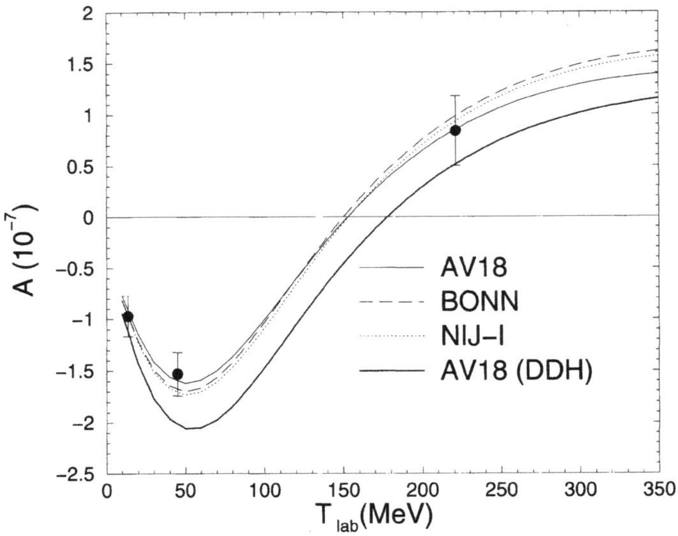

Figure 1. Total nuclear asymmetries obtained with various combinations of strong- and weak-interaction potentials are compared to data, see text.

single-nucleon properties, of the experimental asymmetries, recently measured in the $d(\vec{e},e')pn$ SAMPLE experiments at the MIT-Bates facility [3, 4].

In the present contribution, we review results of recent calculations of the above processes [5, 6, 7], carried out by using the Desplanques, Donoghue, and Holstein (DDH) meson-exchange model of the PV NN interaction [8].

2 Asymmetry in $\vec{p}p$ Elastic Scattering

The longitudinal asymmetry in $\vec{p}p$ elastic scattering is given by [5]

$$A(E) = \frac{\mathrm{Im}\Big[M_{10,00}(E,0) + M_{00,10}(E,0)\Big]}{\mathrm{Im}\Big[\sum_{SM_S} M_{SM_S,SM_S}(E,0)\Big]} \,, \tag{1}$$

where $M_{S'M'_S,SM_S}(E,\theta)$ denotes the amplitude for scattering $SM_S \rightarrow S'M'_S$, SM_S being the pair spin and its z-projection. One should note that, in the absence of PV interactions, the amplitude $M_{00,10}$ would vanish. The expression above for $A(E)$ ignores the contribution of the Coulomb amplitude (a divergent quantity in the limit θ=0), and for this reason $A(E)$ will be referred to as the "nuclear"asymmetry. The effect of including explicitly the amplitude induced by the Coulomb potential is discussed in ref. [5].

The calculated nuclear asymmetries are compared with the available data (see [1] and references therein) in Fig. 1. The curves labeled AV18, BONN, and

NIJ-I correspond to the strong-interaction potentials of refs. [9, 10, 11], respectively, in combination with the DDH potential, with the coupling constants h_ρ^{pp} and h_ω^{pp} determined by a rough fit to data [5]. There is very little sensitivity to the input strong-interaction potential, the larger differences displayed in the earlier study of ref. [12] may be attributable to larger differences in the p-wave phase shifts. The reduced difference is undoubtedly a consequence of the more extended *pp* and *pn* scattering data-base to which present potentials are fitted, as well as of the much higher accuracy achieved in these fits.

We also show in Fig. 1 the AV18 results which correspond to a DDH potential using the "best"estimates for the h_ρ^{pp} and h_ω^{pp} coupling constants [8]. A number of comments are now in order. The data point at 221 MeV essentially determines the value of h_ρ^{pp}. As pointed out by Simonius [13], the dominant contributions to the total asymmetry in the energy range under consideration here are those associated with the $^1S_0$-$^3P_0$ and $^3P_2$-$^1D_2$ partial waves. At energies close to 225 MeV the $^1S_0$-$^3P_0$ contribution vanishes. As a result, the total

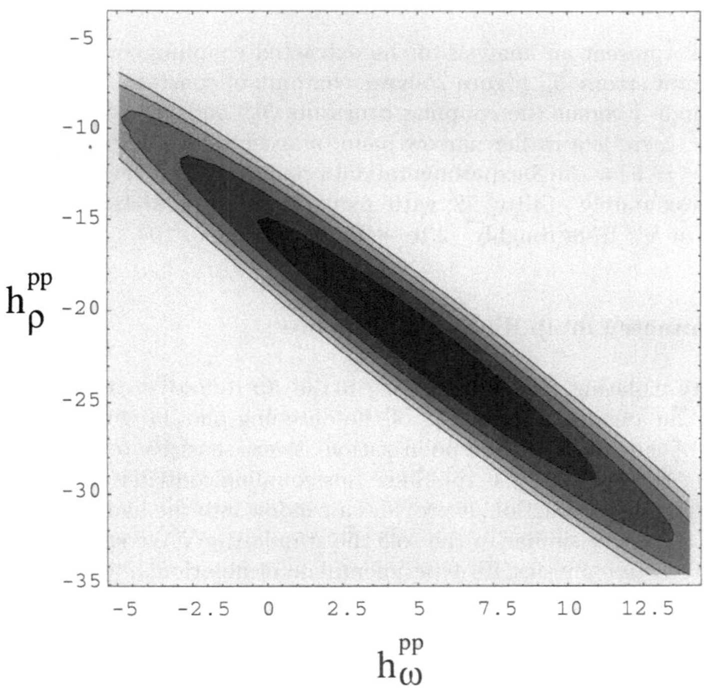

Figure 2. Curves of constant total χ^2 obtained by analyzing the experimental *pp* data with the AV18 model, and ρ- and ω-meson strong-interaction couplings in the DDH potential from the Bonn-2000 model. The curves indicate surfaces of total χ^2 = 1,2,3,4, and 5 for various values of h_ρ^{pp} and h_ω^{pp}.

asymmetry in this energy region is almost entirely due to the $^3P_2$-$^1D_2$ contribution, which is known [13] to be approximately proportional to the following combination of coupling constants, $h_\rho^{pp} g_\rho \kappa_\rho + h_\omega^{pp} g_\omega \kappa_\omega$, where g_ρ, κ_ρ, ... are the vector and tensor coupling constants of the ρ and ω mesons to the nucleon. The ω tensor coupling is much smaller than the ρ tensor coupling, as suggested by vector-meson-dominance arguments. Indeed, in the BONN model, κ_ω is taken to be zero, and hence the data point at 221 MeV fixes h_ρ^{pp} (for given g_ρ, κ_ρ, and Λ_ρ). This is the reason for the $\simeq 44$ % increase (in magnitude) of h_ρ^{pp} with respect to the DDH "best"estimate.

Below 50 MeV, however, the calculated total asymmetry is dominated by the $^1S_0$-$^3P_0$ contribution, approximately proportional to [13] $h_\rho^{pp} g_\rho (2+\kappa_\rho) + h_\omega^{pp} g_\omega (2+\kappa_\omega)$. The increase in magnitude of h_ρ^{pp} required to fit the point at 221 MeV, now leads to a total asymmetry $|A(E)|$ below 50 MeV, which is too large when compared to experiment. Thus, in order to reproduce the 13.6 MeV and 45 MeV data points, the overall strength of the coupling constant combination above needs to be reduced significantly. Since $g_\rho(2+\kappa_\rho)$ and $g_\omega(2+\kappa_\omega) = 2\, g_\omega$ have the same sign, this requires making the sign of h_ω^{pp} opposite to that of h_ρ^{pp}.

We also present an analysis of the extracted coupling constants and their experimental errors [5]. Figure 2 shows contours of constant total χ^2 at levels of 1 through 5 versus the coupling constants h_ρ^{pp} and h_ω^{pp}. As is apparent in the figure, there is a rather narrow band of acceptable values for h_ρ^{pp} and h_ω^{pp} at total $\chi^2 = 1$ for the 3 experimental data points. At this level, h_ρ^{pp} can range from approximately -14 to -28, with a simultaneous (and strongly-correlated) variation in h_ω^{pp} from roughly -2 to $+10$, all in units of 10^{-7}.

3 Asymmetry in $\vec{n}p$ Radiative Capture

The parity-violating (PV) observable in the $\vec{n}p$ radiative capture–the asymmetry in the angular distribution of the outgoing photon with respect to the direction of the initial neutron polarization–is very sensitive to the weak parity-violating πNN coupling [14, 15]. Since this coupling contributes to the longest-range part of the interaction, it provides a window into the hadronic weak interaction in a manner similar to the role the standard πNN coupling provides in strong-interaction physics. Its experimental determination, however, has proven to be very difficult. Measurements of the circular polarization of γ-ray decays in ^{18}F [16] have been interpreted to indicate a very small πNN coupling as compared to theoretical expectations based upon hadronic models [8], while measurements in atomic ^{133}Cs seem to favor a much larger value [17]. The analysis of both these experiments is complicated by the difficulty of handling reliably, from a theoretical stand-point, the many-body aspects of the problem in these complex systems. These issues have been addressed in refs. [18] and [19]. More recently, PV asymmetry measurements of neutron resonances in compound nuclei seem to require even larger values of the weak πNN coupling constant [20].

For these reasons, an experiment measuring the photon asymmetry in $\vec{n}p$ capture has been undertaken at the LANSCE facility [2]. As mentioned above, previous studies of this process [14, 15] showed that this asymmetry is very sensitive to the weak πNN coupling constant, while essentially unaffected by short-range contributions. In ref. [6] we investigated the model dependence of this result by considering several different high-quality interactions fit to the strong-interaction data.

We employed the standard DDH meson-exchange model of the (PV) NN interaction, solving the full Schrödinger equation for the scattering state as well as the deuteron bound state. The equations and relevant mixing parameters in the S-matrix were discussed in considerable detail in ref. [5] for the case of $\vec{p}p$ elastic scattering. While the short-range contributions to the PV interaction should not be viewed as resulting only from the exchange of single mesons, the seven parameters of the DDH model can still be employed to characterize all the low-energy PV mixings. For example, two-pion exchange could play a role [21], however we assumed that its effects could be included, at least at low energy, through the combination of pion- and short-range terms present in the DDH model.

We calculated the asymmetry for the AV18 [9], Bonn-CD [10] and Nijmegen-I (NIJM-I) [11] interactions. Each strong interaction model has associated exchange currents. For the AV18 we included currents from the momentum-independent terms (π- and ρ-currents) as well as terms from the momentum-dependent terms in the interaction, for a review see ref. [22]. For the Bonn-CD and NIJM-I interactions, we only included π- and ρ-currents with the cutoffs from the Bonn-CD model (Λ_π=1.72 GeV and Λ_ρ=1.31 GeV). Contributions from other meson exchanges in the Bonn and Nijmegen models were neglected. In all calculations, the currents associated with Δ excitation and $\omega\pi\gamma$ transition were also considered [22].

The total cross section at thermal neutron energies is due to the well-known M_1 transition connecting the parity-conserving (PC) $^1S_0$ np state to the PC deuteron state. The calculated values for each model are given in Table 1, both for one-body (impulse) currents alone and for the one- and two-body currents. In each case the largest two-body contribution, approximately two-thirds of the total, comes from the currents associated with pion exchange. The total cross section is in good agreement with experimental results, which are variously quoted as 334.2(0.5) mb [23] or 332.6(0.7) mb [24]. It would be possible to adjust, for example, the transition magnetic moment $\mu_{\gamma N\Delta}$ of the Δ-excitation current to precisely fit one of these values, here we simply choose a $\mu_{\gamma N\Delta}$ of 3 n.m., which is consistent with an analysis of γ-N data at resonance.

The PV asymmetry arises from an interference between the M_1 term above and the E_1 transition, connecting the $^3P_1$ PV np state to the PC deuteron state and the $^3S_1$ PC np state to the $^3P_1$ PV deuteron state. The PV components of the wave functions are generated by the DDH potential, including π-, ρ-, and ω-exchange mechanisms. In ref. [6] we took the linear combination of ρ- and ω-weak coupling constants corresponding to pp scattering from an earlier analysis [5] of these experiments. The remaining couplings were taken from the

Table 1. Total np capture cross-sections (in mb) and $\vec{n}p$ radiative capture asymmetries (in units of 10^{-8}) in various models. Asymmetries are reported for pion-exchange only and full DDH interactions.

	Cross Section		Asymmetry	
Interaction	Impulse Current	Full Current	pion-only	full DDH
AV18	304.6	334.2	-4.98	-4.85
NIJM-I	305.4	332.5	-5.11	-4.95
Bonn-CD	306.5	331.6	-4.97	-4.83

DDH "best guess"estimates. Finally, as in the earlier analysis of $\vec{p}p$ scattering, the cutoff values in the meson-exchange interaction were taken from the Bonn-CD potential (Λ_π=1.72 GeV, Λ_ρ=1.31 GeV, and Λ_ω=1.50 GeV).

With these couplings, we obtain the asymmetries also listed in Table 1. The results are consistent with earlier [14] and more recent [25] estimates, and all agree within a few per cent, which is also the magnitude of the contributions from the short-range terms. The E_1 transition has been calculated in the long-wavelength approximation (LWA) with the Siegert form of the E_1 operator, thus eliminating many of the model dependencies and leaving only simple (long-range) matrix elements. In ref. [6] we explicitly calculated corrections beyond the LWA E_1 terms, and found them to be quite small.

4 Asymmetry in $d(\vec{e},e')pn$ Electro-Disintegration

The SAMPLE experiment[3, 4] has measured the parity-violating asymmetry in polarized electron quasi-elastic scattering on the deuteron. This asymmetry has two distinct contributions: one associated with interference of the γ- and Z-exchange amplitudes, and the other induced by PV NN interactions. The first contribution was recently studied in ref. [26], where it was shown that two-body terms in the nuclear electromagnetic and weak neutral currents only produce 1–2% corrections to the asymmetry due to the corresponding single-nucleon currents.

In ref. [6] we investigated the asymmetry originating from hadronic weak interactions, updating and sharpening earlier predictions obtained in Refs. [27, 28]–these studies did not include the effects of PV currents.

The expression for the asymmetry from γ-Z interference was given in ref. [26], while that due to the hadronic weak interaction reads [7]

$$A(q,\omega,\theta_e) = \frac{v'_{\rm T} R'_{\rm T}(q,\omega)}{v_{\rm L} R_{\rm L}(q,\omega) + v_{\rm T} R_{\rm T}(q,\omega)} , \tag{2}$$

where $v_{\rm L}$, $v_{\rm T}$, and $v'_{\rm T}$ denote electron kinematical factors, $R_{\rm L}$ and $R_{\rm T}$ are the standard longitudinal and transverse response functions, and $R'_{\rm T}$ is the PV response function, defined as

$$R'_{\mathrm{T}}(q,\omega) = \overline{\sum_i} \sum_f \delta(\omega + m_i - E_f)\, \mathrm{Im}\left[\mathbf{j}_{fi}(\mathbf{q}) \times \mathbf{j}^*_{fi}(\mathbf{q})\right]_z \ . \tag{3}$$

Here the three-momentum transfer $\mathbf{q}$ has been taken to define the spin-quantization axis (the z-axis), and $\mathbf{j}_{fi}(\mathbf{q}) \equiv \langle f|\mathbf{j}(\mathbf{q})|i\rangle$ are the matrix elements of the electromagnetic current. In terms of electric and magnetic multipole operators, the cross product above is expressed as

$$\mathrm{Im}\left[\mathbf{j}_{fi}(\mathbf{q}) \times \mathbf{j}^*_{fi}(\mathbf{q})\right]_z \propto \sum_{l\geq 1} \mathrm{Re}\Big[\langle J_f||M_l(q)||J_i\rangle\langle J_f||E_l(q)||J_i\rangle^*\Big] \ , \tag{4}$$

and therefore vanishes unless (i) the initial and/or final states do not have definite parity (as is the case here because of the presence of PV NN interactions) and/or (ii) the electric and magnetic multipole operators have unnatural parities $(-)^{l+1}$ and $(-)^l$, respectively, because of PV electromagnetic currents, such as the one-body anapole current [29] and the two-body currents due to PV NN interactions [7].

In the initial study of ref. [6], the one-body anapole current was taken as [29]

$$\mathbf{j}^{(1)}_{\mathrm{PV}}(\mathbf{q}) = \frac{Q^2}{2\,m^2} \sum_i \left[a_{\mathrm{S}}(Q^2) + a_{\mathrm{V}}(Q^2)\tau_{i,z}\right] \sigma_i \, \mathrm{e}^{\mathrm{i}\mathbf{q}\cdot\mathbf{r}_i} \ , \tag{5}$$

where Q^2 is the squared four-momentum transfer ($Q^2{=}q^2 - \omega^2$), and the isoscalar and isovector anapole form factors are normalized as

$$a_{\mathrm{S,V}}(0) = \frac{h_\pi \, g_\pi}{4\sqrt{2}\pi^2}\alpha_{\mathrm{S,V}} \ , \tag{6}$$

with $\alpha_{\mathrm{S}}{=}1.6$ and $\alpha_{\mathrm{V}}{=}0.4$ from a calculation of pion-loop contributions [29]. Here h_π and g_π are weak and strong πNN coupling constants, respectively. More recent estimates of the nucleon anapole form factors give somewhat smaller values for $\alpha_{\mathrm{S,V}}$ [30, 31, 32]. A complete treatment would require estimates of short-distance contributions [33] and electroweak radiative corrections [7].

Only the PV two-body current associated with the π-exchange term in the DDH interaction were included in the calculations of ref. [6] (in addition, of course, to the PC two-body currents discussed in the previous section). The PV currents from ρ and ω exchanges have not yet been considered [7, 19], but are not expected to play a significant role. One should note that at the higher momentum transfers of interest here, 100–300 MeV/c, relevant for the SAMPLE experiments, it is not possible to include the currents through the Siegert theorem, they must be calculated explicitly.

In Fig. 3 we show the asymmetry for one of the SAMPLE kinematics for the AV18 plus full DDH interaction. The total asymmetry is given by the sum of the three contributions shown. The contribution labeled "DDH interaction" is that originating directly from the DDH interaction, while that labeled "PV γ-couplings" is from the (one-body) anapole term and the two-body pion current.

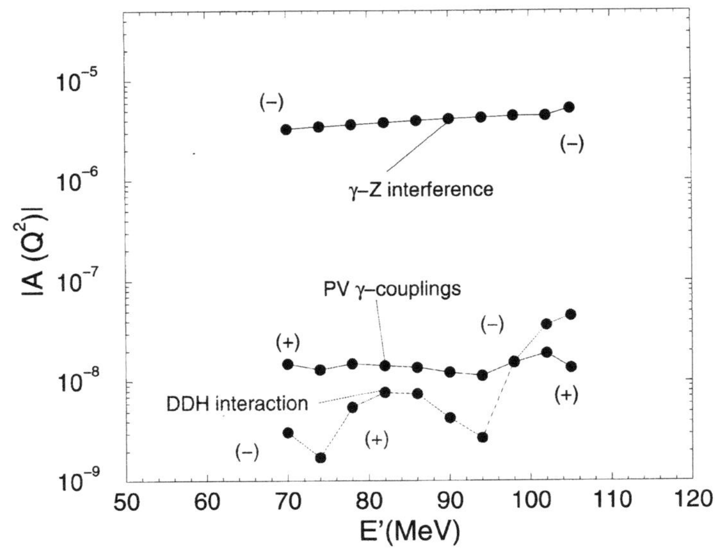

Figure 3. Contributions (in magnitude) to the longitudinal asymmetry in the electrodisintegration of the deuteron as function of the final electron energy. The kinematical setting corresponds to one of those relevant for the SAMPLE experiments: the initial electron energy is 120 MeV and the electron scattering angle is 145.9°. The (−) and (+) symbols denote the signs of the various contributions. See text for further explanations.

Results at higher Q^2 are similar. The Bonn-CD model leads to predictions, for both the asymmetry and inclusive cross section, which are very close to those obtained with the AV18 [7].

In Fig. 4 we plot the asymmetry at the top of the quasi-elastic peak as a function of the four-momentum transfer. SAMPLE kinematics have Q^2 equal to 0.1 $(\mathrm{GeV/c})^2$ and 0.043 $(\mathrm{GeV/c})^2$, respectively. The leading γ-Z term decreases in magnitude with Q^2, as expected. This term does not contribute, of course, for the real photons produced in the $\vec{n}p$ radiative capture. Also shown in the plot is the asymmetry obtained by retaining only the pion term in the DDH interaction, with a coupling given by the "best guess".

These results demonstrate that, in the kinematics of the SAMPLE experiments, the asymmetry from the γ-Z coupling is two-orders of magnitude larger than that associated with the PV hadronic weak interaction. Hence even the largest estimates of the weak πNN coupling constant will not affect extractions of nucleon matrix elements.

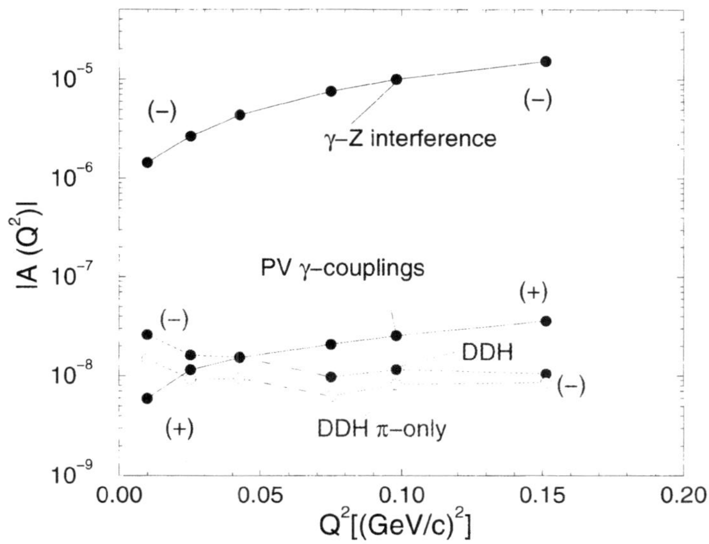

Figure 4. Contributions (in magnitude) to the longitudinal asymmetry in the deuteron electro-disintegration at the top of the quasi-elastic peak, plotted as function of the four-momentum transfer Q^2. The $(-)$ and $(+)$ symbols denote the signs of the various contributions. See text for further explanations.

5 Conclusions

In the present contribution, we have reported recent results obtained for the parity-violating asymmetries in $\vec{p}p$ elastic scattering, $\vec{n}p$ radiative capture, and deuteron electro-disintegration within the framework of the DDH model of the parity-violating hadronic weak interaction.

An analysis of the pp PV longitudinal asymmetry, carried out by using combinations of modern-day strong interaction potentials and the DDH PV potential, shows that the available data provide a strong constraint on allowable linear combinations of ρ and ω PV coupling constants. Combining the statistical and systematic errors in quadrature, h_ρ^{pp} is constrained by present data to approximately 35%, at the level of one standard deviation.

We find that the model-dependence of the np-capture asymmetry is quite small, at a level similar to the expected contributions of the short-range parts of the interaction. This process is in fact dominated by the longer-range pieces of the interaction associated with pion-exchange, and hence this experiment is a clean probe of that physics.

Similarly, we find that the SAMPLE experiment measuring the asymmetry in electron scattering on the deuteron is a very clean probe of nucleon properties. The processes associated with two-nucleons, including parity violation in the deuteron and scattering wave functions, and the currents associated with these interactions, play a very small role at reasonable values of the momentum transfer. These experiments, then, probe distinct aspects of weak-interaction physics.

Acknowledgement. The work of R.S. is supported by the U.S. Department of Energy contract number DE-AC05-84ER40150 under which the Southeastern Universities Research Association (SURA) operates the Thomas Jefferson National Accelerator Facility, while that of J.C. and M.P. was supported by the U.S. Department of Energy under contract W-7405-ENG-36.

References

1. A.R. Berdoz et al.: Phys. Rev. Lett. **87**, 272301 (2001)
2. W.M. Snow et al.: Nucl. Inst. and Meth. A **440**, 729 (2000)
3. R. Hasty et al.: Science **290**, 2117 (2000)
4. D.H. Beck and R.D. McKeown: Ann. Rev. Nucl. Part. Sci. **51**, 189 (2001)
5. J. Carlson et al.: Phys. Rev. C **65**, 035502 (2002)
6. R. Schiavilla, J. Carlson, and M. Paris: submitted to Phys. Rev. C, nucl-th/0212038
7. J. Carlson, M. Paris, and R. Schiavilla: in preparation
8. B. Desplanques, J.F. Donoghue, and B.R. Holstein: Ann. Phys. (N.Y.) **124**, 449 (1980)
9. R.B. Wiringa, V.G.J. Stoks, and R. Schiavilla: Phys. Rev. C **51**, 38 (1995)
10. R. Machleidt: Phys. Rev. C **63**, 024001 (2001)
11. V.G.J. Stoks et al.: Phys. Rev. C **49**, 2950 (1994)
12. D.E. Driscoll and G.A. Miller: Phys. Rev. C **39**, 1951 (1989); Phys. Rev. C **40**, 2159 (1989)
13. M. Simonius: Can. J. Phys. **66**, 548 (1988)
14. B. Desplanques: Nucl. Phys. **A242**, 423 (1975); *ibidem* **A335**, 147 (1980)
15. B.H.J. McKellar: Nucl. Phys. **A254**, 147 (1975)
16. E.G. Adelberger and W.C. Haxton: Ann. Rev. Nucl. Part. Sci. **35**, 501 (1985)
17. V.V. Flambaum and D.W. Murray: Phys. Rev. C **56**, 1641 (1997)
18. W.C. Haxton and C.E. Wieman: Ann. Rev. Nuc. Part. Sci. **51**, 261 (2001)

19. W.C. Haxton, C.-P. Liu, and M.J. Ramsey-Musolf: Phys. Rev. C **65**, 045502 (2002)

20. D. Bowman: private communication.

21. H.J. Pirner and D.O. Riska: Phys. Lett. **B44**, 151 (1973)

22. J. Carlson and R. Schiavilla: Rev. Mod. Phys. **70**, 743 (1998)

23. A.E. Cox, S.A.R. Wynchank, and C.H. Collie: Nucl. Phys. **74**, 497 (1965)

24. S.F. Mughabghab, M. Divadeenam, and N.E. Holden: *Neutron Cross Sections from Neutron Resonance Parameters and Thermal Cross Sections*, London: Academic Press 1981, http://isotopes.lbl.gov/ngdata/sig.htm.

25. B. Desplanques: Phys. Lett. **B512**, 305 (2001)

26. L. Diaconescu, R. Schiavilla, and U. van Kolck: Phys. Rev. C **63**, 044007 (2001)

27. W.-Y. P. Hwang and E.M. Henley: Ann. Phys. (N.Y.) **129**, 47 (1980)

28. W.-Y. P. Hwang, E.M. Henley, and G.A. Miller: Ann. Phys. (N.Y.) **137**, 378 (1981)

29. W.C. Haxton, E.M. Henley, and M.J. Musolf: Phys. Rev. Lett. **63**, 949 (1989)

30. M.J. Musolf and B.R. Holstein: Phys. Rev. D **43**, 2956 (1991)

31. D.O. Riska: Nucl. Phys. A **678**, 79 (2000)

32. C.M. Maekawa and U. van Kolck: Phys. Lett. **B478**, 73 (2000)

33. S.L. Zhu et al.: Phys. Rev. D**62**, 033008 (2000)

Few-Body Systems Suppl. 15, 25–36 (2003)

Few-Body Systems

NN Interaction in Chiral Constituent Quark Models

A. Valcarce[1], F. Fernández[1], P. González[2]

[1] Grupo de Física Nuclear, Universidad de Salamanca, E-37008 Salamanca, Spain

[2] Dpto. de Física Teórica and IFIC, Universidad de Valencia-CSIC, E-46100 Burjassot, Valencia, Spain

Abstract. We review the actual state in the description of the nucleon-nucleon (NN) interaction by means of chiral constituent quark models. We present a series of relevant features that are nicely explained within the quark model framework.

1 Introduction

At the end of the 70's the potentialities of quark model calculations in low-energy hadron physics were established. In a series of pioneering works, Isgur and Karl [1] performed a quite successful exhaustive study of the baryon spectra and at the same time the short-range repulsion of the nuclear force was qualitatively explained [2]. These ideas encouraged several groups to undertake the ambitious project of trying to understand the NN interaction in terms of quark degrees of freedom.

Several authors [3, 4, 5] found a quantitative explanation of the NN short-range repulsion based on dynamical effects induced by one-gluon exchange (OGE). One would say that there exists no repulsive potential in the naive sense. The color-magnetic OGE interaction likes to have as many symmetric pairs in the color-spin space as possible. Due to the connection of the spin-color space with the orbital space as a consequence of the Pauli principle the orbital symmetry $[42]_O$ is preferred. But this symmetry has to have a zero in the S-wave. This zero yields for the phase shift the same results as a hard or a soft core.

Based on this success simple quark models of the nuclear force were constructed. The so-called *hybrid quark models* [4] contained three main ingredients: (i) A short range part given by the quark-exchange interaction, (ii) a long range part by the one-pion exchange (OPE) potential, and (iii) a medium range part by a phenomenological potential or a two-pion exchange potential. These meson exchange potentials were considered to simulate the meson cloud surrounding the quark core. These models allowed to reproduce the NN scattering and bound state data. The next conceptual step in order to understand the NN interaction only based on quark degrees of freedom was the consideration of $(q\bar{q})$ and $(q\bar{q})^2$ excitations generated by off-shell terms of the Fermi-Breit quark-gluon interaction [6]. Although the obtained potential acquired an attractive part in the 0.8–1.5 fm range, it was too weak to bind the deuteron or to fit the extreme low-energy S-wave scattering.

In the middle 80's quark antisymmetry effects on the OPE started to be analyzed [7]. A similar NN short-range repulsion to the one provided by the OGE was obtained. A first quark-model calculation based on gluon and pion exchange at the level of quarks was performed [8]. As the OPE contributes to the $\Delta - N$ mass difference, the rather big quark-gluon coupling constant used in the previous models ($\alpha_s > 1$) got nicely reduced to values around 0.5. However, not enough intermediate range attraction was obtained. To avoid this problem, a phenomenological scalar potential at baryonic level was introduced, returning on this way to a hybrid model. Besides, its coupling constant was fitted to a different value for NN S-waves, than for higher angular momentum partial waves or for the deuteron [8].

2 The Chiral Constituent Quark Model

In the early 90's the constituent quark mass was related to the breaking of chiral symmetry [9, 10]. The underlying idea is that the constituent quark mass is a consequence of spontaneous chiral symmetry breaking. Then, between the chiral symmetry breaking scale ($\Lambda_{\rm CSB} \sim 1$ GeV) and the confinement scale ($\Lambda_{\rm C} \sim 0.2$ GeV), QCD may be simulated in terms of an effective theory of constituent quarks interacting through Goldstone modes. Specifically, in a Nambu-Goldstone realization of chiral symmetry within the linear sigma model, there appear two Goldstone boson fields: the pion and the sigma. The chiral constituent quark model incorporates OPE and one-sigma exchange (OSE) interactions of the form

$$\begin{aligned} V_{\rm OPE}(\boldsymbol{r}_{ij}) &= \frac{1}{3}\alpha_{\rm ch}\frac{\Lambda^2}{\Lambda^2 - m_\pi^2}m_\pi \left\{ \left[Y(m_\pi r_{ij}) - \frac{\Lambda^3}{m_\pi^3} Y(\Lambda r_{ij}) \right] \boldsymbol{\sigma}_i \cdot \boldsymbol{\sigma}_j + \right. \\ & \left. \left[H(m_\pi r_{ij}) - \frac{\Lambda^3}{m_\pi^3} H(\Lambda r_{ij}) \right] S_{ij} \right\} \boldsymbol{\tau}_i \cdot \boldsymbol{\tau}_j \,, \end{aligned} \quad (1)$$

$$V_{\rm OSE}(\boldsymbol{r}_{ij}) = -\alpha_{\rm ch}\frac{4\,m_q^2}{m_\pi^2}\frac{\Lambda^2}{\Lambda^2 - m_\sigma^2}m_\sigma \left[Y(m_\sigma r_{ij}) - \frac{\Lambda}{m_\sigma} Y(\Lambda r_{ij}) \right] , \quad (2)$$

Λ being a cutoff parameter and $Y(x)$ and $H(x)$ the standard Yukawa functions. Perturbative features of QCD are incorporated through the OGE potential,

$$V_{\rm OGE}(\boldsymbol{r}_{ij}) = \frac{1}{4}\alpha_s\,\boldsymbol{\lambda}_i \cdot \boldsymbol{\lambda}_j \left\{ \frac{1}{r_{ij}} - \frac{\pi}{m_q^2}\left[1 + \frac{2}{3}\boldsymbol{\sigma}_i \cdot \boldsymbol{\sigma}_j\right] \delta(\boldsymbol{r}_{ij}) - \frac{3}{4m_q^2 r_{ij}^3} S_{ij} \right\} , \quad (3)$$

where the $\lambda's$ stand for the color SU(3) matrices. Finally, a phenomenological confining potential is introduced. Its detailed radial structure, being fundamental to study the hadron spectra, is expected to play a minor role for the two-baryon interaction. To be consistent with meson spectroscopy it can be taken to be linear

$$V_{\rm CON}(\boldsymbol{r}_{ij}) = -a_c \, \boldsymbol{\lambda_i} \cdot \boldsymbol{\lambda_j} \, r_{ij} \, , \tag{4}$$

though for simplicity harmonic oscillator quark wave functions will be used for the NN interaction.

This model is able to reproduce quite nicely the NN scattering and bound state data [10, 11]. Typical values for the parameters are tabulated in Table 1.

Table 1. Quark-model parameters.

m_q MeV	b fm	α_s	a_c MeV fm^{-1}	$\alpha_{\rm ch}$	r_0 fm	m_σ fm^{-1}	m_π fm^{-1}	Λ fm^{-1}
313	0.518	0.485	67.0	0.0269	0.25	3.42	0.7	4.2

3 Several Reasons to Use Quark Models

3.1 Medium Range Attraction

A new feature of the chiral constituent quark model is that through the OSE potential it incorporates in a natural way the NN medium range attraction without any additional parameter. This fact is illustrated in Fig. 1 where we compare the scalar potential obtained at quark level (solid line) with the baryonic parametrization used in ref. [8] (dashed line) to fit the experimental data. While at the baryonic level two different coupling constants are used: $g^2_{\sigma NN}/4\pi = 3.7$ for (S,T)=(0,1) and $g^2_{\sigma NN}/4\pi = 1.9$ for (S,T)=(1,1), the quark model result is obtained in both cases with the same parameters fixed from the πqq coupling

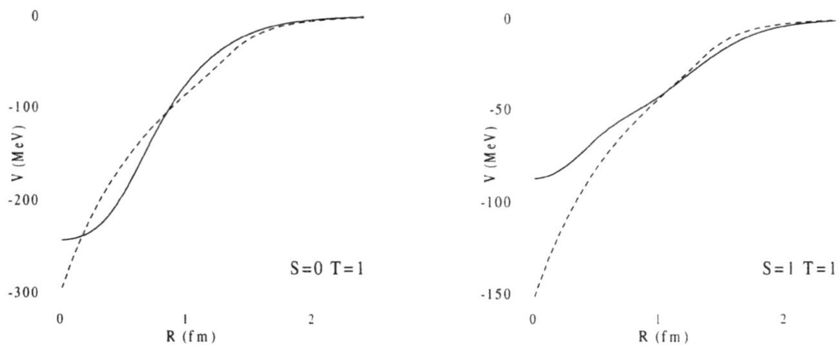

Figure 1. NN σ exchange potential.

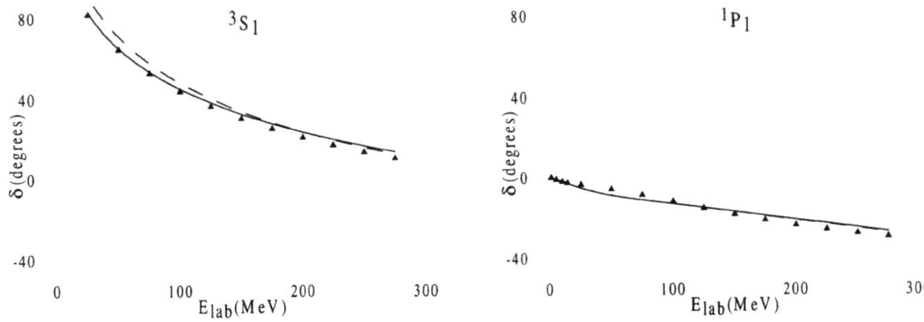

Figure 2. $^3S_1$ and $^1P_1$ NN phase shifts.

constant. In Fig. 2 we compare the phase shifts for both cases and we can see how the predictions are similar.

3.2 NN Short-Range Repulsion

The NN short-range behavior has been explained in terms of an interplay between quark antisymmetry and a particular quark dynamics. In a group theory language, a completely antisymmetric six quark state asymptotically describing two free nucleons in a relative S-wave is given by

$$\Psi_{\rm NN} = \left[\Psi_{\{6\}} + \sqrt{8}\,\Psi_{\{42\}} \right] /3 \,, \tag{5}$$

where

$$\begin{aligned} \Psi_{\{6\}} &= [2^3]_{\rm C}\,[6]_{\rm O}\,[33]_{\rm ST}\,, \\ \Psi_{\{42\}} &= \frac{1}{\sqrt{2}}\,[2^3]_{\rm C}\,[42]_{\rm O}\,([33]_{\rm ST}\;-\;[51]_{\rm ST})\,, \end{aligned} \tag{6}$$

the subindex C, O or ST indicating the color, orbital or spin-isospin representation, respectively. When the distance between the two clusters goes to zero the $\{42\}$ configuration increases in energy. According to our harmonic oscillator wave function ansatz for the quarks, the excitation energy is given by

$$\Delta_{\rm ho} \equiv \Delta \left[E_{\{42\}} - E_{\{6\}} \right]_{\rm ho} = 2\,\hbar\omega = \frac{2\,\hbar^2}{3\,b^2\,m_q}\,. \tag{7}$$

Let us now consider other interactions. Assume first that quarks interact via the OGE used in ref. [4] ($\alpha_s = 1.39$, $m_q = 300$ MeV and $b = 0.6$ fm). We can estimate the energy contribution of the OGE calculating the matrix element $< \{\} \mid \sum_{i<j} V_{\rm OGE}(\boldsymbol{r}_{ij}) \mid \{\} >$, where $\{\}$ stands for the $\Psi_{\{42\}}$ and $\Psi_{\{6\}}$ configurations of Eq. (6). As we are interested in the NN interaction energy we should subtract twice the corresponding nucleon self-energy given by $< {\rm N} \mid \sum_{i<j} V_{\rm OGE}(\boldsymbol{r}_{ij}) \mid {\rm N} >$. For (S,T)=(1,0), the matrix element of the color magnetic potential for the $\{6\}$ configuration is 48.83 MeV, while for the $\{42\}$ configuration it is -235.63 MeV. Since in the energy difference between the $\{42\}$ and $\{6\}$ configurations the subtracted self-energies cancel, we have $\Delta_{\rm OGE} \equiv \Delta\left[E_{\{42\}} - E_{\{6\}}\right]_{\rm OGE} = -284.46$ MeV. This difference compensates the harmonic oscillator energy difference, $\Delta_{\rm ho} = 240.36$ MeV, the $[42]_{\rm O}$ orbital symmetry becoming the lowest in energy. As a consequence of this configuration mixing the S-wave NN relative motion wave function shows an energy independent node, which translates into a hard-core behavior in the phase shift.

In the case of the chiral constituent quark model the value of α_s, which drives the OGE energy gap between the $\{42\}$ and $\{6\}$ configurations, is significantly reduced due to the pseudoscalar contribution to the $\Delta - N$ mass difference and correspondingly the mixing effect. If we recalculate the contribution of the OGE with the parameters of Table 1, we observe again a decrease of the energy difference between the $\{42\}$ and the $\{6\}$ configurations $\Delta_{\rm OGE} = -140.99$ MeV, but it is much smaller than the harmonic oscillator energy difference $\Delta_{\rm ho} = 309.10$ MeV and it is not sufficient to give rise to a significant configuration mixing. It is then convenient to deepen the understanding of the repulsion mechanism.

A similar kind of analysis can be carried out for the OPE and OSE potentials. For (S,T)=(1,0), differently than in the OGE case, the OPE has the same sign for both configurations. Regarding the energy difference one gets $\Delta_{\rm OPE} = -68.62$ MeV. Concerning the OSE one has $\Delta_{\rm OSE} = 133.05$ MeV. Therefore, in the chiral constituent quark model three effects conspire against the symmetry mixing. First, the small value of α_s. Second, the cancellation between the OPE contributions to the $\{42\}$ and $\{6\}$ configurations, and finally the contrary effect of the OSE potential. Putting all the contributions together one obtains, $\Delta\left[E_{\{42\}} - E_{\{6\}}\right]_{\rm OGE+OPE+OSE} = -76.56$ MeV, which is much smaller than the harmonic oscillator energy difference $\Delta_{\rm ho} = 309.10$ MeV. Then, one must conclude that in the chiral constituent quark model there is not enough configuration mixing to account for the hard-core of the NN interaction as a node produced by the $[42]_{\rm O}$ orbital symmetry.

To look for the origin of this repulsive character of the interaction one should go beyond the energy difference between the symmetries and calculate the specific contribution of the interaction for each symmetry. In order to see the repulsive or attractive character of every term of the potential in both configurations one has to subtract twice the corresponding nucleon self-energy, given by $\mathcal{E}_{\rm OGE} = -72.63$ MeV, $\mathcal{E}_{\rm OPE} = -311.40$ MeV, and $\mathcal{E}_{\rm OSE} = -66.90$ MeV. One then obtains, $\rm E_{\{6\}}^{OGE+OPE+OSE} = 455.72$ MeV and $\rm E_{\{42\}}^{OGE+OPE+OSE} = 379.16$ MeV. It is then clear that the chiral potential produces strong repulsion in both configurations, $\{42\}$ and $\{6\}$. The OPE hardly contributes to the mixing between them, however in both symmetries it produces strong repulsion. The OGE reduces the energy difference whereas the OSE increases this energy difference and produces an additional attraction in both configurations. The whole effect is an energy difference of about the same value obtained with the harmonic oscillator model but with an additional repulsion in both symmetries originated mainly by quark antisymmetry on the OPE. Therefore, in this type of models the NN S-wave behavior should be attributed to the strong repulsion generated by the OPE in the $[6]_O$ orbital configuration.

3.3 Universality

A main feature of the quark treatment is its universality in the sense that all the baryon-baryon interactions are treated on an equal footing. Moreover, once the model parameters are fixed from NN data there are no free parameters for any other case. This allows a microscopic understanding and connection of the different baryon-baryon interactions that is beyond the scope of any analysis based only on effective hadronic degrees of freedom. We will illustrate our discussion by means of the recently calculated $NN \to NN^*(1440)$ transition potential [12]. In particular we will determine the $\pi NN^*(1440)$ and $\sigma NN^*(1440)$ coupling constants.

For this purpose, let us realize that asymptotically ($R \geq 4$ fm) the OSE and OPE potentials have at the baryon level the same spin-isospin structure as at the quark level. Hence we can parametrize the asymptotic central interactions as

$$\begin{aligned} V^{OPE}_{NN\to NN^*(1440)}(R) &= \frac{1}{3}\frac{g_{\pi NN}}{\sqrt{4\pi}}\frac{g_{\pi NN^*(1440)}}{\sqrt{4\pi}}\frac{m_\pi}{2M_N}\frac{m_\pi}{2(2M_r)}\frac{\Lambda^2}{\Lambda^2-m_\pi^2} \\ &\quad \times[(\boldsymbol{\sigma}_N.\boldsymbol{\sigma}_N)(\boldsymbol{\tau}_N.\boldsymbol{\tau}_N)]\frac{e^{-m_\pi R}}{R}, \end{aligned} \tag{8}$$

and

$$V^{OSE}_{NN\to NN^*(1440)}(R) = -\frac{g_{\sigma NN}}{\sqrt{4\pi}}\frac{g_{\sigma NN^*(1440)}}{\sqrt{4\pi}}\frac{\Lambda^2}{\Lambda^2-m_\sigma^2}\frac{e^{-m_\sigma R}}{R}, \tag{9}$$

where g_i stands for the coupling constants at the baryon level and M_r is the reduced mass of the $NN^*(1440)$ system.

By comparing these baryonic potentials with the asymptotic behavior of the OPE and OSE obtained from the quark calculation we can extract the $\pi NN^*(1440)$ and $\sigma NN^*(1440)$ coupling constants in terms of the elementary πqq coupling constant and the one-baryon model dependent structure. The sign obtained for the meson-$NN^*(1440)$ coupling constants is determined by the arbitrarily chosen relative sign between the N and $N^*(1440)$ wave functions. Only the ratios between the $\pi NN^*(1440)$ and $\sigma NN^*(1440)$ would be free of this uncertainty.

To get $g_{\pi NN^*(1440)}/\sqrt{4\pi}$ we turn to our numerical results for the $^1S_0$ OPE potential, and fit its asymptotic behavior to Eq. (8). We obtain

$$\frac{g_{\pi NN}}{\sqrt{4\pi}}\frac{g_{\pi NN^*(1440)}}{\sqrt{4\pi}}\frac{\Lambda^2}{\Lambda^2-m_\pi^2} = -3.73, \tag{10}$$

i.e., $g_{\pi NN^*(1440)}/\sqrt{4\pi} = -0.94$. As explained above only the absolute value of this coupling constant is well defined. The coupling scheme dependence can be explicitly eliminated if we compare $g_{\pi NN^*(1440)}$ with $g_{\pi NN}$ extracted from the $NN \to NN$ potential within the same quark model approximation. We get

$$\left| g_{\pi NN^*(1440)} / g_{\pi NN} \right| = 0.25 \,. \tag{11}$$

By proceeding in the same way for the scalar potential we can write

$$\left| g_{\sigma NN^*(1440)} / g_{\sigma NN} \right| = 0.47 \,. \tag{12}$$

The ratio given in Eq. (11) is similar to that obtained in ref. [13]. Nonetheless one can find values for $f_{\pi NN^*(1440)}$ ranging between 0.27–0.47 coming from different experimental analyses. Regarding the ratio obtained in Eq. (12), our result agrees quite well with the only available experimental result, obtained in ref. [14] from the fit of the cross section of the isoscalar Roper excitation in $p(\alpha, \alpha')$ in the 10–15 GeV region, where a value of 0.48 is given. Finally, we can give a very definitive prediction of the magnitude and sign of the ratio of the two ratios,

$$g_{\pi NN^*(1440)} / g_{\pi NN} = 0.53 \; g_{\sigma NN^*(1440)} / g_{\sigma NN} \,. \tag{13}$$

3.4 Pauli Blocking

The Pauli principle plays an essential role to explain the atomic and nuclear structure. For example, the strong repulsion existing in the $\alpha\alpha$ system originates from the nucleonic substructure of the α particle and not from a particular dynamics. The hard-core appears because the saturation of nuclear spin-isospin degrees of freedom allows only four nucleons in the orbital ground state. We shall refer to this saturation phenomenon indicated by the existence of forbidden states at the underlying structure level as *Pauli blocking*. At the level of subnuclear physics the elementary constituents, the quarks, are fermions so a quark Pauli principle is active.

For identical baryons the symmetrization postulate at the baryonic level gives rise to selection rules on the state of the compound system. At the quark level, assuming SU(2) flavor symmetry, this selection rule appears as a particular case of the action of the quark antisymmetrizer, $\mathcal{A}$, on a two cluster state from six identical quarks. Taking into account that any two-baryon state (LST) can be decomposed in a symmetric plus an antisymmetric part under the exchange of the baryon quantum numbers, one can write for a definite symmetry (specified by f)

$$\begin{aligned} \Psi^{LST}_{B_1 B_2}(\boldsymbol{R}) &= \frac{\mathcal{A}}{\sqrt{1+\delta_{B_1 B_2}}} \sqrt{\frac{1}{2}} \Bigg\{ \left[\Phi_{B_1}\left(123; -\frac{\boldsymbol{R}}{2}\right) \Phi_{B_2}\left(456; \frac{\boldsymbol{R}}{2}\right) \right]_{LST} + \\ &+ (-1)^f \left[\Phi_{B_2}\left(123; -\frac{\boldsymbol{R}}{2}\right) \Phi_{B_1}\left(456; \frac{\boldsymbol{R}}{2}\right) \right]_{LST} \Bigg\} , \end{aligned} \tag{14}$$

where B_1 and B_2 denote the baryons, and $-\frac{\boldsymbol{R}}{2}$, $\frac{\boldsymbol{R}}{2}$ are the positions of the clusters. The antisymmetrizer can be decomposed as,

$$\mathcal{A} = \frac{1}{2}\,(1 - 9P_{36})(1 - \mathcal{P}) \,, \tag{15}$$

where P_{36} exchanges quarks 3 and 6, and $\mathcal{P}$ exchanges the two clusters. The $1-\mathcal{P}$ factor implies $L+S_1+S_2-S+T_1+T_2-T+f$ = odd. For non-identical baryons this relation indicates the symmetry associated to a given set of values of (LST). For identical baryons, $B_1 = B_2$, f = even (in order not to have a vanishing wave function) and $L+S+T$ = odd, reproducing the well-known selection rule. Certainly the effect of the quark substructure goes beyond the $(1-\mathcal{P})$ factor in the antisymmetrizer, being also included through the terms containing P_{36}. Let us study the effects of these quark exchanges in a system of two non-identical baryons, in particular the $N\Delta$ system. Assuming the three-quark cluster wave function at position $\boldsymbol{S}$ given by

$$\Phi_B(123;\boldsymbol{S}) = \prod_{i=1}^{3} \eta_{0s}(\boldsymbol{r}_i - \boldsymbol{S}) \otimes [3]_{\mathrm{ST}} \otimes [1]_{\mathrm{C}}, \tag{16}$$

where $\eta_{0s}(\boldsymbol{x})$ is a gaussian with parameter b, and $[3]_{\mathrm{ST}}$ ($[1]_{\mathrm{C}}$) stands for the spin-isospin (color) representation, the norm of the projected $\Psi_{N\Delta}^{\mathrm{LST}}$ state reads

$$N_{N\Delta}^{\mathrm{LST}}(R) = N_L^{\mathrm{di}}(R) - C(S,T,f)\, N_L^{\mathrm{ex}}(R), \tag{17}$$

where $N_L^{\mathrm{di}}(R)$ $[N_L^{\mathrm{ex}}(R)]$ refers to the direct [exchange] radial contribution coming from the $1[P_{36}]$ term in the antisymmetrizer. The spin-isospin coefficient $C(S,T,f)$ is given by

$$\begin{aligned} C(S,T,f) \;=\; & 3\left\{ {}_{\mathrm{ST}}\langle N(123)\,\Delta(456) \mid P_{36}^{ST} \mid N(123)\Delta(456)\rangle_{\mathrm{ST}} + \right. \\ & \left. (-1)^f\; {}_{\mathrm{ST}}\langle N(123)\Delta(456) \mid P_{36}^{ST} \mid \Delta(123)N(456)\rangle_{\mathrm{ST}} \right\}. \end{aligned} \tag{18}$$

Its value for the $N\Delta$ S-waves is given in Table 2 (the NN case is also shown for comparison). In the $R\to 0$ limit and in the dominant order, Eq. (17) transforms

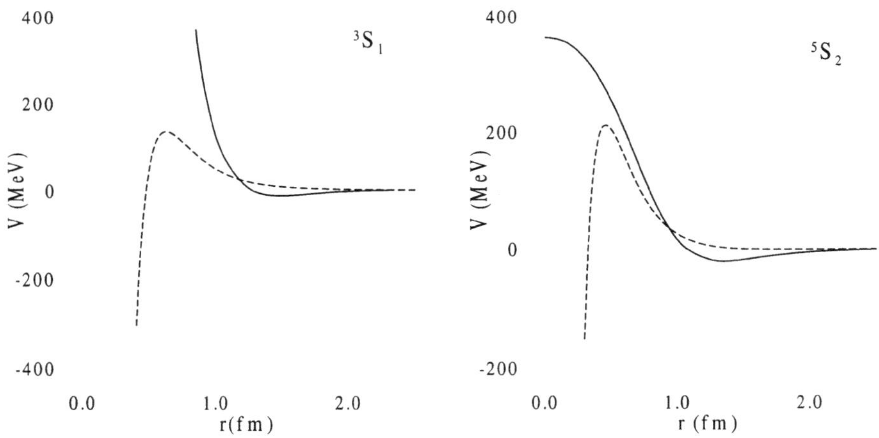

Figure 3. $^3S_1(T=1)$ and $^5S_2(T=1)$ $N\Delta$ potentials.

into

$$N_{N\Delta}^{\mathrm{LST}}(R) \to_{R\to 0} 4\pi \left[1 - \frac{3R^2}{4b^2}\right] \frac{1}{1\cdot 3\cdots(2L+1)} \left[\frac{R^2}{4b^2}\right]^L \times$$
$$\left\{ \left[3^L - C(S,T,f)\right] + \frac{1}{2(2L+3)} \left[\frac{R^2}{4b^2}\right]^2 \left[3^{L+2} - C(S,T,f)\right] + \cdots \right\} . \quad (19)$$

Then, for $3^L = C(S,T,f)$ the overlapping of the two-cluster wave function behaves as R^{2L+4} instead of the centrifugal barrier behavior R^{2L}, indicating that quark Pauli blocking occurs (in other words, a node appears in the relative wave function). This turns out to be the case (see Table 2) for the $N\Delta$ partial waves ${}^3S_1(T=1)$ and ${}^5S_2(T=2)$ giving rise to forbidden states (the $[6]_O$

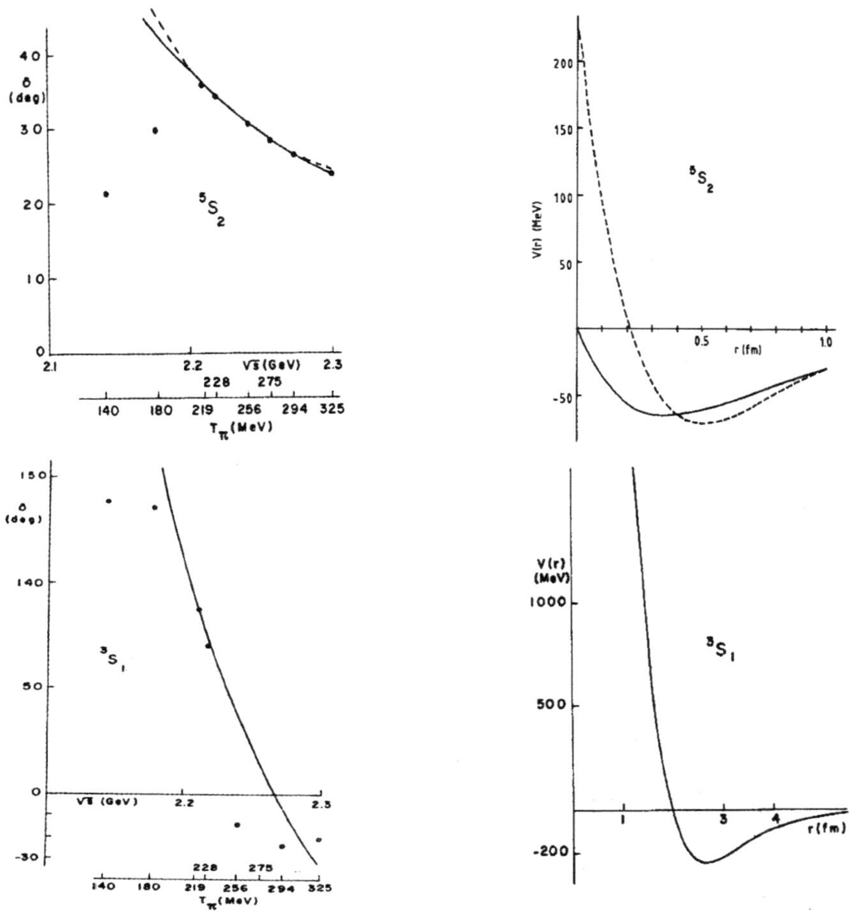

Figure 4. ${}^3S_1(T=1)$ and ${}^5S_2(T=1)$ $N\Delta$ phase shifts and separable potentials.

Table 2. $C(S,T,f)$ coefficients for the NN and $N\Delta$ S-waves.

	NN		$N\Delta$			
	$^1S_0$ $(T=1)$	$^3S_1$ $(T=0)$	$^3S_1$ $(T=1)$	$^3S_1$ $(T=2)$	$^5S_2$ $(T=1)$	$^5S_2$ $(T=2)$
$C(S,T,f)$	$-1/9$	$-1/9$	1	1/9	1/9	1

orbital configuration for $L = 0$). If we now calculate the $N\Delta$ potential, we observe (solid line) the important difference between the potential in a Pauli blocked and a non Pauli blocked channel (see Fig. 3). At the same time we have represented by the dashed line a $N\Delta$ potential obtained from a direct scaling of the NN interaction [15] and we see how the difference between the two channels is just a spin-isospin factor. As shown in Fig. 4, a separable $N\Delta$ potential reproducing the phase shifts obtained from the experimental πd elastic cross section presents the strong repulsion predicted by the quark model in the $^3S_1(T=1)$ partial wave [16].

A similar analysis for the NN system does not show any forbidden configuration, see Table 2. As we have seen, the NN short-range repulsion in the S-waves involves dynamical effects.

3.5 Connection to Other Systems: The Baryon Spectrum

The chiral constituent quark model allows also to study the baryon spectrum in a completely parameter-free way. We have calculated the nonstrange baryon spectrum within a Faddeev approach. We included all the configurations (ℓ, λ, s, t) (ℓ is the orbital angular momentum of a pair, λ is the orbital angular momentum between the pair and the third particle, while s and t are the spin and isospin of the pair) with ℓ and λ up to 5.

For this purpose, the delta function of the OGE interaction has to be regularized. By choosing an exponential regularization

$$\delta(\boldsymbol{r}_{ij}) \Rightarrow \frac{1}{4\pi r_0^2} \frac{e^{-r_{ij}/r_0}}{r_{ij}} , \tag{20}$$

with $r_0 = 0.25$ fm (a typical value for spectroscopic models) we obtain the N and Δ spectra shown in Fig. 5. The predicted spectrum is reasonable except for the fact that all the negative parity states are around 100 MeV below the experimental data. The relative position of the $N(1440)$ and the $N(1535)$ is not very much affected by the modification of α_s (see Fig. 3 of ref. [17]), indicating that the correct level ordering is not connected with the strength of OGE.

A possible mechanism responsible for the inversion of these states can be obtained through the pseudoscalar interaction. To illustrate this point we have calculated the energy of the $N(1440)$ and $N(1535)$ increasing the contribution of the pseudoscalar interaction by letting the cutoff parameter Λ_π of the OPE increase. The results are shown in Fig. 6. As can be seen, the inversion of the

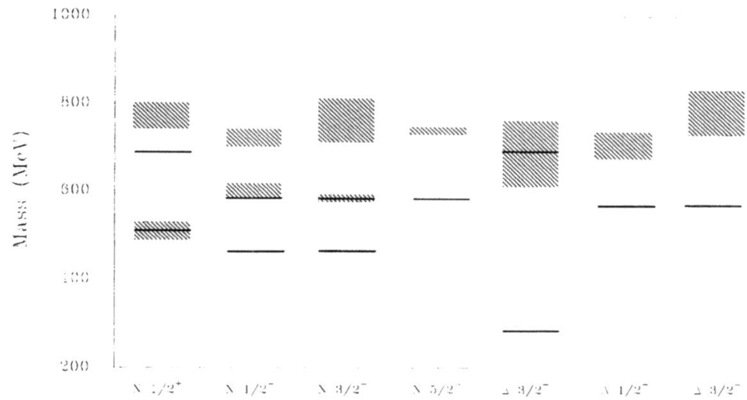

Figure 5. Relative energy nucleon and Δ spectra.

ordering between the positive and negative parity states can be achieved if Λ_π becomes sufficiently large (around 7 fm^{-1} for the set of parameters of Table 2). Such a model would be incompatible with the understanding of the basic features of the two-nucleon system [18].

4 Summary

In this work we have emphasized several physical aspects of the NN interaction that support the results obtained within the chiral constituent quark model. In particular, medium range attraction is satisfactorily explained and the NN short-range repulsion is understood through antisymmetry effects on the OPE interaction. Actually, antisymmetry effects at the level of Goldstone bosons appear as a basic tool to understand the NN phase shifts. Extension of the model to the general baryon-baryon case is straightforward and parameter-free. From it, non-strange meson-baryon-baryon coupling constants can be cal-

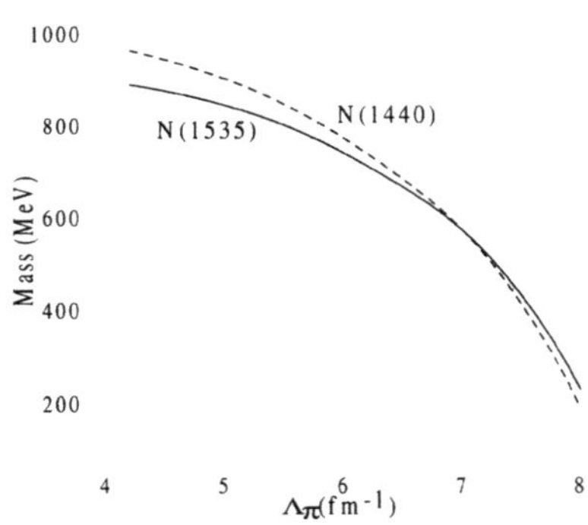

Figure 6. $N(1440)$ and $N(1535)$ masses as a function of Λ_π.

culated. Pauli blocking effects manifest themselves in different partial waves of several systems and find support in indirect experimental data. Such strong repulsion cannot be understood in a meson-exchange model.

Finally, the chiral constituent quark model is able to generate a quite reasonable description of the baryon spectrum with a set of parameters that allows to understand the NN phenomenology. Therefore, the validity of a model where the OGE is combined with Goldstone-boson exchanges is out of question, and it presents advantages with respect to models based only on Goldstone-boson exchanges, as has been recently emphasized in several works [18, 19].

Acknowledgement. This work has been partially funded by Ministerio de Ciencia y Tecnología under Contract No. BFM2001-3563, by Junta de Castilla y León under Contract No. SA-109/01, and by the European Comission IHP program under Contract No. HPRN-CT-2000-00130.

References

1. N. Isgur, G. Karl: Phys. Rev. **D18**, 4187 (1978)
2. V.G. Neudatchin et al.: Prog. Theor. Phys. **58**, 1072 (1977)
3. H. Toki: Z. Phys. **A294**, 173 (1980)
4. M. Oka et al.: Phys. Lett. **90B**, 41 (1980); Nucl. Phys. **A402**, 477 (1983)
5. A. Faessler et al.: Phys. Lett. **112B**, 201 (1982)
6. Y. Fujiwara, K.T. Hetch: Phys. Lett. **171B**, 17 (1986)
7. K. Shimizu: Phys. Lett. **148B**, 418 (1984)
8. K. Bräuer et al.: Z. Phys. **A320**, 609 (1985) ; Nucl. Phys. **A507**, 599 (1990)
9. I.T. Obukhovsky, A.M. Kusainov: Phys. Lett. **238B**, 142 (1990)
10. F. Fernández et al.: J. Phys. **G19**, 2013 (1993)
11. A. Valcarce et al.: Phys. Rev. **C50**, 2246 (1994)
12. B. Juliá-Díaz et al.: Phys. Rev. **C66**, 024005 (2002)
13. D.O. Riska, G.E. Brown: Nucl. Phys. **A679**, 577 (2001)
14. S. Hirenzaki et al.: Phys. Lett. **B378**, 29 (1996)
15. M.T. Peña et al.: Phys. Rev. **C42**, 855 (1990)
16. E. Ferreira et al.: Phys. Rev. **C36**, 1916 (1987)
17. H. Garcilazo et al.: Phys. Rev. **C64**, 058201 (2001)
18. C. Nakamoto, H. Toki: Prog. Theor. Phys. **99**, 1001 (1998)
19. N. Isgur: Phys. Rev. **D62**, 054026 (2000)

Few-Body Systems Suppl. 15, 37–48 (2003)

Few-Body Systems

Near Threshold Meson Production in Nucleon-Nucleon Collisions

J. Haidenbauer,* V. Baru, C. Hanhart, J. Speth

Institut für Kernphysik, Forschungszentrum Jülich GmbH, D-52425 Jülich, Germany

Abstract. A model calculation for pion production in nucleon-nucleon collisions is presented. Direct production, pion re-scattering and contributions from pair diagrams are taken into account. The amplitudes for the elementary processes are based on well established microscopic models of nucleon–nucleon and pion-nucleon scattering. The comparison with experiments focuses on recently measured polarization observables. Furthermore, results of a study of the reaction $NN \to NN\eta$ are presented, performed in the same spirit and the same framework as pion production. Reasonable agreement with available near-threshold cross section data for the reactions $pp \to pp\eta$, $pn \to pn\eta$, and $pn \to d\eta$ is found.

1 Introduction

Meson production in nucleon-nucleon (NN) and nucleon-nucleus collisions is a rather attractive research field in hadron physics at intermediate energies. The revived interest in such reactions is closely connected to recent successful developments in the accelerator technology which enabled experimentalists to perform rather accurate measurements on the production of π, η, η' and other mesons in NN collisions in the near-threshold region [1, 2]. These data in combination with the theoretical analysis open the unique possibility to study the mechanism of the production of different mesons as well as to provide information on resonance properties, baryon–baryon and meson–baryon interactions, short range properties of the NN interaction etc..

The most thoroughly studied reaction is certainly pion production where a large set of accurate near-threshold data is now available for various charge channels ($pp \to pp\pi^0$, $pp \to pn\pi^+$, $pp \to d\pi^+$) [1, 2]. The theoretical interpretation of these data has so far led to rather mixed successes. This concerns in particular the reaction $pp \to pp\pi^0$ where initial investigations based on

* *E-mail address:* j.haidenbauer@fz-juelich.de

the meson-production model proposed by Koltun and Reitan [3] resulted in an under-prediction of the near-threshold cross section by a factor of about 5 [4]. This model includes contributions from direct pion production and (on-shell) πN re-scatting. Subsequently two further mechanisms have been suggested that, in principle, allow to obtain quantitative agreement with the $pp\pi^0$ production data close to threshold. In one of them the pion is produced via an intermediate virtual antinucleon-nucleon state in conjunction with the exchange of a heavy meson like the σ or ω [5, 6]. In the other approach the static πN amplitude of the Koltun-Reitan model (i. e. its on-shell value at threshold) is replaced by the off-shell πN amplitude - as it is required by the structure of the re-scattering process anyway [7, 8]. However, there is still an ongoing debate in the literature about which of the two suggested mechanisms is the physically relevant one [9, 10, 11, 12, 13].

Besides the pion it is primarily the production of η mesons that has been studied rather extensively over the past few years [1, 14, 15, 16, 17]. Among other reasons this reaction is rather interesting because η production near threshold is closely linked with the properties of the S_{11} $N^*(1535)$ resonance. Furthermore, this reaction might provide information about the strength of the ηN (S-wave) interaction. This strength plays a crucial role in the present controversy about the possible existence of η-nuclear quasi-bound states.

Also for η meson production there is already a wealth of rather accurate near-threshold data, providing a sensible testing ground for model calculations. Apart from total cross sections for the reaction $pp \to pp\eta$ [18, 19, 20], there are also some data on the total cross sections of the reactions $pn \to pn\eta$ [21] and $pn \to d\eta$ [22, 23]. Furthermore, there are data on differential cross sections and invariant-mass spectra [20, 24, 25, 26] and even analyzing powers for the reaction $pp \to pp\eta$ [27].

2 Pion Production in Nucleon-Nucleon Collisions

2.1 The Model

Let me first report on a calculation of pion production in NN collisions carried out by the Jülich group [8, 28, 29]. We treat the reaction in distorted wave Born approximation, in the standard fashion. The pion production mechanisms included in our model are shown in Fig. 1. First, we have the usual direct production (Fig. 1a) and re-scattering (Fig. 1b) contributions from NN states. In addition, contributions from pair diagrams involving heavy mesons (Fig. 1c) are also considered. For the distortions in the initial and final NN states we employ the model CCF of Ref. [30]. This potential has been derived from the full Bonn model [31] by means of the folded-diagram expansion. It is a coupled-channels (NN, $N\Delta$, $\Delta\Delta$) model that treats the nucleon and the Δ degrees of freedom on equal footing. Thus, the $NN \leftrightarrow N\Delta$ wave functions that enter in the evaluation of the pion production diagrams involving the Δ isobar (cf. Fig. 1d,e) and the NN wave functions that are used for the other diagrams in Fig. 1 are consistent solutions of the same (coupled-channels)

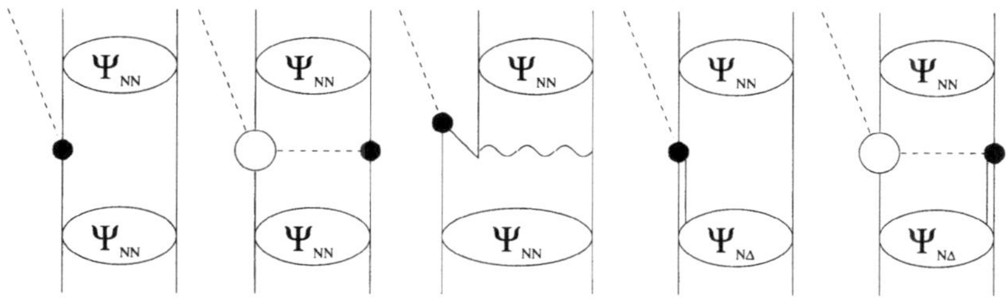

Figure 1. Pion production mechanisms from NN states. (a) direct production; (b) production via πN re-scattering; (c) production via heavy meson exchange; (d),(e) production mechanisms involving the Δ isobar. Ψ_{NN} ($\Psi_{N\Delta}$) denotes the NN ($N\Delta$) wave function in the initial and final state.

Lippmann-Schwinger-like equation. A realistic meson-theoretical model of the πN interaction developed by the Jülich group [32] is utilized for the evaluation of the re-scattering contributions. This model includes the conventional (direct and crossed) pole diagrams involving the nucleon and the Δ-isobar as well as meson exchanges in the scalar (σ) and vector (ρ) channels derived from correlated 2π-exchange (cf. Ref. [33]).

In the actual calculation all NN partial waves up to $L = 2$ and all states with relative orbital angular momentum $l \leq 2$ between the NN system and the pion are considered in the final state.

2.2 Results

Results for the total cross sections (for $pp \to pp\pi^0$, $pp \to d\pi^+$, $pp \to pn\pi^+$, and $pn \to pp\pi^-$) are shown in Fig. 2 as a function of η, where η is the maximum center-of-mass momentum of the produced pion divided by the pion mass. Evidently the predictions of the model are in good overall agreement with the data over a wide energy range. We want to emphasize that the results for the latter three channels do not involve any adjustable parameters and are therefore genuine predictions of our model. In the case of $pp \to pp\pi^0$, however, the basic model (including direct production plus πN re-scattering) yields only about 60% of the measured cross section. Here we have added contributions from the pair diagrams, Fig. 1c, and fixed their "strengths" so that we can reproduce the data (cf. Ref. [28]). The effect of those pair diagrams on the other reaction channels, $pp \to d\pi^+$ and $pp \to pn\pi^+$, is negligible so that the corresponding results remain practically unchanged [28].

As expected, the Δ excitation plays a crucial role for pion production at energies corresponding to η values around or larger than 1. In this energy range it provides the bulk contribution to the total cross section. However, we find

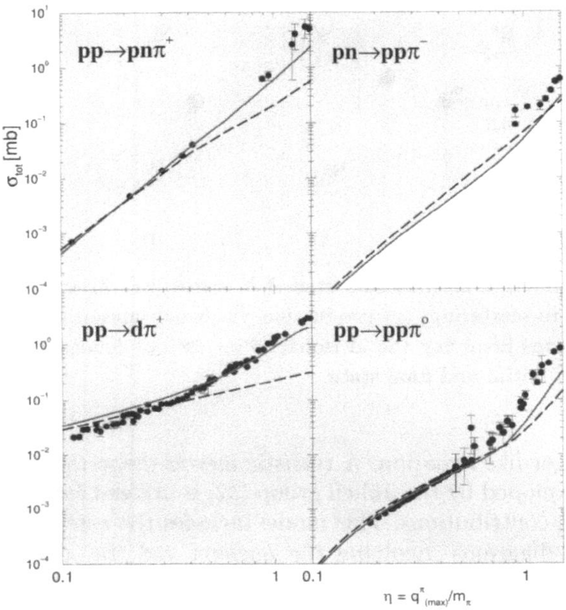

Figure 2. Results for different $NN \to NN\pi$ reaction channels as a function of η, the maximum center-of-mass momentum of the produced pion divided by the pion mass. Solid line: full model; dashed line: contributions of the Δ isobar are omitted.

significant effects of the Δ isobar on the pion production reaction over the whole considered energy range as can be seen from Fig. 2. Specifically differential and spin dependent observables are strongly influenced by contributions involving the Δ - even at energies very close to threshold [34]. Indeed, with Δ degrees of freedom included a nice quantitative agreement of our model predictions with existing data on angular distributions and analyzing powers is found [28, 34].

Very recently high precision data on spin correlation coefficients have become available [35, 36, 37, 38]. These observables can be extracted from measurements where the beam as well as the target are polarized. Predictions for the spin correlation coefficient combinations $A_\Sigma = A_{xx} + A_{yy}$, $A_\Delta = A_{xx} - A_{yy}$ and A_{zz} are shown in Fig. 3 and compared with experimental results (cf. Ref. [35] for a definition of these observables). The figure contains results for $pp \to pp\pi^0$ (left), $pp \to d\pi^+$ (middle), and $pp \to pn\pi^+$ (right) at $T_{lab} = 375$ MeV as a function of the pion angle. Evidently, the data involving charged pions are nicely reproduced by our model. However, there is a serious discrepancy with regard to $pp \to pp\pi^0$ (cf. left panel of Fig. 3). The same situation is

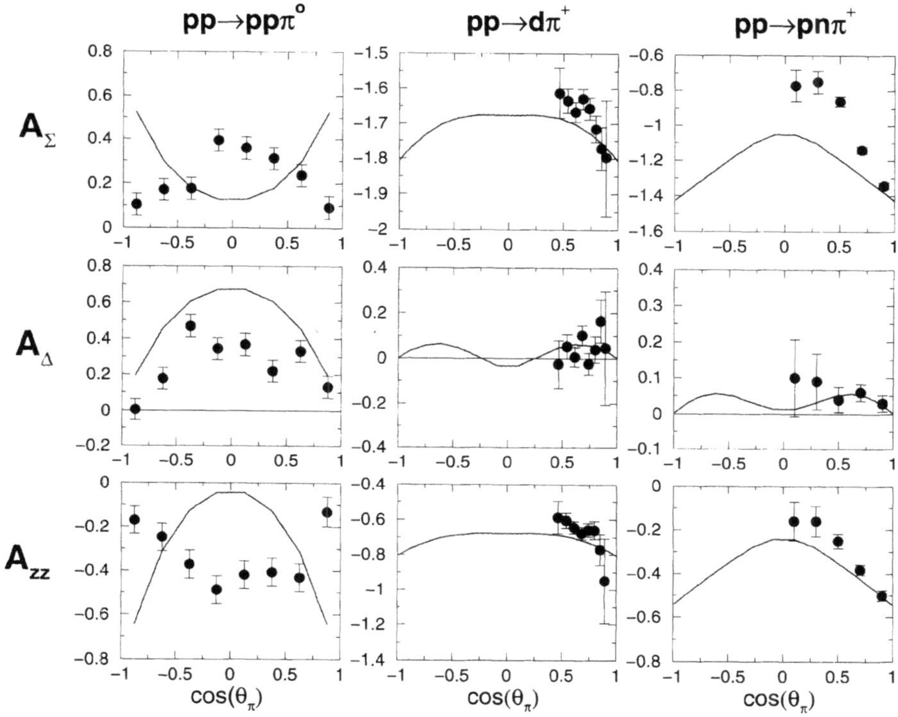

Figure 3. Spin correlation coefficients for $NN \to NN\pi$ at $T_{Lab} = 375$ MeV as a function of the pion angle. The solid line is the full model. The data are taken from Ref. [37] ($pp \to pp\pi^0$), Ref. [36] ($pp \to d\pi^+$) and Ref. [38] ($pp \to pn\pi^+$).

observed for the angular dependence of the spin correlation parameters at other beam energies [36, 37, 38] and also for polar integrals of these observables [29]. Thus, obviously some important contribution to the dynamics of the reaction $pp \to pp\pi^0$ is still missing in our model. The origin of this shortcoming and possibilities for improvements are presently under investigation [39].

3 η Meson Production in Nucleon-Nucleon Collisions

3.1 The Model

Let me now come to our investigation on η production in NN collisions [40]. This study was carried out in the same spirit and in the same framework as the one on pion production reported above. Specifically, again we utilize a microscopic model of meson-nucleon (MN) scattering for the description of the production process and again we take into account consistently the interaction

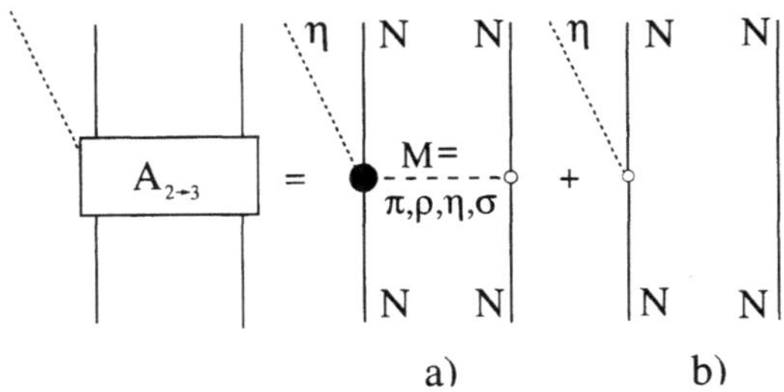

Figure 4. Production mechanisms for the reaction $NN \to NN\eta$ taken into account in our model: (a) η production via $MN \to \eta N$ re-scattering; (b) direct η production.

between the nucleons in the final and in the initial states. The goal is a combined theoretical analysis of all measured channels of the reaction $NN \to NN\eta$, namely $pp \to pp\eta$, $pn \to pn\eta$ and $pn \to d\eta$.

In complete analogy to $NN \to NN\pi$, η-meson production is likewise treated in distorted wave Born approximation. This means that the transition amplitude from the initial to the final state, $M_{2\to3}$, is given by the expression

$$M_{2\to3} = \langle \chi_f \,|\, (1 + T^{FSI}_{NN} G_0) A_{2\to3} (1 + G_0 T^{ISI}_{NN}) \,|\chi_i\rangle. \tag{1}$$

where $A_{2\to3}$ is the (elementary) production operator (cf. Fig. 4) and T^{ISI}_{NN} and T^{FSI}_{NN} are the NN reaction amplitudes in the initial and final states. (Note that there is evidence that the ηN final state interaction plays also a role, at least for excess energies up to 10-15 MeV [18, 22, 41]. Those effects are neglected in the present study.) The production amplitude $A_{2\to3}$ consists of re-scattering diagrams with π, ρ, η, and σ meson exchanges (Fig. 4a) plus the direct η emission (Fig. 4b). The elementary re-scattering amplitudes $MN \to \eta N$ entering in $A_{2\to3}$ (Fig. 4a) are derived from a microscopic interaction model. Specifically, they are taken from a coupled-channels model for πN scattering that has been developed recently by the Jülich group [42]. This model is an extension of the πN interaction [32], that has been used for pion production (cf. Sect. 2), to higher energies. The interactions in and between the various channels (πN, ηN, ρN, σN, and $\pi\Delta$) are derived in the meson-exchange picture starting out from effective chiral Lagrangians. The model includes the $N^*(1535)$ resonance as essential contribution but in addition also various (t–channel) meson and (u–channel) baryon exchange diagrams. The parameters of the model are fixed by requiring a simultaneous description of the πN phase shifts and inelasticities as well as of the transition cross sections for the reactions $\pi N \to \eta N$ and $\pi N \to \rho N$ over an energy range that extends well beyond the ηN threshold. The reaction amplitudes are obtained from solving a coupled-channels relativistic scattering equation of Lippmann-Schwinger type. Clearly, in such a model not only the $\pi N \to \eta N$ amplitude is determined by empirical data, but also the transition amplitudes involving heavier mesons are to a large extend constrained by the phase shifts and inelasticity parameters of πN scattering. This is also true for the relative phases between the various amplitudes.

Vertex parameters appearing in the MNN vertices of the diagrams in Fig. 4 are taken over from the full Bonn NN model [31] (π, ρ, σ) or from the πN model [42] (η), cf. [40]. Thus, we do not have any free and/or adjustable parameters in the production amplitude $A_{2\to 3}$. Accordingly, our results for the reaction $NN \to NN\eta$ are genuine model predictions.

Effects from the NN FSI as well as ISI are fully taken into account. Here we employ the same NN model (CCF [30]) that has been already used in our study of pion production. This model describes the relevant NN phase shifts reasonably well up to energies around the η production threshold, i.e. up to $T_{lab} \approx 1300$ MeV [40, 30].

We focus on the description of η-meson production in the near-threshold region, i.e. for excess energies Q up to around 50 MeV. Therefore, we restrict our investigation to S-wave contributions.

3.2 Results

Results for the reactions $pp \to pp\eta$, $pn \to pn\eta$ and $pn \to d\eta$ are presented in Fig. 5 and compared with empirical information from the Uppsala/Celsius [18, 21, 22, 23] and Jülich/COSY [19, 20] accelerator facilities. The dashed curves correspond to the calculation based on the original Jülich MN model [42] for the elementary η-production amplitude and the CCF NN model [30] for the ISI and FSI.

Evidently, the calculation based on the Jülich MN model [42] yields a qualitative overall description of the experimental data. This has to be certainly considered as success because there are no adjustable parameters in our calculation of $NN \to NN\eta$. As one can observe from Fig. 5, for the $pp\eta$ case we overestimate the cross section by approximately 30% whereas for $pn \to pn\eta$ the model calculation underestimates the experimental data by about 50%. The situation for $pn \to d\eta$ is very similar to $pn \to pn\eta$ since both processes are governed by the $I = 0$ isospin channel. As is known from earlier investigations [16, 17] for the latter reaction the contribution of the $I = 1$ channel is much smaller than the one for $I = 0$.

It is interesting to investigate the sensitivity of our results to the MN amplitude and, in particular, to see whether the model predictions for $NN \to NN\eta$ can be improved by modifications in the MN model. Here specifically the $\rho N \to \eta N$ transition amplitude is of interest because (i) in the original model [42] no direct $\rho N \to \eta N$ transition potential at all was included and (ii) this particular transition amplitude plays a rather crucial role in the analysis of $NN \to NN\eta$ by Fäldt and Wilkin [16]. For this purpose we created a variant of the MN model where we explicitly include a coupling of the ρN channel to the ηN channel via the S_{11} $N^*(1535)$ resonance, cf. Ref. [40] for details. Varying the bare coupling constants ($g_{\rho NN^*}$, etc.) we explored the variation in the amplitude for $\rho N \to \eta N$ while staying as close as possible to the experimental phase shift and inelasticity of the S_{11} πN partial wave

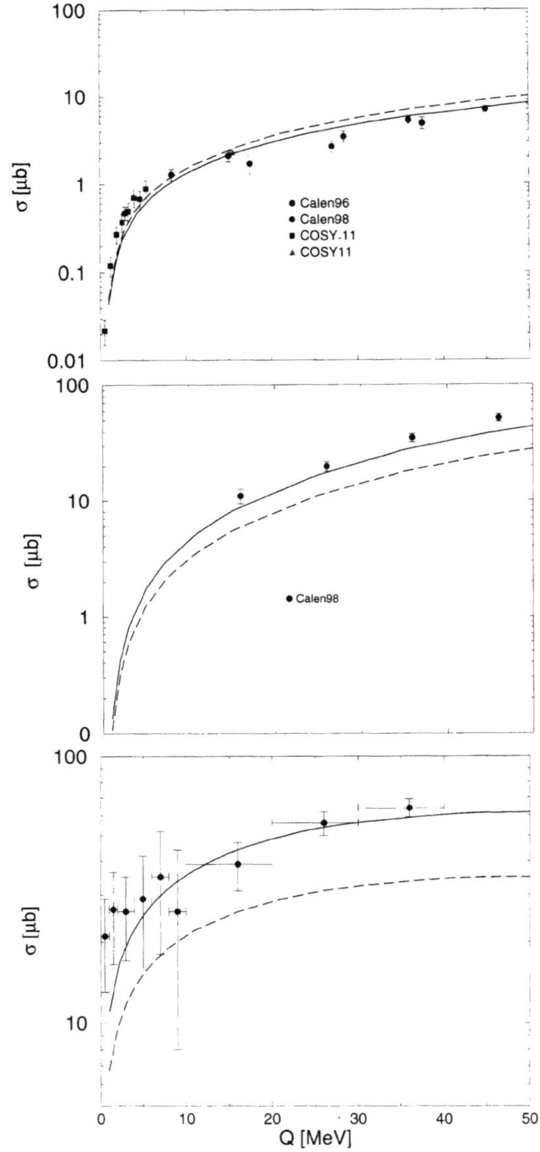

Figure 5. Total cross sections of the reactions $pp \to pp\eta$ (top), $pn \to pn\eta$ (middle) and $pn \to d\eta$ (bottom) as a function of the excess energy Q. The dashed lines represent the results of our calculation with the original MN model of Ref. [42], whereas the solid lines are based on the extended MN model described in the text. Data are taken from Refs. [18, 19, 20, 21, 22, 23].

and the $\pi N \to \eta N$ transition cross section produced by the original model. Thereby it turned out that there is not much room for variations within our MN model. Obviously, the requirement of reproducing the πN phase shifts as well as the $\pi N \to \eta N$ transition cross section constrains also other transition amplitudes of the various coupled channels to a large extent. In particular, the magnitude of the amplitudes is basically fixed. Only the relative phase between the T-matrices could be changed within certain limits. For the original MN model the amplitudes of the π- and ρ-exchange contributions to η production are almost orthogonal to each other. Thus, there is practically no interference between these contributions. The additional diagrams which we introduced into the MN model allow a slight modification of the orientation of these amplitudes in the complex plane and, as a consequence, now interference effects do occur. Specifically, it is possible to generate a destructive interference in the isotriplet channel ($I = 1$) and a constructive interference in the $I = 0$ case. This leads to a slight reduction of the cross section in the reaction $pp \to pp\eta$ and to a significant enhancement for $pn \to pn\eta$, cf. the solid lines in Figs. 5, bringing now the results quite close to the experiment. Also the prediction for the reaction $pn \to d\eta$ is now in good agreement with the experimental data.

Let us analyze the contributions of the individual exchange diagrams. We do this for the calculation based on the extended MN model. Corresponding results are presented in Fig. 6 for $pp \to pp\eta$ (top) as well as for $pn \to pn\eta$ ($I = 0$) (bottom). Evidently, in our model π exchange plays the dominant role in both reaction channels. All other meson exchanges lead to contributions that are significantly smaller. Nevertheless, some of the other contributions, especially those of the η and ρ exchanges and the direct term are still important due to interference with the dominant π exchange amplitude. For example, in the reaction $pp \to pp\eta$ the η exchange and the direct term add up constructively with the dominant π exchange piece. Those three processes together yield a cross section that would overshoot the data. However, the destructive interference from the ρ and (to some extend also) the σ mesons reduce the results slightly so that they finally agree roughly with the data.

The cross section for $pn \to pn\eta$ is primarily determined by the isovector particles (π and ρ) in the $I = 0$ channel, because their contributions are weighted by the large isospin coefficient (9 versus 1 for $I = 1$). As mentioned already, also in this reaction channel the dominant role belongs to the pion exchange, though now the contribution from ρ exchange is much more important as in the $I = 1$ case (cf. Fig. 6). The individual contributions from the other production mechanisms are again small. Like in case of $pp \to pp\eta$ interferences play a role once we add all contributions coherently. Specifically, the η- and the ρ-meson exchange exhibit opposite features as in the $pp \to pp\eta$ case, i.e. the η is destructive while the ρ is constructive. The direct term acts again constructively and its contribution nearly cancels with the one resulting from η exchange. Thus, it is primarily the constructive interference of the π- and the ρ-meson exchange contributions that is responsible for bringing our predictions close to the experimental data.

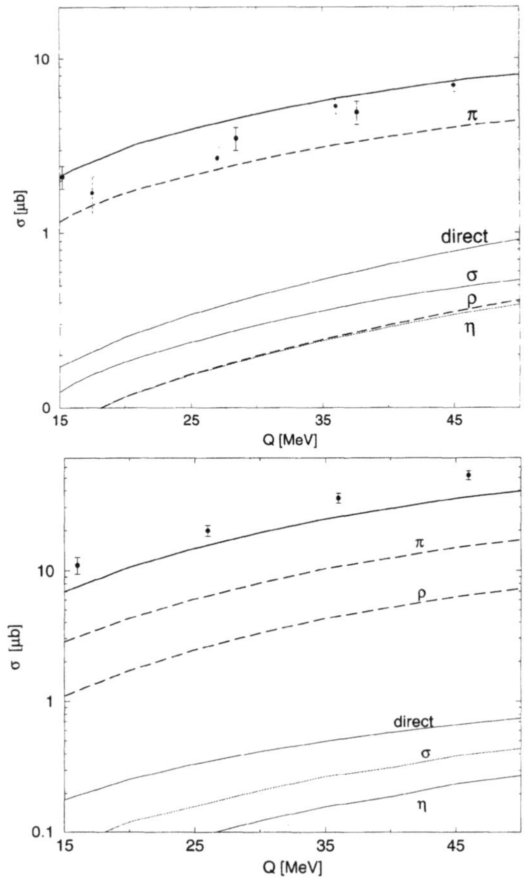

Figure 6. Contributions of the individual meson exchanges to the reactions $pp \to pp\eta$ (top) and $pn \to pn\eta$ ($I = 0$) (bottom). The presented results are based on the extended MN model. The solid line is our full result, i.e. when all contributions are summed up coherently.

4 Summary

We presented a microscopic calculation of near-threshold production of the π as well as η mesons in nucleon-nucleon collisions, performed in an unified framework. For both reactions the elementary production mechanism is assumed to consist of direct emission and of production via elementary re-scattering processes ($\pi N \to \pi N$ or $MN \to \eta N$ with $M = \pi$, ρ, η, and σ). The amplitudes for those re-scattering processes are taken from a (coupled-channels) meson-exchange model of the πN system developed by the Jülich group. Furthermore, in both production reactions effects of the NN interaction in the final as well as in the initial state are taken into account microscopically.

With regard to pion production a good quantitative agreement with practically all existing data (total cross sections, angular distributions, analyzing powers, spin correlation parameters) from the pion production threshold up to the Delta resonance region is achieved. The only significant discrepancies with experiments occur in the spin correlation parameters of the reaction $pp \to pp\pi^0$. The origin of this shortcoming and possibilities for improvements are presently under investigation.

In case of η production reasonable agreement with available data on total cross section is found for the reactions $pp \to pp\eta$, $pn \to pn\eta$, and $pn \to d\eta$ within the considered energy range of 50 MeV. Here higher partial waves need be included in the calculation in order to facilitate an extension of the model to higher excess energies and also to make a comparison with measured angular distributions, polarizations and invariant mass spectra meaningful.

Acknowledgement. We want to thank A.M. Gasparyan, O. Krehl, and A.E. Kudryavtsev for collaboration on the topics presented here.

References

1. H. Machner, J. Haidenbauer: J. Phys. **G25** (1999) 231
2. P. Moskal et al.: Prog. Part. Nucl. Phys **49** (2002) 1
3. D. Koltun, A. Reitan: Phys. Rev. **141** (1966) 1413
4. G. Miller, P. Sauer: Phys. Rev. **C44** (1991) 1725
5. T.-S. Lee, D. Riska: Phys. Rev. Lett. **70** (1993) 2237
6. C.J. Horowitz, H.O. Meyer, D.K. Griegel: Phys. Rev. **C49** (1994) 1337
7. E. Hernández, E. Oset: Phys. Lett. **B350** (1995) 158
8. C. Hanhart, J. Haidenbauer, A. Reuber, C. Schütz, J. Speth: Phys. Lett. **B358** (1995) 21
9. T.Y. Park et al.: Phys. Rev. **C53** (1996) 1519
10. T.D. Cohen et al.: Phys. Rev. **C53** (1996) 2661
11. T. Sato, T.-S.H. Lee, F. Myhrer, K. Kubodera: Phys. Rev. **C56** (1997) 1246
12. J. Adam, A. Stadler, M.T. Peña, F. Gross: Phys. Lett. **B407** (1997) 97
13. C. Hanhart, J. Haidenbauer, M. Hoffmann, U.-G. Meißner, J. Speth: Phys. Lett. **B424** (1998) 8

14. E. Gedalin et al.: Nucl. Phys. **A634** (1998) 368

15. M.T. Peña, H. Garcilazo, D.O. Riska: Nucl.Phys. **A683** (2001) 322

16. G. Fäldt, C. Wilkin: Phys. Scr. **64** (2001) 427

17. K. Nakayama, J. Speth, T.-S. H. Lee: Phys. Rev. **C65** (2002) 045210

18. H. Calén et al.: Phys. Lett. **B366** (1996) 366

19. J. Smyrski et al.: Phys. Lett. **B474** 2000) 182

20. P. Moskal et al.: πN Newsletter **16** (2002) 367

21. H. Calén et al.: Phys. Rev. **C58** (1998) 2667

22. H. Calén et al.: Phys. Rev. Lett. **80** (1998) 2069

23. H. Calén et al.: Phys. Rev. Lett.**79** (1997) 2642

24. H. Calén et al.: Phys. Lett.**B 458** (1999) 190

25. M. Abdel-Bary et al.: `nucl-ex/0205016`

26. P. Moskal et al.: `nucl-ex/0208004`, `nucl-ex/0210019`

27. P. Winter et al.: Phys. Lett. **B544** (2002) 251

28. C. Hanhart et al.: Phys. Lett. **B444** (1998) 25

29. C. Hanhart et al.: Phys. Rev. **C61** (2000) 064008

30. J. Haidenbauer, K. Holinde, M.B. Johnson: Phys. Rev. **C48** (1993) 2190

31. R. Machleidt, K. Holinde, Ch. Elster: Phys. Report **149** (1987) 1.

32. C. Schütz et al.: Phys. Rev. **C51** (1995) 1374

33. C. Schütz et al: Phys. Rev. **C49** (1994) 2671

34. C. Hanhart: Ph.D. thesis: Jül-3476, Jülich (1997)

35. H.O. Meyer et al.: Phys. Rev. Lett. **81** (1998) 3096

36. B. von Przewoski et al.: Phys. Rev. **C58** (1998) 1897

37. H.O. Meyer et al.: Phys. Rev. **C63** (2001) 064002

38. W.W. Daehnick et al.: Phys. Rev. **C65** (2002) 024003

39. C. Hanhart: `nucl-th/0207015`; C. Hanhart, N. Kaiser: `nucl-th/0208050`

40. V. Baru et al., in preparation

41. H. Garcilazo, M.T. Peña: Phys. Rev. **C66** (2002) 034606

42. O. Krehl et al.: Phys. Rev. **C62** (2000) 025207

Few-Body Systems Suppl. 15, 49–60 (2003)

Few-Body Systems

The Importance of the Final-State Interaction in $np \to \eta d$ near Threshold

H. Garcilazo,[1,*] M. T. Peña[2]

[1] Escuela Superior de Física y Matemáticas, Instituto Politécnico Nacional, Edificio 9, 07738 México D.F., Mexico

[2] Centro de Física das Interacções Fundamentais, IST, Av. Rovisco Pais, P-1049-001 Lisboa, Portugal

Abstract. We have studied the process $np \to \eta d$ in the region near threshold, within the framework of a three-body model. Although the magnitude of the cross section is highly dependent on the reaction mechanism, its shape is essentially determined by the ηd final-state interaction. This strong signature provides information about the ηN scattering length. We found that the $np \to \eta d$ experimental data can only be explained if the ηN scattering length has a small real part.

1 Introduction

The reaction $np \to \eta d$ has been measured recently with high accuracy in the region near threshold [1, 2]. Its cross section shows a strong enhancement at threshold, as compared with the predictions based on a two-body phase space, which based the speculation on the existence of an ηNN quasi-bound state [3]. However, the solutions of the Faddeev equations for ηd elastic scattering do not support the existence of a ηNN quasi-bound state [4, 5]. Nevertheless, the strong ηN interaction in the resonant S_{11} channel is indeed responsible for an enhancement of the elastic scattering amplitude near threshold. Accordingly, it is natural to expect that in the reaction $np \to \eta d$ near threshold the ηd final-state interaction will produce a similar enhancement. In this talk we report on very recent work [6] where such a threshold effect was investigated.

2 Separable Two-Body Interactions

We consider a non-relativistic three-body model of the ηNN system, where the elementary ηN interaction is obtained from a separable potential model of the

* *E-mail address:* humberto@esfm.ipn.mx

coupled ηN-πN-σN system. Here, σN represents effectively the $\pi\pi N$ channel, a decay mode of the S_{11} resonance, corresponding to a 10 % branching ratio. Therefore, we will take $m_\sigma = 2m_\pi$.

We have constructed six models based on the analysis [7, 8, 9, 10] which will be labeled 1-6, while model 0 is based in the analysis of ref. [11].

The potentials describing the meson-nucleon transitions in all the models built are of separable form

$$< p|V_{ii}|p' >= \lambda_i \, g_i(p)g_i(p'), (i = \eta, \pi, \sigma) \tag{1}$$

$$< p|V_{ij}|p' >= \pm\sqrt{\lambda_i\lambda_j} \, g_i(p)g_j(p'), (i, j = \eta, \pi, \sigma) \tag{2}$$

with

$$g_\eta(p) = \frac{A_\eta + p^2}{(\alpha_\eta^2 + p^2)^2}, \tag{3}$$

$$g_\pi(p) = \frac{A_\pi + p^2}{(\alpha_\pi^2 + p^2)^2}, \tag{4}$$

$$g_\sigma(p) = \frac{p}{(\alpha_\sigma^2 + p^2)^2}, \tag{5}$$

where the ηN and πN form factors correspond to S waves, while the σN form factor corresponds to a P wave, as required by parity conservation.

In the fitting procedure we used the $\eta N \to \eta N$ transition amplitude obtained from the corresponding data analysis and the $\pi^- p \to \eta n$ total cross-section data. We show in Fig. 1 the comparison of the predictions of our model with the $\pi^- p \to \eta n$ cross-section data.

The separable potential for the NN subsystem (in the $^3S_1$ channel) is the so-called PEST model [12],

$$< p|V_{NN}|p' >= -g_N(p)g_N(p'), \tag{6}$$

with form factor given by

$$g_N(p) = \sum_{n=1}^{6} \frac{c_n}{\beta_n^2 + p^2}. \tag{7}$$

This model yields the same deuteron wave function as the Paris potential.

3 Formalism and Results for ηd Scattering

We consider the system of three particles (η, N, and N), where two of them are identical, and all three interact through pairwise interactions. We represent these interactions with separable potentials. As for the possible three-body channels, we restrict the orbital states between spectator particles and the center of mass of interacting pairs to S waves ($\lambda = 0$). In particular, for the

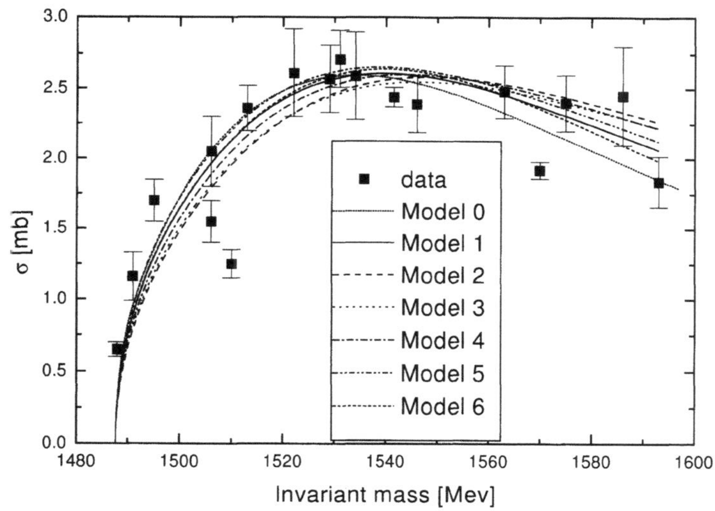

Figure 1. Comparison of the predictions of the models for the $\pi_{-}p \to \eta n$ cross section, where the experimental data have been compiled in ref. [7].

channels with an η being a spectator, this truncation is a good approximation for energies near threshold, where the effect of the centrifugal barrier has to be small. As for the relative orbital momentum of the states for the different pairs, we also restrict these states to S-waves, to the exception of the $\sigma - N$ pair case: due to the intrinsic parity of the σ (two-pions), the contribution to the S wave of the $\eta N \to \eta N$ amplitude yielded by a $\sigma - N$ intermediate state comes necessarily from a P wave. For the ηd system, the relative orbital S-wave states for the interacting pairs are the S_{11} channel for the eta-nucleon pair, and the ${}^3S_1$ channel for the nucleon-nucleon pair.

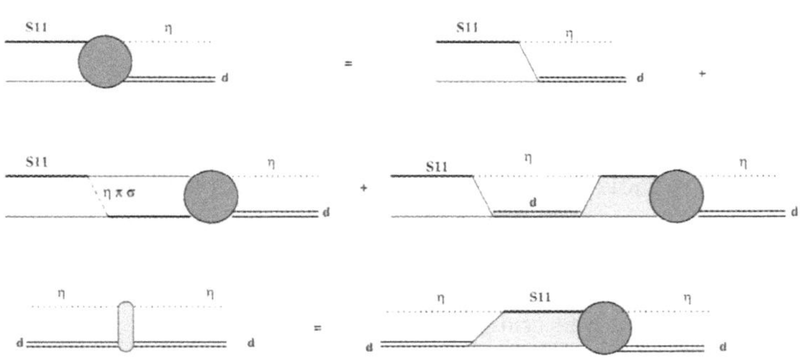

Figure 2. Faddeev equations for ηd elastic scattering.

We denote the η by the label 1 and the two nucleons by labels 2 and 3. The

Faddeev equations for ηd elastic scattering are shown diagrammatically in Fig. 2. In the second equation of this figure, there is a term with the nucleon-nucleon interaction proceeding while a meson is a spectator. That meson can only be the η. Indeed, the term with a pion as the spectator meson would imply an intermediate state (formed by a pion and a NN state in the $^3S_1$ channel) of isospin 1, whereas the ηd final system has isospin 0. Moreover, the intermediate state where a pion is the spectator and the NN state is in the $^1S_0$ scattering state can not proceed either, due to the fact that this state has total spin 0, while the ηd system has total spin 1. As for the σ-meson, it also does not contribute as a spectator to the scattering series due to conservation of parity, since we are restricting the orbital angular momentum between the spectator and the pair to be an S wave, and the final system has a total negative parity.

The integral equations corresponding to Fig. 2 are written analytically as

$$\begin{aligned} T_2(q_2;E) &= K_{21}(q_2,q_{10};E) \\ &+ \int_0^\infty {q_2'}^2 dq_2' K_{23}(q_2,q_2';E)\tau_2(E-{q_2'}^2/2\nu_2)T_2(q_2';E) \\ &+ \int_0^\infty q_1^2 dq_1 K_{21}(q_2,q_1;E)\tau_1(E-q_1^2/2\nu_1)T_1(q_1;E), \end{aligned} \quad (8)$$

$$T_1(q_1;E) = 2\int_0^\infty {q_2}^2 dq_2 K_{12}(q_1,q_2;E)\tau_2(E-{q_2}^2/2\nu_2)T_2(q_2;E). \quad (9)$$

The function $T_2(q_2;E)$ represents the antisymmetric (with respect to nucleons 2 and 3) transition amplitude from the ηd state to a state with a nucleon and a S_{11} isobar while the function $T_1(q_1;E)$ represents the transition amplitude from the ηd state to the ηd state. The kernels $K_{12}(q_1,q_2)$, $K_{21}(q_2,q_1)$, and $K_{23}(q_2,q_2')$ correspond to the various one-particle-exchange transitions depicted in Fig. 2, while $\tau_1(E-q_1^2/2\nu_1)$ and $\tau_2(E-q_2^2/2\nu_2)$ are, respectively, the propagator of a NN isobar with a η as spectator and the propagator of a ηN isobar with a N as spectator.

We present in Table 1 the predictions for the ηd scattering length of the seven separable models for the coupled ηN-πN-σN system. We show the results obtained considering: a) η-exchange only, b) η- and π-exchange, and c) η-π- and σ-exchange. From Table 1 we conclude that the η contribution is the dominant one, and the π- and σ-contributions have a very small effect on the ηd scattering length. Moreover, these results do not imply the existence of a ηNN quasi-bound state (whose signature is that the real part of the ηd scattering length becomes negative while the imaginary part becomes large).

We show in Fig. 3 the ηd elastic cross section predicted by the 7 models. One realizes how the cross section at threshold increases by one order of magnitude, as one goes from model 0 to model 6, i.e., from Re $a_{\eta N}$ = 0.42 fm to Re $a_{\eta N}$ = 1.07 fm, so that the strength of the ηd final-state interaction in the reaction $np \to \eta d$ depends crucially on the value of Re $a_{\eta N}$.

Table 1. ηd scattering length (in fm) predicted by the seven separable potential models of the coupled ηN-πN-σN system. We give the results obtained including only η-exchange, η- and π-exchange, and η-π- and σ-exchange in the driving terms of Fig. 2.

Model	$a_{\eta N}$	η	$\eta + \pi$	$\eta + \pi + \sigma$
0	0.42+i0.34	1.01+i1.24	1.00+i1.28	0.99+i1.28
1	0.72+i0.26	2.53+i1.51	2.56+i1.51	2.57+i1.51
2	0.75+i0.27	2.75+i1.64	2.75+i1.62	2.76+i1.62
3	0.83+i0.27	3.28+i1.93	3.28+i1.91	3.30+i1.91
4	0.87+i0.27	3.55+i2.07	3.56+i2.05	3.57+i2.04
5	1.05+i0.27	4.91+i2.72	4.92+i2.70	4.93+i2.70
6	1.07+i0.26	4.77+i2.25	4.79+i2.25	4.79+i2.24

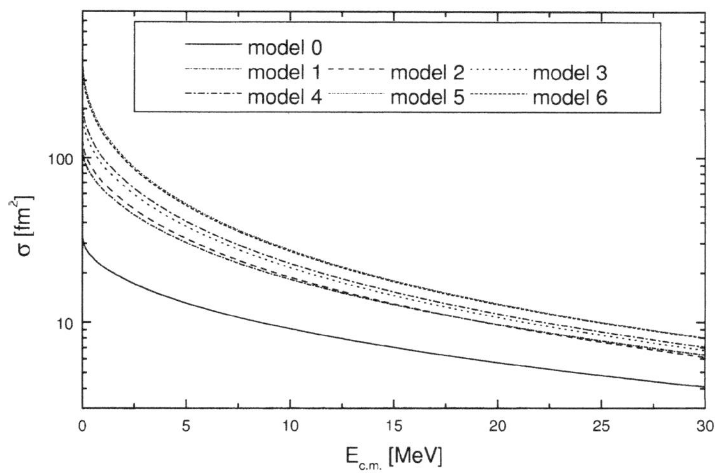

Figure 3. ηd elastic scattering cross section obtained using the seven models of the coupled ηN-πN-σN system.

4 Modeling the Production Mechanism

In Fig. 4 we represent the model considered in this work for the mechanism of the $np \to \eta d$ reaction. Fig. 4.a shows the driving terms for the meson production reaction, which consist of the direct-production diagram and of the box diagram, containing the re-scattering of the η, π and σ mesons plus other heavy meson exchanges (generically labeled by M) through excitation of the S_{11} resonance. Fig. 4.b shows the distortion of the driving terms by the initial- and final-state interactions.

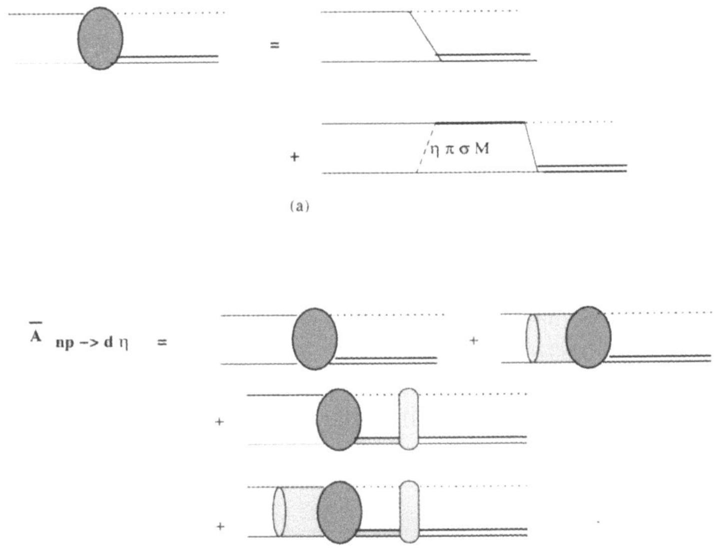

(b)

Figure 4. Model of the $np \to \eta d$ process. a) Direct-production diagram and box diagram including η, π, σ and heavy meson exchanges. b) NN initial-state interaction and ηd final-state-interaction diagrams.

The initial-state interaction is calculated using a two-body NN distorted wavefunction

$$\Psi q_{N0}(q_N) = \frac{1}{q_N^2}\delta(q_N - q_{N0}) + \frac{m_N}{q_{N0}^2 - q_N^2 + i\epsilon} T_{NN}(q_N; E), \tag{10}$$

where q_{N0} is the NN on-shell momentum, $m_N/(q_{N0}^2 - q_N^2 + i\epsilon)$ is the Lippmann-Schwinger propagator of the NN system, and $T_{NN}(q_N; E)$ is the half-shell nucleon-nucleon t matrix which will be discussed in the next section.

As for the final-state interaction, it is treated using a three-body distorted wave involving $\tau_1(E - q_1^2/2\nu_1)$, the propagator of a NN isobar and a spectator η, and $T_1(q_1; E)$ which is the $\eta d \to \eta d$ three-body scattering amplitude obtained from the solution of the integral equations (8) and (9).

The effect of the NN initial- and ηd final-state interactions is thus given by the two-loop calculation

$$\begin{aligned}\bar{A}_{np\to\eta d} &= \int_0^\infty q_1^2 dq_1 \int_0^\infty q_N^2 dq_N \left[\frac{1}{q_1^2}\delta(q_1 - q_{10}) + \tau_1(E - q_1^2/2\nu_1) T_1(q_1; E)\right] \\ &\quad \times A_{np\to\eta d}(q_1, q_N)\end{aligned}$$

$$\times \left[\frac{1}{q_N^2} \delta(q_N - q_{N0}) + \frac{m_N}{q_{N0}^2 - q_N^2 + i\epsilon} T_{NN}(q_N; E) \right] . \tag{11}$$

The production amplitude without initial- and final-state interactions $A_{np \to \eta d}(q_1, q_N)$ which appears in Eq.(11) is shown diagrammatically in Fig. 4.a. In order to make the Feynman diagrams on this figure consistent with the Faddeev diagrams of Fig. 2 their non-relativistic limit is taken and their projection on the initial $NN\ {}^1P_1$ channel and final $\eta d\ {}^3S_1$ channel is performed.

For the calculation of $A_{np \to \eta d}(q_1, q_N)$, the driving terms (direct production and the box diagrams) include a πNN and ηNN vertex with pseudovector coupling, For the η case in the non-relativistic limit the vertex takes the galilean-invariant form

$$V_{\eta NN} = \frac{f_\eta}{m_\eta} \left(1 + \frac{m_\eta}{m_N} \right) \boldsymbol{\sigma} \cdot \boldsymbol{p}, \tag{12}$$

where $\boldsymbol{p}$ is the relative momentum between the meson and the nucleon. Similarly, the πNN and σNN vertices are

$$V_{\pi NN} = \frac{f_\pi}{m_\pi} \left(1 + \frac{m_\pi}{m_N} \right) \boldsymbol{\sigma} \cdot \boldsymbol{p}\ \boldsymbol{\tau} \cdot \boldsymbol{\phi}_\pi, \tag{13}$$

$$V_{\sigma NN} = f_\sigma. \tag{14}$$

If instead we start with pseudoscalar coupling we would get

$$V_{\eta NN} = \frac{f_\eta}{m_\eta} \boldsymbol{\sigma} \cdot \boldsymbol{k}_\eta, \tag{15}$$

where $\boldsymbol{k}_\eta$ is the momentum of the η, and similarly for the pion. We will introduce a form factor that regularizes the vertices at large momenta as

$$V_{mNN} \to V_{mNN} \frac{\Lambda^2 + p_0^2}{\Lambda^2 + p^2}, \tag{16}$$

where p_0 is the on-shell momentum. We took for the cutoff parameter Λ =1800 MeV/c which is a typical value for meson-exchange models [13, 14]. As for the S_{11} isobar, the meson-nucleon couplings are specified by the separable potential models of the coupled ηN-πN-σN system described above. We also included in the driving terms the contribution of the heavy mesons η' and σ' with masses of 958 and 980 MeV, and nucleon vertex functions similar to Eqs. (12) and (14), respectively.

5 Distortion from the NN Initial-State Interaction

The initial-state nucleon-nucleon interaction for this problem is needed at energies well above the pion-production threshold, and no realistic NN potential built so far is valid for this energy range, describing satisfactorily the non-vanishing nucleonic inelasticities. Therefore, we constructed a purely phenomenological extension of the Paris NN potential, to those energies, by including an energy dependent adjustable imaginary term

$$V(p,p') = V_{PARIS}(p,p') - i\gamma \frac{pp'}{(\alpha^2+p^2)(\alpha^2+p'^2)}. \tag{17}$$

The parameters $\gamma = 0.6$ and $\alpha = 0.75$ fm^{-1} were obtained by fitting the experimental amplitude of Arndt et al [16] very near the η production energy threshold. The NN amplitude in the $^1P_1$ initial nucleonic channel could then be calculated from the Lippmann-Schwinger equation with the complex potential above tuned this way. As a result, the NN initial state interaction yields a suppression of the cross section of a factor 5 (i.e., the cross section gets multiplied by a factor 0.2), approximately a constant reduction factor, as usually assumed for the nucleon-nucleon interaction at high energies [17, 18].

6 Results for the Reaction $np \to \eta d$

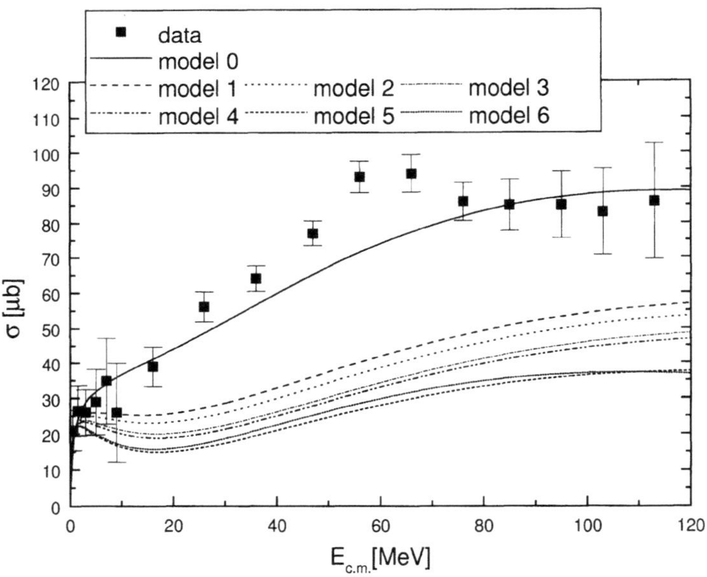

Figure 5. Cross section of the reaction $np \to \eta d$ for the seven models described in the text as compared with the data of refs. [1,2].

The model of the $np \to \eta d$ reaction considered here and depicted in Fig. 4, has the ηNN coupling constant as an input. However this coupling is poorly known. The situation is no better for the coupling constants of the heavy mesons with the nucleon and with the S_{11} isobar. Therefore, at the first stage of our study we consider the model of Fig. 4 without heavy meson exchanges and use the data of $np \to \eta d$ to try to fix the ηNN coupling constant. In this attempt we noticed that the results differ considerably whether one uses the pseudovector vertex (12) or the pseudoscalar one (15). This vertex dependence comes exclusively form the direct production diagram of Fig. 4.a, since the box diagram gives almost exactly the same results with either type of coupling. The reason is that in the first case there are significant contributions from kinematics where the exchanged nucleon is off-mass-shell. We found that at this stage the pseudoscalar vertex (which leads to a negligible contribution for the direct term) allows us to obtain a better description of the experimental data. We show this result in Fig. 5, where after having fitted for each model the very-near-threshold region (see Fig. 6) we predicted from there the cross section at higher energies.

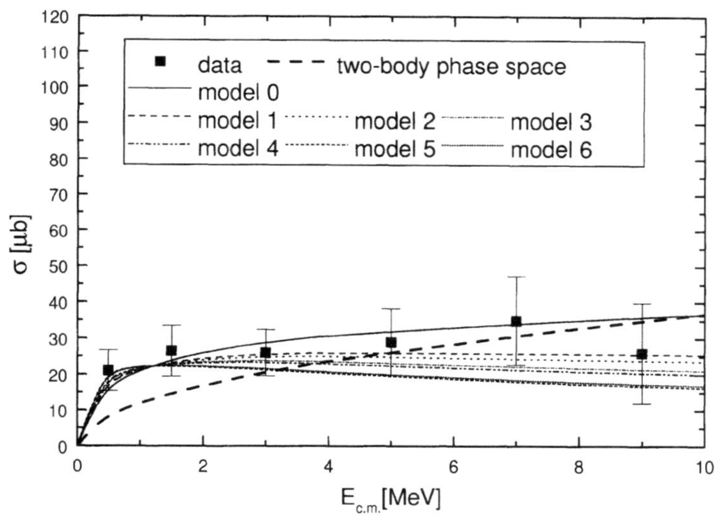

Figure 6. Same as Fig. 5 but for the region very near threshold.

It is clear from Figs. 5 and 6 that

a) the very low-energy region (< 10 MeV in the CM) is not the suitable kinematic domain for the $np \to \eta d$ reaction to probe the several ηN dynamical models, since the data does not distinguish and select between them.

b) however, the higher energy region clearly favors model 0, the only one which describes the data reasonably well throughout the full energy range.

Besides, as we will see below, this preference of the data for a model with small Re $a_{\eta N}$ is to a large extent independent of the production mechanism, indicating that the characteristic shape of the experimental cross section is a signature of the ηd final-state interaction alone, whereas its absolute value involves the contribution of several different processes. This result is not so surprising since, as mentioned in the introduction, this system has a very strong

ηd interaction in the final state, Therefore, it follows from Watson's theorem [19] that the shape of the cross section will be determined basically by this final-state interaction.

The values of the ηNN coupling constant that we extract from this analysis are very large since if we consider $g_\eta = (2m_N/m_\eta)f_\eta$ we will get for model 0 $g_\eta^2/4\pi = 93.2$. This quantity is not well known but certainly it must be much smaller. For example, the Bonn potential [20], requires $2 < g_\eta^2/4\pi < 7$ while the Nijmegen potential [21] uses $g_\eta^2/4\pi = 0.25$. The photoproduction process, on the other hand [22, 23], gives $g_\eta 2/4\pi \approx 1$. Thus, it seems that other processes have to contribute substantially to the production mechanism, in order to allow for a smaller ηNN coupling constant as dictated by several evidences.

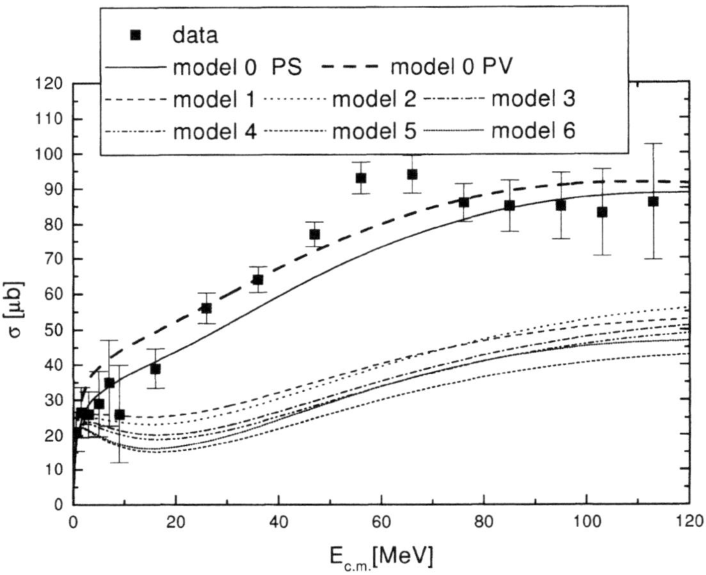

Figure 7. Same as Fig. 5 but with the inclusion of the heavy pseudoscalar meson η'.

Accordingly, we include the exchange of a heavy meson (isoscalar, scalar, or vector) in the production mechanism, but used the fixed value $f_\eta = 1.5$, which corresponds to $g_\eta^2/4\pi = 2.1$, i.e., the lower limit determined from the Bonn potential. We show in Fig. 7 the results that are obtained by allowing in the production mechanism a pseudoscalar exchange corresponding to the η' meson. Once more, the shape of the cross section can only be reproduced at all energies by model 0, and the theoretical cross sections are actually very similar to those of Fig. 5. We adjusted the product of the coupling constants of the η' with the nucleon and with the S_{11} for models 0 to 6. Since the ηNN coupling constant is now much smaller, the relative importance of the direct term in the production amplitude is reduced and the difference between pseudovector and pseudoscalar coupling becomes less dramatic. This is shown by the thick-dashed line in Fig. 7, which corresponds to model 0 and the coupling constants values of the solid line, but where the pseudovector vertex (12) is used instead.

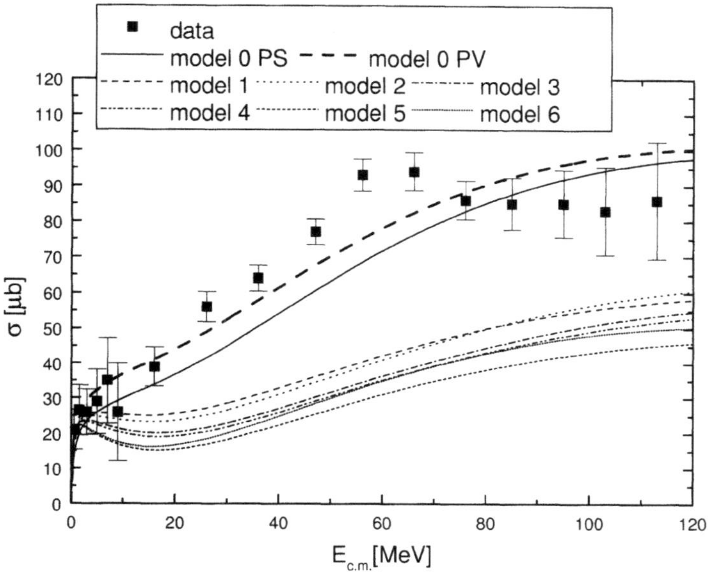

Figure 8. Same as Fig. 5 but with the inclusion of the heavy scalar meson σ'.

We show in Fig. 8 the results obtained when instead of the η' the mechanism of meson production includes the exchange of a σ' [the $f_0(980)$]. The product of the coupling constants of the σ' with the nucleon and with the S_{11} were adjusted again for models 0 to 6. Once again, only model 0 gives more or less reasonable results and there is not much difference between the results of pseudovector and pseudoscalar coupling. In the case of vector-meson exchange, the dominant contribution, which corresponds to total orbital angular momentum $L = 0$, has exactly the same structure as the one of the pseudoscalar-meson exchanges considered and therefore the results would be similar to those of Fig. 7.

Thus, we conclude from these results that different short-range mechanism can be adjusted to the magnitude of the cross-section. Most importantly is that, nevertheless, the good description provided by model 0 for all energies, and essential for the ηd dynamics in the final state, is independent of the production mechanism considered.

Acknowledgement. This work was supported in part by COFAA-IPN (México) and by Fundação para a Ciência e a Tecnologia, MCT, under contracts and PRAXIS/P/FIS/10031/1998, SAPIENS/36291/99, CERN/P/FIS/40135/2000 (Portugal).

References

1. H. Calén et al.: Phys. Rev. Lett. **79**, 2642 (1997)
2. H. Calén et al.: Phys. Rev. Lett. **80**, 2069 (1998)
3. T. Ueda: Phys. Rev. Lett. **66**, 297 (1991)
4. H. Garcilazo, M. T. Peña: Phys. Rev. **C61**, 064010 (2000)
5. H. Garcilazo, M. T. Peña: Phys. Rev. **C63**, 021001 (2001)
6. H. Garcilazo, M. T. Peña: Phys. Rev. **C66**, 034606 (2002)
7. M. Batinić, I. Šlaus, A. Švarc, B. M. K. Nefkens: Phys. Rev. **C51**, 2310 (1995)
8. M. Batinić, I. Dadić, I. Šlaus, A. Švarc, B. M. K. Nefkens, T.-S. H. Lee: Physica Scripta **58**, 15 (1998)
9. A. M. Green, S. Wycech: Phys. Rev. **C55**, R2167 (1997)
10. A. M. Green, S. Wycech: Phys. Rev. **C60**, 035208 (1999)
11. A. Sibirtsev, S. Schneider, Ch. Elster, J. Haidenbauer, S. Krewald, J. Speth: Phys. Rev. **C65**, 044007 (2002)
12. H. Zankel, W. Plessas, J. Haidenbauer: Phys. Rev. **C28**, 538 (1983)
13. F. Gross, J. W. Van Orden, Karl Holinde: Phys. Rev. **C45**, 1094 (1992)
14. R. Machleidt, K. Holinde, C. Elster: Phys. Rep. **149**, 1 (1987)
15. K. Holinde: Phys. Reports **68**, 121 (1981)
16. R. A. Arndt, C. H. Oh, I. I. Strakovsky, R. L. Workman, F. Dohrmann: Phys. Rev. **C56** 3005 (1997)
17. M. Batinć, A. Švarc, T-S. H. Lee: Physica Scripta **56**, 321, (1997)
18. V. Yu. Grishina, L. A. Kondratyuk, M. Buscher, J. Haidenbauer, C. Hanhart, J. Speth: Phys. Lett. **B475**, 9 (2000)
19. Kenneth M. Watson: Phys. Rev. **88**, 1163 (1952)
20. R. Machleidt: Adv. Nucl. Part. Phys. **19**, 189 (1989)
21. T. Rijken, V. G. J. Stoks: Phys. Rev. **C59**, 21 (1999)
22. M. Benmerrouche, N. C. Mukhopadhyay: Phys. Rev. Lett. **67**, 1070 (1991)
23. L. Tiator, C. Bennhold, S. S. Kamalov: Nucl. Phys. **A580**, 455 (1994)

Few-Body Systems Suppl. 15, 61–66 (2003)

Isospin Symmetry Breaking: Experimental Observation

B.J. Roy for the GEM Collaboration

J. Bojowald[1], A. Budzanowski[3], A. Chatterjee[8], J. Ernst[6], L. Freindl[3], D. Frekers[7], P. Hawranek[1,2], J. Ilieva[4], L. Jarczyk[2], V. Jha[8], K. Kilian[1], D. Kirillov[12], S. Kliczewski[3], W. Klimala[2], D. Kolev[5], M. Kravčiková[11], T. Kutsarova[4], J. Lieb[9], H. Machner[1], A.Magiera[2], G. Martinská[10], H. Nann[13], L. Pentchev[4], N. Piskunov[12], D. Protič[1], P. von Rossen[1], B.J. Roy[1,8], I. Sitnik[12], R. Siudak[3], A. Strzałkowski[2], R. Tsenov[5], M. Uličný[1,10], J. Urbán[6,10], A. Wrońska[1,2]

[1] Institut für Kernphysik, Forschungszentrum Jülich, Germany
[2] Institute of Physics, Jagellonian University, Kraków, Poland
[3] Institute of Nuclear Physics, Kraków, Poland
[4] Institute of Nuclear Physics and Nuclear Energy, Sofia, Bulgaria
[5] Physics Faculty, University of Sofia, Sofia, Bulgaria
[6] Institut für Kernphysik der Universität Bonn, Bonn, Germany
[7] Institut für Kernphysik, Universität Münster, Münster, Germany
[8] Nuclear Physics Division, BARC, Bombay, India
[9] Physics Department, George Mason University, Fairfax, Virginia, USA
[10] Department of Nuclear Physics, University Košice, Slovakia
[11] Department of Physics, Technical University Košice, Slovakia
[12] Laboratory for high energy physics, JINR Dubna, Russia
[13] IUCF, Indiana University, Bloomington, USA

Abstract. Experiments have been performed at COSY, Juelich to investigate isospin symmetry breaking effects in nuclear reactions by studying simultaneously the two reactions $p+d \to {}^3\mathrm{H}+\pi^+$ and $p+d \to {}^3\mathrm{He}+\pi^0$ at beam energies close to corresponding η production threshold. The beam momentum dependence of the ratio of cross section of these two reactions would be indicative of isospin symmetry breaking, if any. The present measurement reports observation of isospin symmetry breaking effect. The results are presented.

1 Introduction

The isospin symmetry breaking (ISB) of strong interaction is one of the most interesting problems of nuclear and particle physics. At quark level this symmetry breaking is believed to be associated with the mass difference between up and down quarks [1]. In the pionic sector, isospin symmetry violation could be responsible for the mixing of π^0 and η and also that of ρ^0 and ω. However, there is little direct evidence for such violation in nuclear reactions. A good example of isospin symmetry violation would be the measurement of $dd \rightarrow ^4\mathrm{He}\pi^0$ reaction which is forbidden by isospin symmetry. The mere detection of a non-zero cross section of this reaction would be clear evidence for ISB. The reaction was measured [2] at an incident deuteron energy of 1100 MeV and a cross section (at $\theta_{\mathrm{cm}} = 107$ deg) $\sim$ 1 pb/sr was reported. However, the result was questioned by some participants of the experiment, also a recent analysis [3] indicates that the signal, under this experimental condition, may be confused with the back ground reaction $dd \rightarrow ^4\mathrm{He}\gamma\gamma$.

A strong indication of ISB can also be obtained in the reactions $pd \rightarrow ^3\mathrm{H}\pi^+$ and $pd \rightarrow ^3\mathrm{He}\pi^0$. Isospin symmetry predicts that the ratio of these two reactions should be 2 apart from some deviation mainly due to the difference in the three-nucleon wave function and Coulomb correction. It was suggested by Wilkin [4] and later by Magiera and Machner [5] that the isospin symmetry breaking effect, due to π-η mixing, can be very pronounced in these reactions at large $\pi - p$ angles and at beam energies close to the η threshold. The advantage of this reaction is that a high measurement accuracy can be achieved by detecting these two reactions simultaneously, since in a simultaneous measurement various systematical errors get cancelled. Experiments have been performed at COSY, Juelich, for the simultaneous detection of the two reactions $p + d \rightarrow$ $^3\mathrm{H}+\pi^+$ and $p + d \rightarrow$ $^3\mathrm{He}+\pi^0$, at beam energies close to the corresponding η production threshold as detailed below.

2 Experiment

The experiment was performed at COSY using a proton beam of intensity $5 \cdot 10^8$ particles/s and at five different beam momenta: 1.56, 1.57, 1.571, 1.59 and 1.70 GeV/c. The beam momentum accuracy was better than 0.1%. The target was a cell of liquid deuteron with a thickness of 1 cm. The reaction products were detected using the Big Karl magnetic spectrometer shown schematically in Fig. 1. The experimental details are reported in a previous communication [6], here only some major features are mentioned.

The outgoing ^{3}H and ^{3}He particles have almost the same momenta, but their charges differ by factor of 2. Therefore,the ^{3}H magnetic rigidity is twice that of ^{3}He. In order to detect simultaneously both these particles, the Big Karl spectrometer was upgraded. The standard focal plane is equipped with two drift chambers for particle tracking and scintillation hodoscopes for particle identification. This set-up is used for ^{3}He detection. In the first dipole D1 there is a special hole in the magnet outer yoke originally foreseen for beam dump.

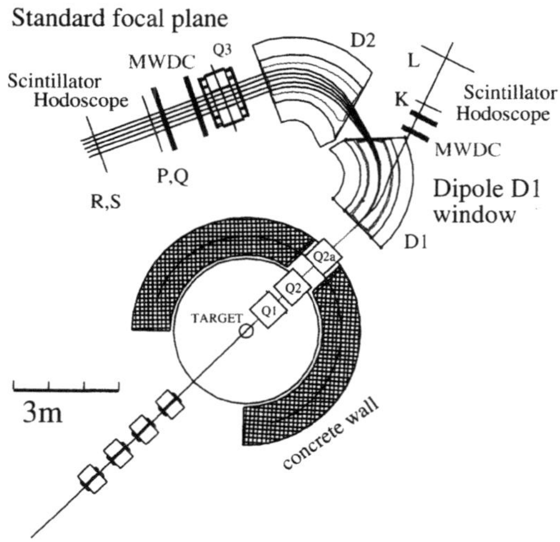

Figure 1. Schematic view of the Big Karl magnetic spectrograph.

This place is used for ^{3}H detection. It was equipped (Fig. 1) with a newly built but similar detection system as that for the standard focal plane.

The magnetic rigidities of ^{3}H and beam protons are very similar. Therefore the beam hits the dipole D1 yoke very close to the detection system, thereby, producing a large background. In order to reduce this background, a small beam incidence angle at the target was introduced. It was achieved by an additional steering magnet placed close to the target [6]. The beam incident angle was calibrated using the reaction $pp \to d\pi^+$ measured at the normal focal plane. Further background reduction was achieved by demanding a correlation trigger between appropriate paddles of the hodoscopes, as detailed in ref. [6].

The crucial point for reaching a high accuracy in the ratio R is the determination of the relative acceptance of the ^{3}He/^{3}H detection systems. We have tried two methods of measuring the acceptance ratio. The first method is based on registering the kinematical coincidences of the particles emitted from the $pp \to d\pi^+$ reaction. This, however, delivered only an upper limit for the relative acceptance. The second method is based on the equality of the center of mass cross sections for the $pp \to d\pi^+$ reaction. At beam momenta close to 1.57 GeV/c the ratio of maximum and minimum kinematically allowed deuteron momenta is about 2. A simultaneous detection of these deuterons simulates almost exactly registering of ^{3}He/^{3}H in the goal experiment. This relative acceptance includes not only the geometry of the detection systems but also the absolute efficiency of the detectors. However, since the ratio of the maximum and minimum kinematically allowed deuteron momenta is not exactly equal 2, some corrections were needed to apply which resulted in a larger systematical error.

3 Results

A very good ^3He separation was achieved in the standard focal plane as seen in Fig. 2a. The example of missing mass spectrum used for ^3He events integration is shown in Fig. 2b and clearly identifies the reaction of interest. The situation is somewhat worse for ^3H where the background is much higher. The

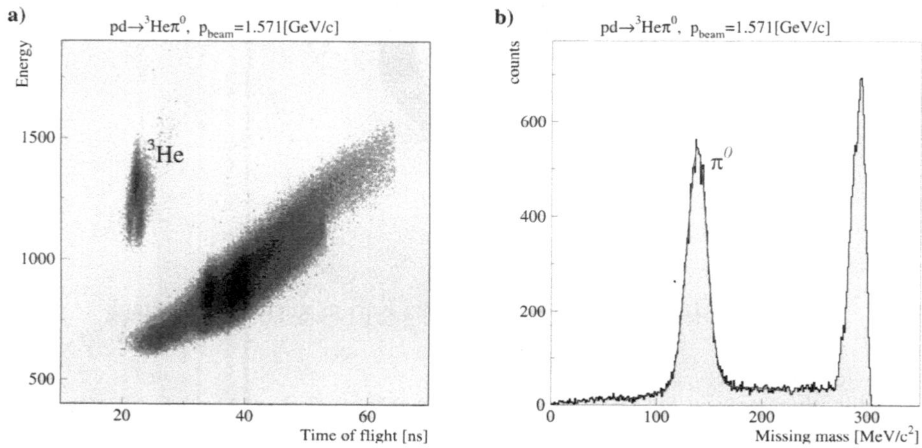

Figure 2. Spectra measured at the standard focal plane

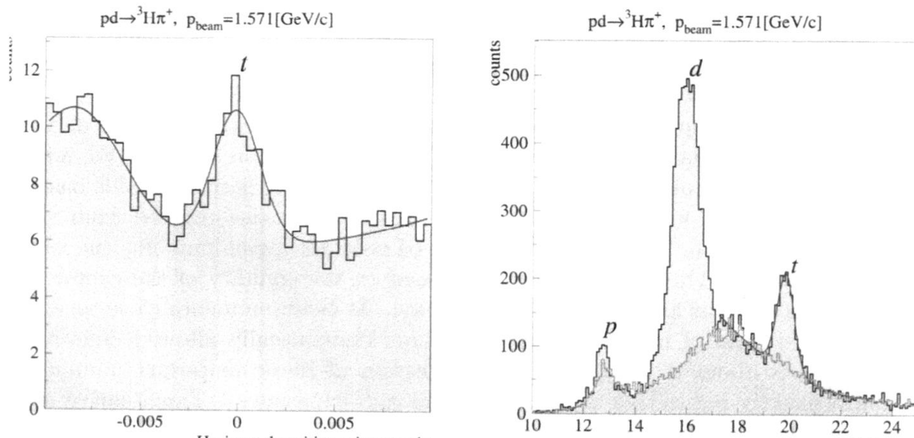

Figure 3. Spectra obtained with the D1 exit detection system. Tritons are clearly visible.

position and angle information is used for gating. An example of the horizontal position vs. horizontal angle (corresponding to the particle momentum) spectrum is presented in Fig. 3. The peak for ^{3}H from the reaction $pd \rightarrow ^3\mathrm{H}\pi^+$ is clearly visible in the middle. The time of flight spectrum is also shown in Fig. 3 which was used, after proper back-ground substraction, for calculating the cross section. The preliminary results for the measured cross section ratio of ^{3}H/^{3}He are plotted in Fig. 4. Due to the large errors in the relative acceptance the data at the highest beam momentum, where the mixing effect is expected to be minimal, are normalized to $R = 2$. The values near the η threshold are very different from the isospin predicted value of 2.

4 Summary

The ratio of the differential cross sections of the two isospin related reactions $pd \rightarrow ^3\mathrm{H}\pi^+$ and $^3\mathrm{He}\pi^0$ are measured at few energies near the η threshold and far away from η threshold. The $p - \pi$ opening angle was taken as $\sim$ 180 deg. Due to the large errors in the relative acceptance measurement, the ratio R of these two reaction cross sections at the highest beam momentum is normalized to $R = 2$. The present results are indicative of isospin symmetry violation. When interpreted in a meson-mixing model [5], the π-η mixing angle(θ_m) can be deduced from the data. A value of $\theta_m = 0.026$ rad is obtained from the fit to the present data (Fig. 4). However, the relavite acceptance measurement

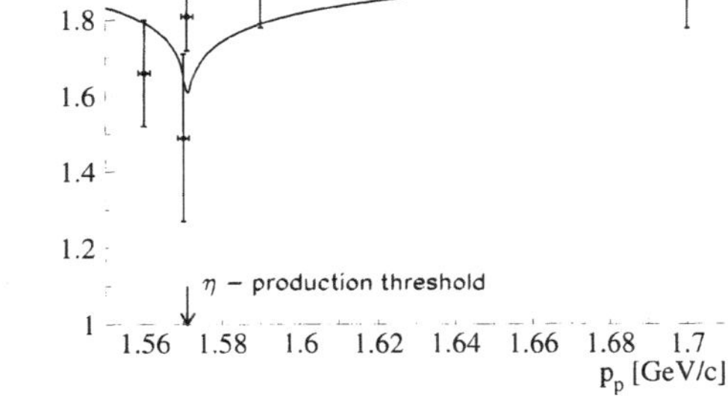

Figure 4. Preliminary results for the observed ratio of cross section of the reactions $pd \rightarrow ^3\mathrm{H}\pi^+/^3\mathrm{He}\pi^0$. The model calculation [5] for mixing angle θ=0.026 rad are also shown.

of the two detection system needs further improvement. Measurements in this respect are in progress.

References

1. G.A. Miller, B.M.K. Nefkens, I. Slaus: Phys. Rep. **194**, 1 (1990)
2. L. Goldzahl, J. Banaigs, J. Berger et al.: Nucl. Phys. **A533** 675 (1991)
3. D. Dobrokhotov, G. Faldt, A. Gardestig et al.: Phys. Rev. Lett. **83**, 5246 (1999)
4. C. Wilkin: Phys. Lett. **B331**, 275 (1993)
5. A. Magiera, H. Machner: Nucl. Phys. **A674**, 515 (2000)
6. J. Bojowald, A. Budzanowski, A. Chatterjee et al.: Nucl. Inst. Meth. **A487**, 74 (2002)

Few-Body Systems Suppl. 15, 67–78 (2003)

Few-Body Systems

Relativistic Analysis of Proton-Proton Bremsstrahlung

J.A. Tjon*

Theory Group, Jefferson Laboratory, Newport News VA 23606, USA

Abstract. A microscopic calculation of pp bremsstrahlung based on a relativistic one boson exchange model for the nucleon-nucleon interaction is reviewed. For small photon momenta, gauge invariance imposes an important constraint on the single nucleon current contribution. Within the dynamical model it is found, that the pair term is compensated to a large extent by the rescattering contribution. Polarization predictions are in good agreement with the experimental data.

Another approach to $pp\gamma$ is the soft photon approximation. It is based on a low energy expansion. Current conservation is satisfied by the introduction of an e.m. contact term. For one kinematics a large difference is found between the two approaches. We discuss this in some detail. In particular, it shown that the dynamical model predictions are very sensitive to the low energy NN interaction. Fitting to the pp data agreement is found with the observed $pp\gamma$ cross section.

1 Introduction

One of the major motivations to study the nucleon-nucleon (NN) bremsstrahlung process has been the exploration of possible sensitivities to model dependence of the nuclear force. Similarly as in the case of electron scattering there exists the complication of the presence of two-body e.m. currents. Moreover, relativistic spin corrections are found to be important in $NN\gamma$. All these different aspects have been in the past the subject of theoretical studies [1]-[3].

More recently, polarized and unpolarized $pp\gamma$ experiments [4, 5] have been performed with high precision, stimulating new theoretical calculations [6]-[13]. Both real and virtual bremsstrahlung in pp scattering have experimentally been studied at KVI. In this talk I will describe some of the theoretical approaches. The so-called soft photon approximation (SPA) [2, 6, 7] and a microscopic analysis based on a relativistic one-boson-exchange (OBE) model will be discussed [10, 11]. Here we confine ourself to the case that the photon momentum $q^2 = 0$.

* *E-mail address:* tjon@jlab.org

2 General Remarks

The dominant terms in the description of the $pp\gamma$ process are given by the external bremsstrahlung contributions corresponding to emission of the photon from one of the initial or final nucleon legs. Rough estimates of the relative size of the relativistic Z-graph contributions can be made. Consider as an example the graph, where the photon is emitted with momentum q from the initial proton 1, carrying momentum p_1. The contribution can be written as

$$M_\mu = T \frac{1}{\not{p}_1 - \not{q} - m} J_\mu u(p_1), \tag{1}$$

where the current operator J_μ is given by

$$J_\mu = \gamma_\mu + \mathrm{i}\frac{\kappa}{2m}\sigma_{\mu\nu}q_\nu. \tag{2}$$

In Eq. (1) T is the NN T-matrix. It is an operator in the Dirac space of the two nucleons, characterized by their ρ-spin: $T(\rho_1' \; \rho_2', \rho_1 \; \rho_2)$. Moreover, it depends on the off-shell momentum $p_1 - q$. Because of this we expect that the bremsstrahlungs process may lead to sensitivity on the microscopic models used for the NN interaction. However, as already pointed out 30 years ago by Sauer and collaborators [14]-[16] the latter may be strongly constrained by the effective range behaviour.

We can readily make an estimate of the magnitude of the current matrix element at low photon momenta. Since $pp\gamma$ is predominantly dipole radiation we may consider here only the 3-current matrix element and confine ourself to the γ_k term in the current. The graph can be split into two pieces $M(+)$ and $M(-)$, corresponding to the contributions from the positive and negative energy state propagation of the intermediate proton. We get

$$M_k(+) = T(++,++)\frac{\bar{u}(p_1 - q)\gamma_k u(p_1)}{E - E_{p_1-q} - \omega}, \tag{3}$$

while

$$\begin{aligned} M_k(-) &= T(-+,++)\frac{\bar{v}(p_1 - q)\gamma_k u(p_1)}{E + E_{p_1-q} + \omega} \\ &\approx \frac{1}{2M}T(-+,++). \end{aligned} \tag{4}$$

The latter contribution is nothing else than the well-known Z-graph contribution. Since the matrix element of γ_k between positive spinors vanishes for $q \to 0$, the Z-graph contribution is expected not to be substantially smaller than Eq. (3). In the case of Compton scattering it is the dominant contribution at low energy, yielding the classical Thompson scattering result.

Since the T-matrix elements for different ρ-spin sectors are roughly of the same order of magnitude, the ratio of Eqs. (4) and (3) will in general not to be

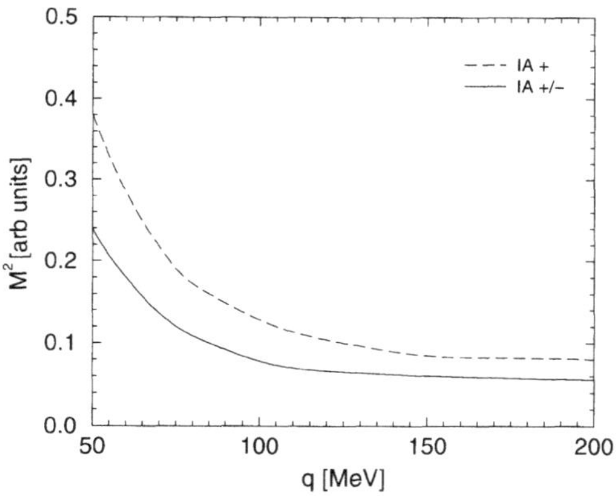

Figure 1. The external bremsstrahlung contribution using the full (+/-) or keeping only the positive positive energy state propagator (+) of the intermediate nucleon. Kinematics is for fixed c.m. proton angles $\theta_1 = 36^0$, $\theta_2 = 69^0$ and photon angle $\theta_\gamma = 124^0$. The photon momentum was varied by varying the energy of the incoming proton. Calculations are from the dynamical model described in section 3.

small. In the relativistic OBE model one indeed finds sizable corrections due to these negative energy nucleon state contributions. This is shown in Fig. 1.

A more quantitative treatment of the external photon radiation contribution has been proposed by Liou et al. [2, 6, 7]. It is based on the low energy expansion of Low [17]. For small enough q the T-matrix is close to the on-shell value. The off-shell T-matrix is a function of the Lorentz invariants s, t and the invariant masses of the external nucleons. The only off-shell dependence in the external radiation graph resides in the invariant mass of the nucleon leg emitting the photon. Expanding the T-matrix up to linear terms in q around the on-shell point, the NN amplitude is assumed to be expressed in terms of the on-shell one and its derivative. Consequently, the T-matrix and its derivative can be determined from the experimental NN phaseshifts.

Usually one makes use of the five β-decay covariants as the representation for the NN amplitude. The on-shell invariant amplitudes are functions of only two independent invariants. A specific choice of the two Lorentz invariants (s, t or u) has to be assumed in the parameterization of the amplitudes. In general sizable sensitivities are found on the specific choice made. See for details refs. [7, 18].

The obtained expression for the external radiation contributions forms the basis of the so-called soft photon approximation (SPA). To satisfy current conservation an e.m. contact term is introduced. It should be noted, that the choice of the covariants to characterize the Dirac structure of the NN amplitude is not unique [19]. Depending on which covariants are used the contact term can be very different in magnitude.

Imposing further the Pauli exclusion principle a unique contact current operator has been constructed by Liou et al. [6] for their specific choice of Lorentz invariants to parameterize the NN amplitude. In Fig. 2 is shown the prediction of the SPA for the KVI kinematics at 190 MeV.

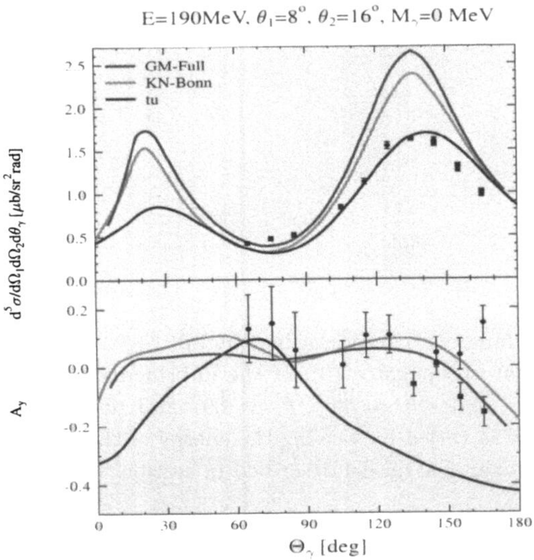

Figure 2. The unpolarized cross section and polarization in the KVI kinematics. Data are from ref. [4]. The two upper curves in the upper panel are microscopic calculations from refs. [9, 10], while the lower curve is the SPA prediction. In the A_y plot the SPA result [7] is the curve deviating the most from the experimental data.

From this we see that the agreement of the SPA with the experimental unpolarized cross-section data is remarkably good. The SPA predictions for the polarization observables are on the other hand poor. These results should be contrasted to the theoretical predictions obtained from dynamical models [9, 10]. For these models the cross section is substantially overestimated near the peak, whereas the polarization is well in agreement with the experimental data. We will discuss this in more detail in the next section.

3 Microscopic Model of $pp\gamma$

In this section we present some results for the relativistic OBE model of refs. [20]-[22]. The model has been applied with success to the description of elastic and inelastic electron scattering off the deuteron [21, 22]. The case of the real bremsstrahlung process has been studied by Martinus et al. [10]. Starting from a relativistic quasi-potential framework, the NN force is characterized by a standard set of exchanges of $\pi, \sigma, \rho, \omega, \eta$ and δ mesons.

The BSLT [23] prescription is used for the two-nucleon free Greens' function, including both positive and negative energy intermediate spinor states. No retardation effects are included in the NN interaction, that is the relative energy variable is set to zero in the NN driving force. To get a reasonable fit to the NN phase shifts we had to use a pseudo-vector πNN coupling, leading to pair suppression in the NN interaction. Moreover, a monopole meson-nucleon form factor is adopted with a cut-off mass of around 1200 MeV.

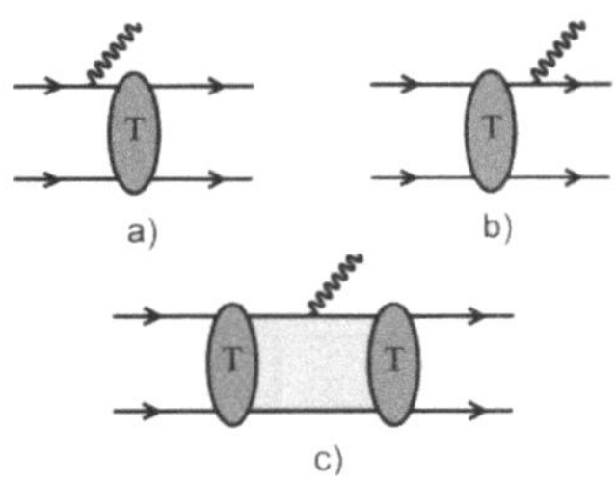

Figure 3. The external radiation (a-b) and rescattering (c) graphs for $pp\gamma$ bremsstrahlung.

With the described nucleon force model, the pp bremsstrahlung reaction can be calculated. Using the single-nucleon current (2), the external radiation graphs (IA) and the rescattering (RSC) contributions, shown in Fig. 3, have been determined. The RSC has clearly to be included if one wants to satisfy gauge invariance. In the present dynamical model it essentially plays the role of the contact term to warrant current conservation. Due to the emission of the photon, the NN T-matrix in general has to be boosted. At the considered energies the boost effects are found not to be large.

The effects of two-body currents, the $\pi\rho\gamma$-like mesonic exchange (MEC) and isobar currents, Fig. 4, can be significant [10, 24]. It should be noted, that the pion-in-flight and corresponding contact term do not contribute to $pp\gamma$. In the Martinus et al. calculation [10] it is done in a fully relativistic way including initial and final state interaction. Consequently, to calculate the 2-body current matrix elements two-loop calculations have to be carried out.

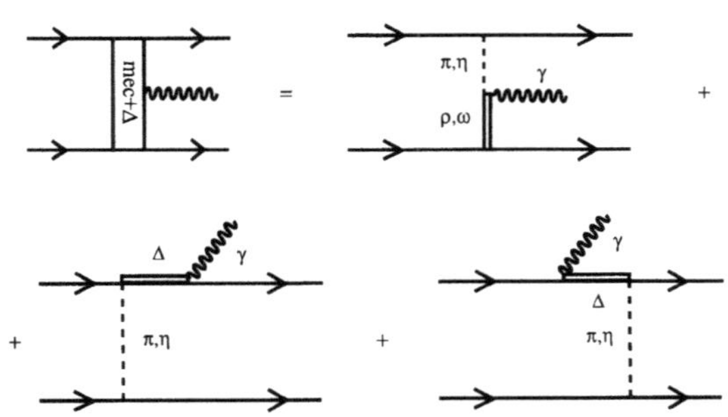

Figure 4. The $\pi\rho\gamma$-like mesonic exchange currents and isobar currents.

In Fig. 5 we show the predictions in the case of the TRIUMF kinematics [25]. The overall agreement with the TRIUMF data is reasonable except for an overall normalization factor of 2/3. This has also been found in the nonrelativistic calculations of Workman and Fearing [3]. Moreover, as can been seen from the figure, of the two-body currents the Δ contribution is the dominant one for the considered kinematics. For the Δ propagator the Rarita-Schwinger propagator has been used with an energy-dependent complex mass [26]. Only a small sensitivity has been found on the various choices of the Δ couplings used in the literature.

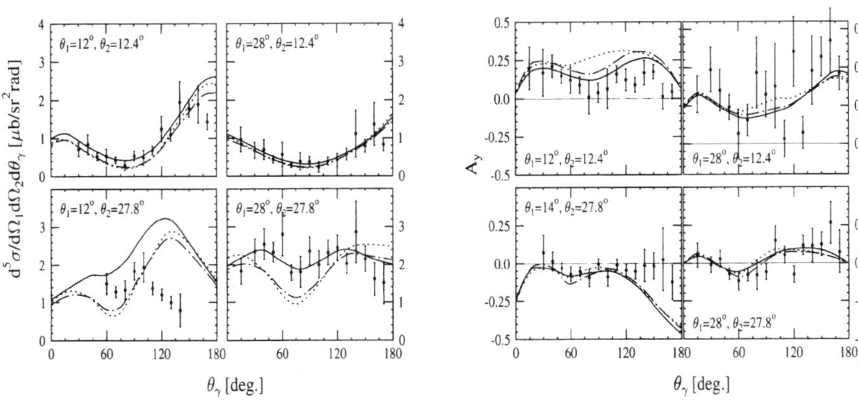

Figure 5. The cross section and analyzing power at TRIUMF kinematics, $T_{\mathrm{Lab}} = 280$ MeV and various sets of outgoing proton angles θ_1, θ_2. Predictions using only the single nucleon current are shown as dot-dashed lines. The full calculation with MEC and isobar currents corresponds to the solid lines. Also are shown separately the inclusion of only MEC (dotted) and of the isobar current (dashed) to the one-body contribution. The experimental cross sections have been multiplied by a normalization factor of 2/3.

As we discussed in the previous section, there is a sizable contribution from the negative energy state nucleon propagator if one considers only the relativistic IA graphs. In Fig. 6 are shown the predictions if one also includes the RSC. From this we see, that the negative energy state contribution in the IA is almost completely cancelled by the rescattering graph, at least at lower photon momentum.

Since this Z-graph contribution corresponds essentially to a contact term, we see that the total contact term arising from the Z-graph and the rescattering graphs cancel each other, so that an analysis based on the IA and only positive energies are close to the full result. At higher photon energies the cancellation is not complete and some sensitivity is found, as can be seen in the enlarged plot in the righthand panel of Fig. 6.

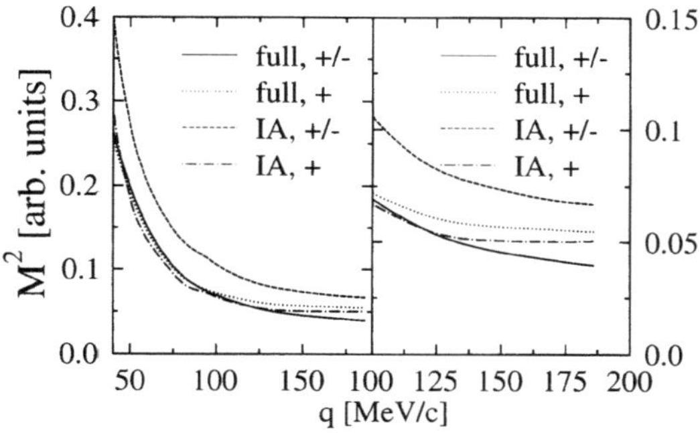

Figure 6. Cancellation of the negative energy state IA contribution by the rescattering graph. Kinematics is the same as in Fig. 1.

4 The KVI Puzzle

As we have shown in Fig. 2, a rather surprising descrepancy in the differential $pp\gamma$ cross section occurs in the KVI kinematics. The microscopic model discussed in the previous section predicts a substantial higher value near the peak, whereas there is good agreement with the SPA calculation.

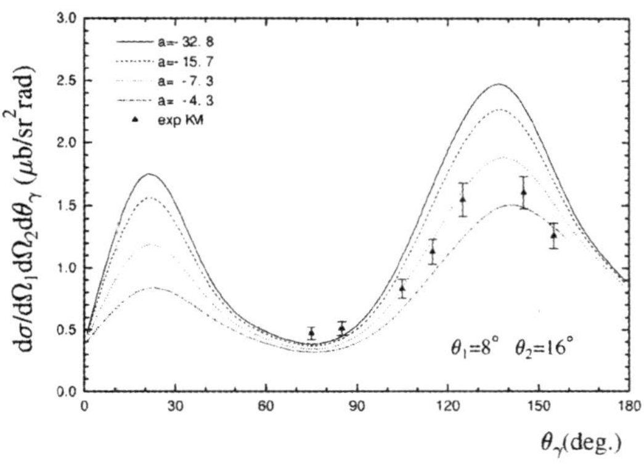

Figure 7. The dependence of the $pp\gamma$ cross section on the $^1S_0$ scattering length a (in fermi) in the KVI kinematics for OBE models with various a. Proton angles are 8^0 and 16^0. The models were obtained by simply varying the σNN coupling constant to yield a given a.

We now discuss a possible reason for the too large strength found in the dynamical model [12, 13]. Considering the KVI kinematics, we find that the contributions from the two-body currents can safely be neglected. So the main difference in the two approaches should come from the single nucleon current matrix elements. Moreover, the negative energy state contributions are small in this case. An interesting property of the KVI kinematics is, that in the diagram, where the photon is externally radiated in the initial state the remaining final state NN system has a rather low lab energy (about 10 MeV) at the photon angle of 140 degrees, where the peak in the cross section is located. This suggests, that the result may well depend on the low energy behaviour of the NN interaction. In particular, at these energies the pp s-wave scattering deviates significantly from the np one. Since the NN interaction has been fitted to the np data, the given OBE model gives a scattering length of a=-23.7 fm in the ${}^1S_0$ channel, whereas for the pp channel it should be a=-7.8 fm. To get some idea how the $pp\gamma$ is affected by changes in the ${}^1S_0$ channel, we have varied in this channel the σNN coupling constant leading to NN interactions with different scattering lengths. In Fig. 7 are shown the dependence of the $pp\gamma$ cross section on a. From this we see, that the $pp\gamma$ in this region is indeed very sensitive to the changes of the NN interaction at low energy. The predictions agree remarkably well with the data for the case that the scattering length of the model corresponds to the empirical pp one.

From this study we see that in the considered KVI kinematics the $pp\gamma$ is sensitive to the details of the NN force. Although the SPA leads to a good description, the NN information going into that analysis are the empirical phase shifts at considerable higher energies. This suggests, that the agreement may be accidental.

To treat pp scattering properly the long range Coulomb interaction has to be included. Mostly this is done by solving the equations in coordinate space. It is known that the effect of Coulomb is very important at energies, say, below 10 MeV, but becomes substantially less at higher energies. Therefore we may hope, that an improved $pp\gamma$ analysis can be carried out without accounting for the full complexity of the Coulomb interaction, but only by fitting the OBE model to the empirical strong pp phaseshifts like those of the Nijmegen phaseshift analysis PWA93 [27]. We have carried out a χ^2 fit of the OBE model to the PWA93 pp phaseshifts between 10 MeV and 220 MeV. Considering the ${}^1S_0$ channel we find with the refitted OBE model a scattering length of about -12 fm.

In Fig. 8 are shown the $pp\gamma$ predictions using the refitted OBE model. We see that around the peak the cross section is reduced substantially as compared to the predictions using the original np fitted OBE model, getting closer to the experimental results. The strength at the peak is still too high. Given the higher value of the scattering length a of the model than the PWA93 one, the result is consistent with our a-dependent study.

To investigate how large the Coulomb effect is for the KVI kinematics, we have carried out a model study, where the Coulomb is exactly accounted for. The model consists in addition to the Coulomb potential of a strong s-wave NN potential of a separable type

$$V(p,p') = \lambda g(p) g(p'), \tag{5}$$

where p is the relative NN momentum.

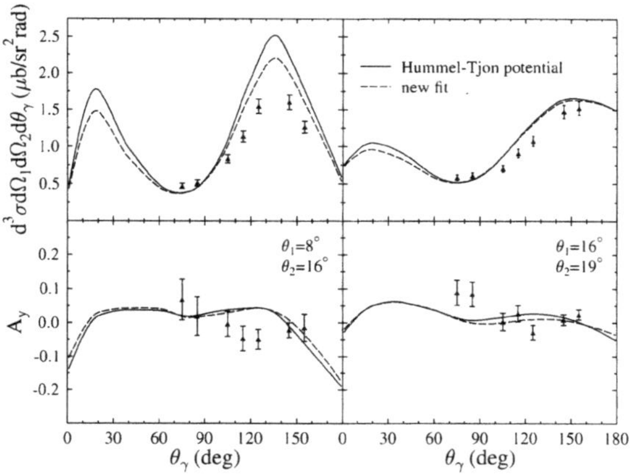

Figure 8. The $pp\gamma$ cross section in the KVI kinematics using a refitted OBE model to the strong PWA93 pp phaseshifts. The pp scattering length of the refitted OBE model a=-12 fm.

For the form factor $g(p)$ various forms have been used. The parameters of $g(p)$ are fitted to describe the phaseshift of the $^1S_0$ channel over an energy region up to 150 MeV. The pp T-matrix can be determined exactly for this model using a two-potential approach. In Fig. 9 is shown the effect of including the Coulomb on the $pp\gamma$ cross section at 190 MeV as a function of the angle of proton 1. From this we see, that the Coulomb effect at small angles becomes substantial, but is only of the order of a few percent at the existing KVI kinematics. It should be interesting to see such an enhancement experimentally.

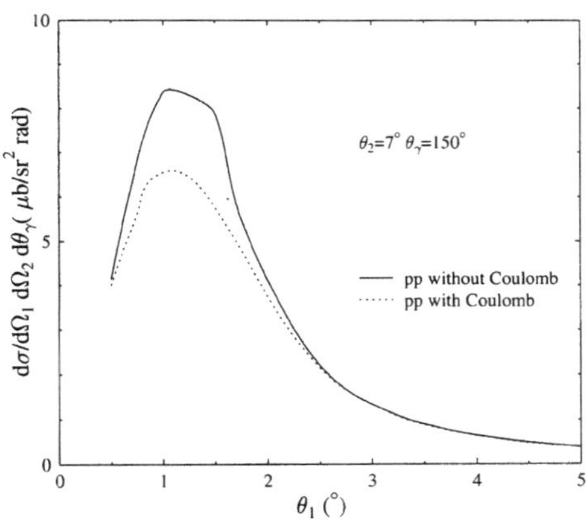

Figure 9. The $pp\gamma$ cross section in the KVI kinematics as a function of the angle of proton 1 with or without Coulomb.

Since the pp fitted OBE model is not exactly phase equivalent with the PWA93 it is interesting to examine how sensitive the predictions are for the on-shell amplitude. For this purpose we have modified the OBE half-off-shell T-matrix to reproduce the PWA93 on-shell by

$$T(\hat{p}, p'; s) \rightarrow T_{\mathrm{OBE}}(\hat{p}, p'; s) \frac{T_{SC}(\hat{p})}{\hat{T}_{\mathrm{OBE}}}, \tag{6}$$

where $\hat{p}$ is the on-shell relative momentum. The T_{SC} is the Coulomb modified amplitude obtained using the PWA93 phases. It is given by [28]

$$T_{SC}(\hat{p}) = \frac{e^{2\mathrm{i}\sigma_0}}{\hat{p}\cot(\delta_{\mathrm{PWA93}}) - \mathrm{i}\hat{p}}, \tag{7}$$

where $\sigma_0 = \arg\Gamma(1 + \mathrm{i}\eta)$ is the pure Coulomb s-wave phaseshift with $\eta = \alpha m/2\hat{p}$. The on-shell $\hat{T}_{\mathrm{OBE}}$ is expressed in terms of the OBE phaseshifts as

$$\hat{T}_{\mathrm{OBE}} = \frac{1}{\hat{p}\cot(\delta_{\mathrm{OBE}}) - \mathrm{i}\hat{p}}. \tag{8}$$

Using this modification we have calculated the bremsstrahlung process. Preliminary results [13] show that the $pp\gamma$ cross section did not change visibly.

5 Concluding Remarks

We have discussed two approaches to describe the bremsstrahlung in pp scattering. One is based on an expansion for low photon momenta and makes use of the low energy theorem. It has the simplicity that only the on-shell NN amplitude is needed, but relies on certain approximations. The other is a field theory based analysis of the $pp\gamma$ process employing a relativistic OBE model for the NN interaction. The actual calculations are more complicated due to the rescattering contribution, necessary to satisfy gauge invariance. Since the dynamics in this approach is given by the mesonic degrees of freedom also the contributions of the e.m. mesonic and isobar two-body currents can be included.

Comparing the predictions with existing experiments, the overall agreement of both approaches is reasonable at low photon momenta. We have paid extensively attention to a kinematics in the KVI experiment, where the predictions of the two approaches are very different. It is shown that in this case the predictions of the dynamical model are very sensitive to the low energy NN interaction. We find in particular if the NN model fits the pp data, the difference between the OBE model and the experimental data has become substantially smaller due to the proper low energy behaviour of the pp interaction. The relativistic model can be compared with the soft photon approximation. Although the kinematics demands low energy NN information in the microscopic model, in the SPA due to the choice of the invariants only on-shell NN data around 100 MeV are needed in the calculation, suggesting that the agreement of the latter model with the experimental data may be accidental.

In the theoretical analysis we have found rather small sensitivities to the off-shell dependence of the $pp\gamma$ at low photon energies. This can be attributed to the constraints imposed by gauge invariance. At higher energies more sensitivity exists. It should be interesting to study the bremstrahlung process at higher energies. This applies also to the case of np bremsstrahlung. The latter has the advantage that the Coulomb long-range problem is absent.

Acknowledgement. This work is supported by the U.S. Department of Energy contract number DE-AC05-84ER40150 under which the Southeastern Universities Research Association (SURA) operates the Thomas Jefferson National Accelerator Facility.

References

1. V. Brown: Phys. Rev. **177**, 1498 (1969)
2. M.K. Liou, M.I. Sobel: Ann. Phys. (NY) **72**, 323 (1972)
3. R. Workman, H. Fearing: Phys. Rev. **C34**, 780 (1986)
4. H. Huisman et al.: Phys. Rev. Lett. **83**, 4017 (1999); Phys. Rev. **C65**, 031001 (2002)
5. J.G. Messchendorp et al.: Phys. Rev. **C61**, 064007 (2000)
6. M.K. Liou et al.: Phys. Lett. **B345**, 372 (1995)
7. M.K. Liou et al.: Phys. Rev. **C54**, 1574 (1996); Phys. Rev. **C65**, 014001 (2002)
8. J. Eden, M. Gari: Phys. Rev. **C53**, 1102 (1996)
9. F. de Jong et al.: Phys. Lett. **B333**, 1 (1994); V. Herrmann et al.: Phys. Rev. **C46**, 2199 (1992); Nucl. Phys. **A582**, 568 (1995)
10. G. Martinus et al.: Phys. Lett. **B402**, 7 (1997); Phys. Rev. **C56**, 2945 (1997); ibid. **C58**, 686 (1998)
11. G. Martinus et al.: Few-Body Syst. **26**, 197 (1999)
12. D. Cozma et al.: Phys. Rev. **C65**, 024001 (2002)
13. D. Cozma et al.: in preparation
14. M. Baranger et al.: Nucl. Phys. **A138**, 1 (1969)
15. P.U. Sauer: Nucl. Phys. **A170**, 497 (1971)
16. P.U. Sauer, J.A. Tjon: Nucl. Phys. **A216**, 541 (1973)
17. F. Low: Phys. Rev. **96**, 1428 (1954)

18. A. Korchin et al.: Nucl. Phys. **A581**, 493 (1995)

19. J.A. Tjon, S.J. Wallace: Phys. Rev. **C32**, 1667 (1985); ibid. **C36**, 1085 (1987)

20. J. Fleischer, J.A. Tjon: Nucl. Phys. **B84**, 375 (1974); Phys. Rev. **D15**, 2337 (1977); ibid. **D21**, 87 (1980)

21. M.J. Zuilhof, J.A. Tjon: Phys. Rev. **C22**, 2369 (1980)

22. E. Hummel, J.A. Tjon: Phys. Rev. **C42**, 423 (1990); ibid. **C49**, 21 (1994)

23. A.A. Logunov, A.N. Tavkhelidze: Nuovo Cim. **29**, 380 (1963); R. Blankenbecler, R. Sugar: Phys. Rev. **142**, 1051 (1966)

24. M. Jetter, H.W. Fearing: Phys. Rev. **C51**, 1666 (1995)

25. K. Michaelian et al.: Phys. Rev. **D41**, 2689 (1990)

26. E.E. van Faassen, J.A. Tjon: Phys. Rev. **C28**, 2354 (1983); ibid. **C30**, 285 (1984)

27. V.G.J. Stoks et al.: Phys. Rev. **C48**, 792 (1993); M.C.M. Rentmeester et al.: Phys. Rev. Lett. **82**, 4992 (1999)

28. D.R. Harrington: Phys. Rev. **139**, B691 (1965); R.G. Newton: *Scattering theory of waves and particles*. New York: McGraw-Hill (1966)

Few-Body Systems Suppl. 15, 79–85 (2003)

Few-Body Systems

Relativity in Polarized Electron Scattering Observables

J.A. Caballero[1,*], M.C. Martínez[1], T.W. Donnelly[2], E. Moya de Guerra[3], J.M. Udías[4], J.R. Vignote[4]

[1] Departamento de Física Atómica, Molecular y Nuclear, Universidad de Sevilla, E-41080 Sevilla, Spain

[2] Center for Theoretical Physics, Massachusetts Institute of Technology, Cambridge, MA 02139, USA

[3] Instituto de Estructura de la Materia, CSIC, E-28006 Madrid, Spain

[4] Departamento de Física Atómica, Molecular y Nuclear, Universidad Complutense de Madrid, E-28040 Madrid, Spain

Abstract. Coincidence scattering of polarized electrons from nuclei with polarization transfer to outgoing nucleons is studied within the context of relativistic mean field theory. Effects introduced by the dynamical enhancement of the lower components of the bound and scattered nucleon wave functions are analyzed for the polarized response functions and transferred polarization asymmetries. Results obtained by projecting out the negative-energy components are compared with the fully-relativistic calculation. The crucial role played by the relativistic dynamics in some spin-dependent observables is clearly manifested even for low/medium values of the missing momentum. Kinematical relativistic effects are also analyzed. A discussion of the factorization approach and the mechanism for its breakdown is also briefly presented.

1 Introduction

Quasielastic coincidence electron scattering reactions have provided over the years important insight into single-particle properties of nuclei. This is so because at quasielastic kinematics the reaction mechanism underlying $(e, e'N)$ reactions can be treated with confidence in the impulse approximation (IA), i.e., assuming the virtual photon attached to a single bound nucleon that absorbs the whole momentum (q) and energy (ω) (see [1, 2] for details).

A large fraction of the theoretical analyses of $(e, e'N)$ reactions in past years was carried out on the basis of non-relativistic calculations. Within this scheme,

* *E-mail address:* juan@nucle.us.es

the bound and ejected nucleons are described by non-relativistic wave functions which are solutions of the Schrödinger equation with phenomenological potentials. Moreover, the current operator is also described by a non-relativistic expression derived directly from a Pauli reduction. Such non-relativistic reductions constitute the basis for the standard distorted-wave impulse approximation (DWIA) that has been widely used to describe $(e, e'N)$ experiments performed at intermediate energies [1, 2].

In the last decade some experiments performed have involved momenta and energies high enough to invalidate the non-relativistic expansions assumed in DWIA. A consistent description of these processes requires one to incorporate relativistic degrees of freedom wherever possible. Within this context, nuclear responses and cross sections have been investigated recently by our group using the relativistic mean field approach [3]. This constitutes the basis of the relativistic distorted-wave impulse approximation (RDWIA), where bound and scattered wave functions are described as Dirac solutions with scalar and vector potentials, and the relativistic free nucleon current operator is assumed.

Relativistic contributions can be cast into two general categories, kinematical and dynamical relativistic effects. The former are directly connected with the structure of the 4-vector current operator, compared with the non-relativistic one that usually involves p/M_N, q/M_N and ω/M_N expansions. The latter, dynamical relativistic effects, come from the difference between the relativistic and non-relativistic nucleon (bound and ejected) wave functions involved. Within these dynamical relativistic effects one may distinguish effects associated with the Darwin term (dynamical depression of the upper component of the scattered nucleon wave function in the nuclear interior that mainly affects the determination of spectroscopic factors at low missing momenta) and effects due to the dynamical enhancement of the lower components of the relativistic wave functions (expected to be especially relevant at high missing momenta, although they have proven to play an important role for some particular observables even at low/medium p values).

So far, fully-relativistic analyses of $(e, e'p)$ reactions have clearly improved the comparison with experimental data [3, 4]. In recent work we have undertaken a systematic study of the relativistic effects in different observables. We started with the relativistic plane-wave impulse approximation (RPWIA), i.e., neglecting final-state interactions (FSI) between the outgoing nucleon and the residual nucleus. Although being an oversimplication, the RPWIA approach allows one to simplify the analysis, disentangling the relativistic effects from other distortion effects. The presence of negative-energy components in the relativistic bound nucleon wave function was shown to be very important for some observables even at low/moderate values of the missing momentum. In particular, the interference TL and TT responses were proved to be very sensitive to dynamical effects of relativity affecting the lower components. These results persist also when FSI are included.

Following similar arguments, we have presented in [5] a systematic study of the new response functions that enter in the description of $\mathrm{A}(\vec{e}, e'\vec{N})\mathrm{B}$ processes. Since spin and relativity go hand in hand, one may *a priori* consider the relativistic approach to be better suited to describe nucleon polarization observables. Kinematical and dynamical relativistic effects for responses and polarization observables have been analyzed in detail within the plane wave limit for the outgoing nucleon. In work in progress effects in the final state are also being incorporated through relativistic FSI. They are briefly summarized in this work.

2 Formalism

In the case of $\mathrm{A}(\vec{e},e'\vec{N})\mathrm{B}$ reactions, the hadronic response functions are usually given by referring the recoil nucleon polarization to the coordinate system defined by the axes: $\mathbf{l}$ (parallel to the momentum $\mathbf{p}_N$ of the outgoing nucleon), $\mathbf{n}$ (perpendicular to the plane containing $\mathbf{p}_N$ and the momentum transfer $\mathbf{q}$), and $\mathbf{s}$ (determined by $\mathbf{n}\times\mathbf{l}$). In terms of the polarization asymmetries, the differential cross section can be expressed in the form

$$\frac{d\sigma}{d\varepsilon' d\Omega_e d\Omega_N} = \frac{\sigma_0}{2}\left[1+\mathbf{P}\cdot\boldsymbol{\sigma}+h\left(A+\mathbf{P}'\cdot\boldsymbol{\sigma}\right)\right] , \tag{1}$$

where σ_0 is the unpolarized cross section, A denotes the electron analyzing power, and $\mathbf{P}$ ($\mathbf{P}'$) represents the induced (transferred) polarization. A general study of the properties and symmetries of all of these responses and polarizations can be found in [6]. In the plane wave limit for the outgoing nucleon, the induced polarization $\mathbf{P}$ and the analyzing power A are zero. In terms of nuclear responses, from the total of eighteen response functions, only nine survive within RPWIA. Four, R_0^L, R_0^T, R_0^{TL} and R_0^{TT}, represent the unpolarized responses and the five remaining, $R_l^{T'}$, $R_s^{T'}$, $R_l^{TL'}$, $R_s^{TL'}$ and $R_n^{TL'}$ (this one enters only for out-of-plane kinematics), depend explicitly on the recoil nucleon polarization and only enter when the electron beam is also polarized. In this work, we restrict our discussion to these observables known as transferred polarization responses and/or transferred asymmetries.

First, we consider the case of the plane wave limit for the outgoing nucleon wave function (RPWIA) The role played by the negative-energy projection components in the polarized responses was discussed in [5]. There we show that in two responses, $R_l^{T'}$ and $R_s^{TL'}$, the contribution of the negative-energy projections is almost negligible, that is, dynamical relativistic effects from the bound nucleon wave function do not significantly affect these responses. On the contrary, $R_s^{T'}$ and $R_l^{TL'}$ are much more sensitive. This result resembles what appeared for the unpolarized interference TL response [7]. Hence there exists a strong discrepancy between RPWIA results and those corresponding to the standard factorized PWIA.

Dynamical enhancement of the lower components in the bound nucleon wave function is even more clearly seen when analyzing the transferred polarization asymmetries (Fig. 1). Here the fully-relativistic RPWIA results (dashed lines) corresponding to the Coulomb gauge with the $CC1$ and $CC2$ choices of the current (see ref. [5] for details on the current operators), are compared with the

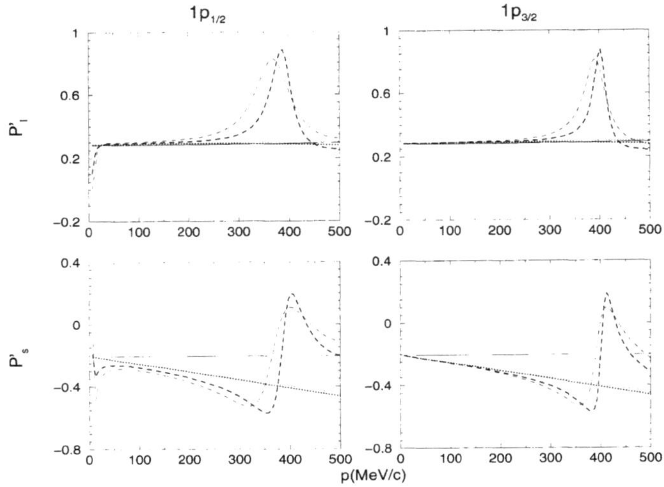

Figure 1. Transferred polarization asymmetries P_l' and P_s' in the plane wave limit for the outgoing nucleon. Fully-relativistic results (dashed lines) are compared with their positive-energy projection contributions (dotted lines). Thin lines correspond to the $CC1$ current operator and thick lines to $CC2$. We also show for comparison the static limit result (solid line).

transferred polarizations obtained by projecting out the negative-energy components (dotted lines). Results for $p_{1/2}$ (left-hand panels) and $p_{3/2}$ (right-hand panels) are shown. Kinematics corresponds to $q = 1$ GeV/c, $\omega = 440$ MeV/c and forward ($\theta_e = 23^0$) scattering angle. First, note the difference between the relativistic and projected results observed at very small missing momentum values for the $p_{1/2}$ shell. This effect comes directly from the quantum number $\overline{\ell}$ involved in the lower component of the bound state wave function ($\overline{\ell} = 0$ for $p_{1/2}$). Moreover, it is also important to point out that fully-relativistic and positive-energy projected results typically do not differ appreciably for p-values up to ~ 300 MeV/c. For $p > 300$ MeV/c relativistic and projected results start to deviate from each other. This general behaviour is what one expects because of the clear dominance of the positive-energy projection component of the momentum distribution in the region $p \leq 300$ MeV/c [7]. On the contrary, for $p > 300$ MeV/c the negative-energy components are similar to or even larger than the positive ones, and hence the effects of the dynamical enhancement of the lower components in the bound relativistic wave functions are clearly visible in the transfer polarization asymmetries.

Obviously, final state interactions (FSI) between the ejected nucleon and the residual nucleus should be included in the analysis of $A(\vec{e}, e'\vec{N})$ processes in order to compare with data. In fact, FSI introduce significant modifications in the responses and transferred polarization asymmetries. However, the high sensi-

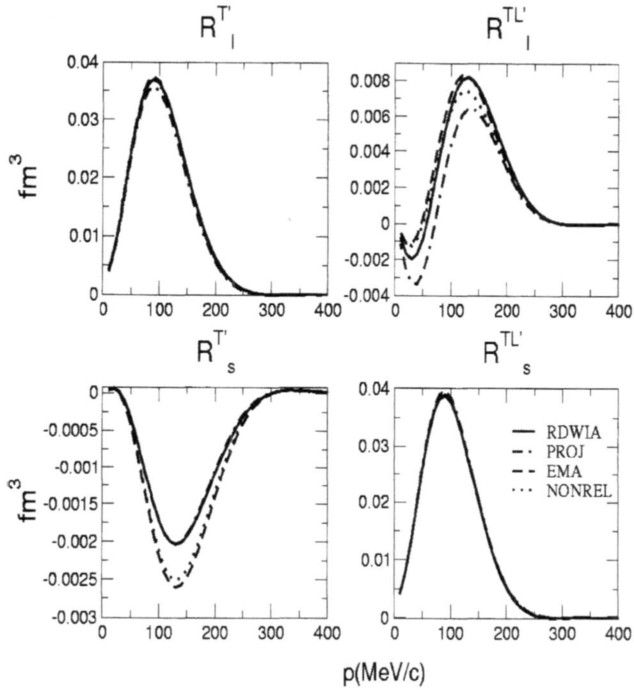

Figure 2. Polarized response functions for the $1p_{1/2}$ shell. Fully-relativistic response (solid line) is compared with the projected one (dot-dashed), EMA (dashed) and non-relativistic approach (dotted).

tivity of polarization-related-observables to negative-energy projections shown within RPWIA is maintained in the relativistic distorted-wave impulse approximation (RDWIA). This was already the case for the unpolarized interference longitudinal-transverse response R^{TL} and A_{TL}-asymmetry. The analysis presented in [8] proves the crucial role played by both kinematical and dynamical relativistic effects in order to fit the experiment. In particular, the richness shown by the structure of the left-right asymmetry (A_{TL}) is only consistent with predictions of relativistic calculations that include dynamic enhancement of the lower components of Dirac spinors.

Similar comments can be also applied to the transferred polarization observables. In particular, the different sensitivity to relativistic effects shown by the transferred polarized responses within RPWIA, persists when RDWIA calculations are involved. This is illustrated in Fig. 2 where we present the polarized responses corresponding to the $p_{1/2}$-shell in ^{16}O. Kinematics has been selected as in the previous figure. We compare the fully distorted relativistic calculation using the Coulomb gauge and the current operator CC2 (solid line) with the results after projecting the bound and scattered proton wave

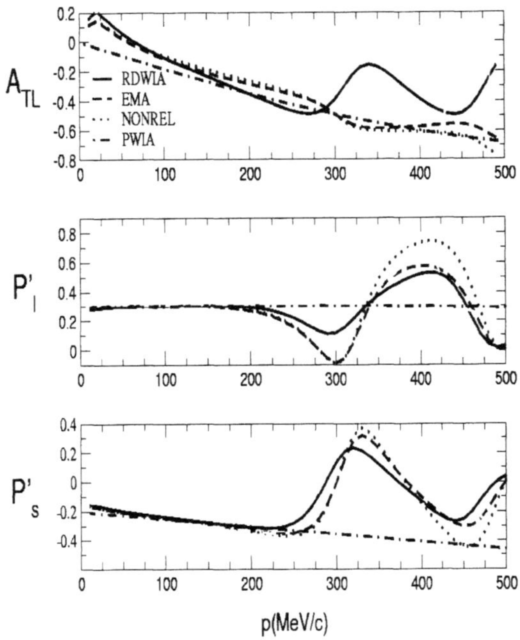

Figure 3. Asymmetry A_{TL} and transferred polarization ratios P_l' and P_s'. RDWIA results (solid) are compared with EMA (dashed), non-relativistic (dotted) and factorized PWIA (dot-dashed) calculations.

functions over positive-energy states (dot-dashed) and using the asymptotic momenta (dashed). This last approach is equivalent to the effective momentum approximation (EMA). Also for comparison we show a non-relativistic calculation where the non-relativistic current operator (obtained from a Pauli reduction) corresponds to the expression given in [9]. Note that two responses, $R_s^{TL'}$ and $R_l^{T'}$, are very insensitive to dynamical and kinematical relativistic effects. On the contrary, effects due to relativity are clearly visible for $R_l^{TL'}$ and $R_s^{T'}$. This result agrees with the findings within RPWIA [5].

Finally, in Fig. 3 we show the RDWIA results for the left-right asymmetry A_{TL} and transferred polarization asymmetries P_l' and P_s'. The curves are labelled as in the previous figure. It is particularly interesting the comparison between the fully relativistic results and the ones corresponding to the asymptotic projection (equivalent to EMA) and the non-relativistic calculation. Note that the richer structure shown by the relativistic calculation in A_{TL} is basically lost within the EMA and non-relativistic approaches. Moreover, these results get closer to the factorized PWIA calculation. This modest deviation from the factorized (free) result is mainly due to the spin-orbit coupling in the final state. The behaviour shown by the transferred polarization asymmetries is clearly different. Here, the EMA and non-relativistic calculations do no change substantially the structure shown by the fully relativistic result, differing clearly from the factorized (free) one. This means that the spin-orbit coupling in the final wave function breaks down completely factorization in the polarized observables.

Summarizing, in this work we have analyzed outgoing nucleon polarized observables within a relativistic mean field approach. We have shown that dynamical relativistic effects, namely the enhancement of the lower components of Dirac wave functions, affect very significantly the observables. This was already the case for the unpolarized observables, R^{TL} and A_{TL}. However, here we also show that the behaviour presented by the transferred polarization asymmetries is clearly different from the one corresponding to the left-right asymmetry.

Acknowledgement. This work was supported by funds provided by DGICYT (Spain) under Contracts Nos. PB/98-1111, PB/98-0676 and the Junta de Andalucía (Spain) and by the U.S. Department of Energy under Cooperative Research Agreement No. DE-FC02-94ER40818. M.C.M acknowledges support from a fellowship from the Fundación Cámara (University of Sevilla). J.A.C. also acknowledges MEC (Spain) for a sabbatical stay at MIT (PR2001-0185).

References

1. S. Boffi, C. Giusti, F.D. Pacati, M. Radici: *"Electromagnetic Response of Atomic Nuclei"*, Oxford-Clarendom Press 1996; Phys. Rep. **226**, 1 (1993)
2. J.J. Kelly: Adv. Nucl. Phys. **23**, 77 (1996)
3. J.M Udías, P. Sarriguren, E. Moya de Guerra, E. Garrido, J.A. Caballero: Phys. Rev. **C48**, 2731 (1993); **51**, 3246 (1995); **53**, R1488 (1996)
4. J.M. Udías, J.A. Caballero, E. Moya de Guerra, J.R. Vignote, A. Escuderos: Phys. Rev. **C64**, 024614 (2001)
5. M.C. Martínez, J.A. Caballero, T.W. Donnelly: Nucl. Phys. **A707**, 83 (2002); **A707**, 121 (2002)
6. A. Picklesimer, J.W. Van Orden: Phys. Rev. **C35** 266 (1987); **40**, 290 (1989)
7. J.A. Caballero, T.W. Donnelly, E. Moya de Guerra, J.M. Udías: Nucl. Phys. **A632**, 323 (1998); **A643** 189 (1998)
8. J.M. Udías, J.A. Caballero, E. Moya de Guerra, J.E. Amaro, T.W. Donnelly: Phys. Rev. Lett. **83**, 5451 (1999)
9. J.E. Amaro, J.A. Caballero, T.W. Donnelly, A.M. Lallena, E. Moya de Guerra, J.M. Udías: Nucl. Phys. **A602**, 263 (1996); **A611**, 163 (1996)

Few-Body Systems Suppl. 15, 87–98 (2003)

Few-Body Systems

Three-Nucleon Electroweak Capture Reactions

L.E. Marcucci[1,2], M. Viviani[2,1], A. Kievsky[2,1], S. Rosati[1,2], R. Schiavilla[3,4]

[1] Physics Department, University of Pisa, Via Buonarroti 2, I-56100 Pisa, Italy

[2] INFN, Sezione di Pisa, Via Buonarroti 2, I-56100 Pisa, Italia

[3] Jefferson Laboratory, Newport News, Virginia 23606, USA

[4] Physics Department, Old Dominion University, Norfolk, VA 23529, USA

Abstract. Recent advances in the study of the $p-d$ radiative and $\mu-{}^3$He weak capture processes are presented and discussed. The three-nucleon bound and scattering states are obtained using the correlated-hyperspherical-harmonics method, with realistic Hamiltonians consisting of the Argonne v_{14} or Argonne v_{18} two-nucleon and Tucson-Melbourne or Urbana IX three-nucleon interactions. The electromagnetic and weak transition operators include one- and two-body contributions. The theoretical accuracy achieved in these calculations allows for interesting comparisons with experimental data.

1 Introduction

A number of electromagnetic (EM) and weak transitions in light nuclei have interesting astrophysical implications. They are also very important for an understanding of nuclear structure and dynamics. The theoretical description of these processes requires the knowledge of the initial and final nuclear states, which can be either bound or in the continuum. Furthermore, it is important to use a realistic model for the EM and weak current operators. The three-nucleon systems, in particular, provide a unique "laboratory" due to the capability, achieved in the last few years, of obtaining very accurate bound and continuum wave functions. The accuracy of the calculated three-nucleon wave functions has been verified by comparing results for a variety of observables obtained by a number of groups using different techniques [1]. At present, good overall agreement exists between the theoretical and experimental $N-d$ elastic and inelastic scattering observables (a notable exception, however, is the A_y analyzing power at low energies) [2, 3]. Therefore the study of EM and weak transitions in the three-nucleon system does not suffer of uncertainties related to the computation of the nuclear wave functions and it is a direct test of the nuclear Hamiltonian H from which these wave functions are obtained, and of the model used to describe the nuclear currents. Since the nuclear EM current is related to H through current conservation, it is clear that the two topics are inter-related. Furthermore, it is interesting to understand whether relativistic corrections as well as Δ-isobar and additional sub-nucleonic degrees of freedom play a role in these processes.

The model for the nuclear EM and weak current considered here has been recently reviewed in refs. [4, 5, 6] and has been tested in numerous few-nucleon processes. It includes one- and two-body operators. The one-body operators are obtained directly from the non-relativistic limit of the covariant single-nucleon vector and axial currents. In the study of the muon capture, the contribution coming from the induced pseudo-scalar term of the nucleon axial current has to be included (it gives a negligible contribution to β-decay processes). However, the experimental value of the corresponding form factor $G_{\rm PS}(q_\sigma^2)$ is rather uncertain. Assuming pion-pole dominance, the partially conserved axial current (PCAC) hypothesis, and the Goldberger-Treiman relation, $G_{\rm PS}$ is predicted to be [7, 8, 9, 10, 11]

$$G_{\rm PS}^{\rm PCAC}(q_\sigma^2) = -\frac{2\, m_\mu m_N}{m_\pi^2 + q_\sigma^2}\, G_{\rm A}(q_\sigma^2) \ , \qquad (1)$$

where q_σ is the four-momentum transferred to the nuclear system, m_N, m_μ and m_π indicates the nucleon, muon and pion mass, respectively, and $G_{\rm A}$ is the axial form factor. In our calculation, we have assumed

$$G_{\rm PS}(q_\sigma^2) = R_{\rm PS}\, G_{\rm PS}^{\rm PCAC}(q_\sigma^2) \ , \qquad (2)$$

where $R_{\rm PS}$ is a parameter which has been varied to study the sensibility of our results to this form factor and to investigate to which extent $G_{\rm PS}^{\rm expt} = G_{\rm PS}^{\rm PCAC}$. However, most of the calculations have been performed with the choice $R_{\rm PS} = 1$.

The two-body EM current is separated in two terms. There is a "model-independent" part which is constructed consistently with the nucleon-nucleon interaction, in order to satisfy the current conservation relation [12]. Current conservation is only approximate for the momentum dependent terms. The second part includes "model-dependent" contributions which come from the $\rho\pi\gamma$ and $\omega\pi\gamma$ processes and the Δ degrees of freedom. The latter contribution is included in the current and in the nuclear medium in an approximate way, by following the procedure described in ref. [13]. The two-body weak vector current is then obtained from the isovector part of the EM current, in accordance with the conserved-vector-current hypothesis.

Two-body terms have been taken into account in both the axial charge and current operators. The two-body axial charge operator has been obtained consistently with the two-nucleon interaction model, following the methods of ref. [14]. The two-body axial current operators are derived from a meson-exchange model, including π- and ρ-exchanges and the $\rho\pi$-transition processes, as well as Δ-isobar excitation [7, 15]. The latter process gives the dominant contribution. However, its magnitude depends critically on the value adopted for the $N\Delta$ axial coupling constant g_A^*. In the quark model, g_A^* is related in a simple way to the axial coupling constant of the nucleon g_A ($g_A \approx 1.26$). However, given the uncertainties inherent in quark model predictions, a more reliable estimate for g_A^* is obtained by adjusting its value to reproduce the experimental value of the Gamow-Teller matrix element in tritium β-decay [7, 15]. In this way, the model dependence of the weak axial current is significantly reduced, as shown by previous studies of proton weak captures on ^{1}H [15] and ^{3}He [7]. It is important to note that the value of g_A^* depends on how the Δ-isobar degrees of freedom are treated. Moreover, given the procedure used to determine g_A^*, the latter cannot be naively interpreted as the $N\Delta$ axial coupling constant, since the contributions of additional resonances not included in the present study will contaminate its value.

The ^{3}He, ^{3}H bound and the $p-d$ continuum wave functions have been calculated by expanding on a basis of pair-correlated hyperspherical harmonic (PHH) functions [16]. Such a technique has been proven to be very accurate. Various $n-d$ elastic scattering observables calculated by solving the Faddeev equations [2] are in remarkable agreement with the corresponding results obtained with the PHH technique. For example, the phase shift and mixing angle parameters calculated by the two methods at the center-of-mass (c.m.) energy $E_{\rm c.m.} = 2$ MeV have been found to differ at level of 0.1%, at most [1]. It should be pointed out that with the PHH technique the inclusion of the Coulomb potential, clearly very important in the energy range considered here, does not present any difficulty.

This paper is organized as follows. The $p-d$ radiative capture and the muon capture on ^{3}He are discussed in Sect. 2 and 3, respectively. The last section contains a summary and some concluding remarks.

2 Radiative $p-d$ Capture

The applications of our formalism for c.m. energies ranging from zero to 2 MeV, namely below the deuteron breakup threshold (DBT), were already presented in refs. [17] and [18]. Recently we have extended the PHH technique, in order to compute also $p-d$ scattering wave functions above the DBT [19]. We can therefore compute $p-d$ capture observables at higher energies than previously published. We present here in Fig. 1 a preliminary study at $E_{\rm c.m.} = 3.33$ MeV, where high-quality data, including differential cross sections, vector and tensor analyzing powers [20] exist. Results for differential cross sections at $E_{\rm c.m.} = 6.67$ and 9.87 MeV and deuteron tensor analyzing powers at $E_{\rm c.m.} = 5.83$ MeV are given in Figs. 2 and 3 respectively. The corresponding experimental data are taken from refs. [21] and [22].

All the theoretical calculations reported in this section have been obtained using the Argonne v_{18} (AV18) [23] two-nucleon and Urbana IX (UIX) [24] three-nucleon interactions. By inspection of Fig. 1 we can conclude that our results

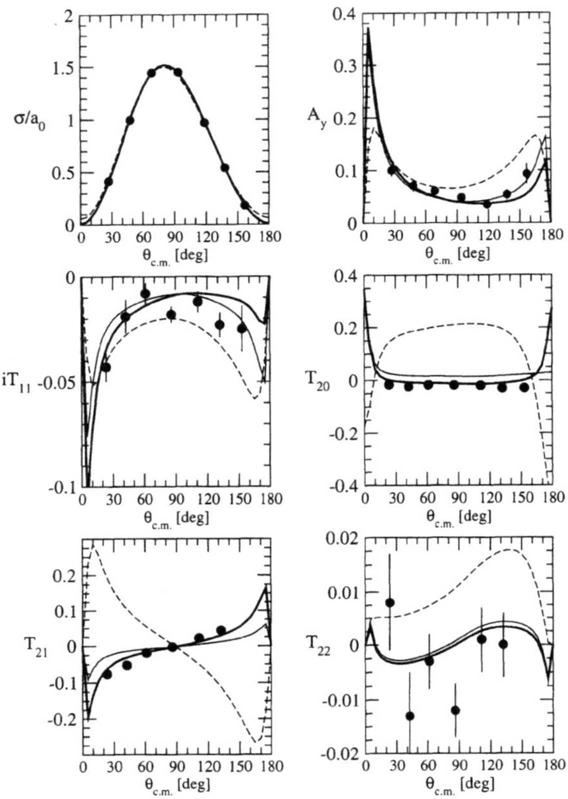

Figure 1. Differential cross section, proton vector analyzing power, and the four deuteron tensor analyzing powers for $p-d$ capture at $E_{\text{c.m.}} = 3.33$ MeV, obtained with the AV18/UIX Hamiltonian model and one-body only (thick dashed lines) or both one- and two-body currents (thin solid lines), are compared with the experimental values of ref. [20]. The results obtained in the LWA approximation for the spin-flip E_1 RME's are also shown (thick solid lines). In the first panel, $a_0 = \int d\Omega\ \sigma/(4\pi)$.

(thin solid lines) for the differential cross section and the observables A_y and iT_{11} are in good agreement with the experimental data. On the contrary, for the observables T_{20} and T_{21}, significant discrepancies can be observed. A similar situation was found for the $E_{\text{c.m.}} = 2$ MeV observables [18]. The same situation is present also at higher energies. In fact the differential cross sections in Fig. 2 are well reproduced by our full calculation, as well as the $A_y(d)$ observable in Fig. 3, which is basically proportional to iT_{11}. Also at $E_{\text{c.m.}} = 5.83$ MeV, as can be seen from Fig. 3, discrepancies are present for the deuteron tensor observables A_{xx}, A_{yy} and A_{zz}, which are a linear combination of T_{20}, T_{21} and

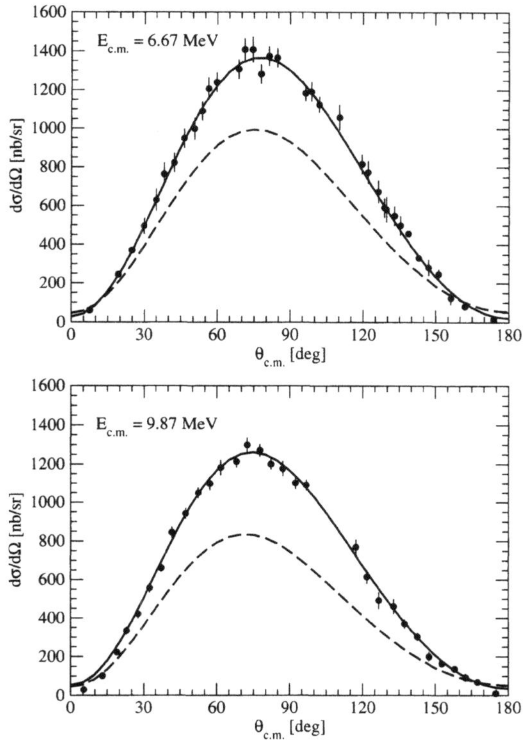

Figure 2. Differential cross section for $p-d$ capture at $E_{\text{c.m.}} = 6.67$ MeV (top panel) and $E_{\text{c.m.}} = 9.87$ MeV (lower panel). Notation as in Fig. 1. Thick and thin solid lines are on top of each other. The data are from ref. [21].

T_{22} (the latter observable is however rather small).

The analysis of the $E_{\text{c.m.}} = 2$ MeV results in ref. [18] traced back the problem to an overprediction of the spin-flip electric dipole E_1 reduced matrix elements (RME's) (namely, those arising from the transitions where the $p-d$ spins in the incident channel are coupled to $S = 3/2$). In fact, when the same RME's are computed in the long wavelength approximation (LWA) at leading order (thick solid line), the observables T_{20} and T_{21} are better reproduced (see Figs. 1 and 3). Interestingly, an analysis of the next-to-leading order terms in the LWA performed in ref. [18], has shown that they give a sizeable contribution to the spin-flip E_1 RME's, showing the inadequacy of the use of the leading order only for the calculation of this small spin-flip transition matrix elements. The origin of the discrepancies observed in the deuteron tensor polarization observables is currently under investigation.

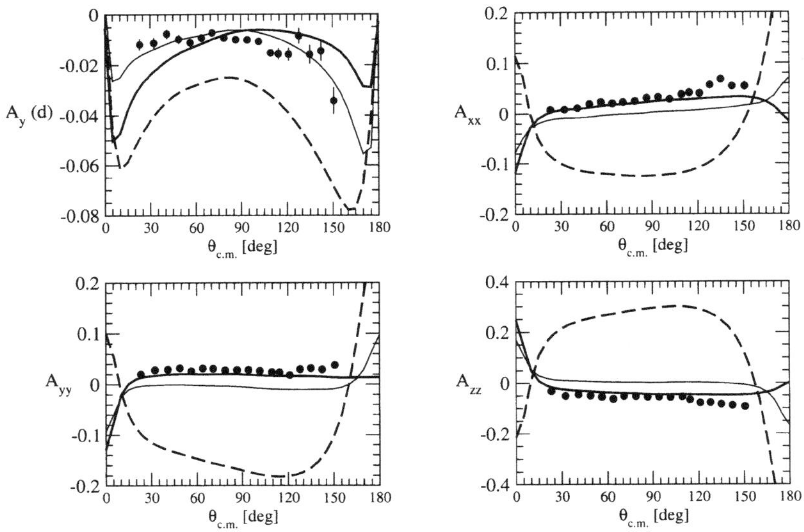

Figure 3. Deuteron tensor analyzing powers for $p - d$ capture at $E_{\mathrm{c.m.}} = 5.83$ MeV. Notation as in Fig. 1. The data are from ref. [22].

3 Muon Capture

The μ^- weak capture on ^{3}He can occur through three different hadronic channels:

$$\mu^- + ^3\mathrm{He} \rightarrow \ ^3\mathrm{H} + \nu_\mu \quad (70\%) \ , \tag{3}$$

$$\mu^- + ^3\mathrm{He} \rightarrow n + d + \nu_\mu \quad (20\%) \ , \tag{4}$$

$$\mu^- + ^3\mathrm{He} \rightarrow n + n + p + \nu_\mu \quad (10\%) \ . \tag{5}$$

The focus of the present section is on the first process [8]. Some of the nuclear physics issues in muon capture have been reviewed recently in ref. [25]. When the triton polarization is not detected, the differential capture rate for the reaction (3) is given by

$$\frac{\mathrm{d}\Gamma}{\mathrm{d}(\cos\theta)} = \frac{1}{2}\Gamma_0 \left[1 + A_\mathrm{v} P_\mathrm{v} \cos\theta + A_\mathrm{t} P_\mathrm{t} \left(\frac{3}{2}\cos^2\theta - \frac{1}{2} \right) + A_\Delta P_\Delta \right] \ , \tag{6}$$

where Γ_0 is the total capture rate, A_v, A_t and A_Δ angular correlation parameters and θ is the angle between the muon polarization and the leptonic momentum transfer. The coefficients P_v, P_t and P_Δ are linear combinations of the probabilities $P(f, f_z)$ of finding the μ^- ^{3}He system in the total-spin state

$|f\, f_z\rangle$, and are defined as [26, 27]

$$\begin{aligned} P_{\rm v} &= P(1,1) - P(1,-1) \ , \\ P_{\rm t} &= P(1,1) + P(1,-1) - 2\,P(1,0) \ , \\ P_{\Delta} &= P(1,1) + P(1,-1) + P(1,0) - 3\,P(0,0) \ . \end{aligned} \tag{7}$$

Therefore, $P_{\rm v}$ and $P_{\rm t}$ are proportional to the vector and tensor polarizations of the f=1 state, respectively, while P_{Δ} indicates the deviation of the f=0 population density from its statistical factor 1/4. Because of the small energy splitting between the f=0 and f=1 hyperfine states (1.5 eV) compared to the $\mu^-\,{}^3$He binding energy, and hence small deviation of $P(f,\, f_z)$ from its statistical value, direct measurements of the angular correlation parameters are rather difficult [25, 26, 28].

The results reported here have been obtained using either the AV18 and AV18/UIX interactions or the older Argonne v_{14} (AV14) two-nucleon interaction [29] in conjunction with the the Tucson-Melbourne (TM) three-nucleon interaction [30]. Note that both three-nucleon interactions have been adjusted to reproduce the triton binding energy.

Particular care has been put in the calculation of the 1s wave function $\psi_{1s}(x)$ of the muon in the electric field of the ^{3}He nucleus. Since the Bohr radius of the muonic atom is about 130 fm, i.e. much larger than the nuclear radius, it is well justified to approximate $\psi_{1s}(x)$ in the computation of the weak transition matrix elements with an average value $\psi_{1s}^{\rm av}$, defined as:

$$\psi_{1s}^{\rm av} = \frac{\int d\mathbf{x}\ {\rm e}^{{\rm i}\mathbf{q}\cdot\mathbf{x}}\psi_{1s}(x)\,\rho(x)}{\int d\mathbf{x}\ {\rm e}^{{\rm i}\mathbf{q}\cdot\mathbf{x}}\rho(x)} \ , \tag{8}$$

where $\rho(x)$ is the ^{3}He nucleus charge distribution and $\mathbf{q}$ is the leptonic momentum transfer. Finally, it is common to introduce the factor $\mathcal{R}$ defined as [10, 11]:

$$|\psi_{1s}^{\rm av}|^2 \equiv \mathcal{R}|\psi_{1s}(0)|^2 = \mathcal{R}\,\frac{(2\,\alpha\, m_{\rm r})^3}{\pi} \ , \tag{9}$$

where $\psi_{1s}(0)$ denotes the Bohr wave function evaluated at the origin for a point charge $2e$, and $m_{\rm r}$ is the reduced mass of the $\mu^-\,{}^3$He system. The factor $\mathcal{R}$ has been calculated with the following procedure: for a given Hamiltonian model we have calculated $\rho(x)$. Knowing $\rho(x)$, the Poisson equation has been solved to obtain the static electric potential. Finally we have solved the Dirac equation and found $\psi_{1s}(x)$. Using Eqs. (8) and (9), we have calculated $\mathcal{R}$ for the three nuclear Hamiltonian models considered here. The results are given in Table 1. They are in rather good agreement with the values given in literature [26].

Results for the capture rate Γ_0 and angular correlation parameters $A_{\rm v}$, $A_{\rm t}$, and A_{Δ} are presented in Table 2. The uncertainty (in parentheses) in the predicted values is due to the uncertainty in the determination of the $N\Delta$ transition parameter $g_{\rm A}^*$, as discussed in the Introduction. The latter reflects the experimental error in the Gamow-Teller matrix element of tritium β-decay.

Table 1. Values for $\mathcal{R}$ as defined in Eq. (9) for three different Hamiltonian models H.

H	$\mathcal{R}$
AV18	0.979
AV18/UIX	0.980
AV14/TM	0.980

Inspection of Table 2 shows that the theoretical determination of the total capture rate Γ_0, when the AV18/UIX and AV14/TM Hamiltonian models are used, is within 1 % of the recent experimental result [31]. Furthermore, the model dependence in the calculated observables is very weak: the AV18/UIX and AV14/TM results differ by less than 0.5 %. The agreement between theory and experiment and the weak model dependence mentioned above reflect, to a large extent, the fact that both the AV18/UIX and AV14/TM Hamiltonian models reproduce: i) the experimental binding energies as well as the charge and magnetic radii [32] of the three-nucleon systems; ii) the Gamow-Teller matrix element in tritium β-decay. In this respect, it is interesting to note that the capture rate predicted by the AV18 Hamiltonian model is about 4 % smaller than the experimental value. The same result has been found with the two-nucleon AV14 Hamiltonian model [8].

The value for the angular correlation parameter $A_{\rm v}$ listed in Table 2 is also in reasonable agreement with the corresponding experimental result which has however a rather large error. Note that the polarization observables are not sensitive to the inclusion of the three-nucleon force, as can be seen comparing the AV18 and AV18/UIX lines of Table 2.

The contributions of the different components of the weak current and charge operators to the observables are reported for the AV18/UIX model in Table 3. The column labelled "One-body" lists the contributions associated

Table 2. Capture rate Γ_0 (sec^{-1}) and angular correlation parameters $A_{\rm v}$, $A_{\rm t}$, and A_Δ calculated using PHH wave functions corresponding to the AV18, AV18/UIX and AV14/TM Hamiltonian models are compared with the experimental results. The theoretical uncertainties, shown in parentheses, reflect the uncertainty in the determination of the $N\Delta$ transition axial coupling constant $g_{\rm A}^*$. The experimental values of Γ_0 and $A_{\rm v}$ have been taken from ref. [31] and [28], respectively. Here, we have assumed $R_{\rm PS} = 1$ in Eq. (2).

Observable	AV18	AV18/UIX	AV14/TM	Expt.
Γ_0	1441(7)	1484(8)	1486(8)	1496(4)
$A_{\rm v}$	0.5341(14)	0.5350(14)	0.5336(14)	0.63(15)
$A_{\rm t}$	−0.3642(9)	−0.3650(9)	−0.3659(9)	
A_Δ	−0.1017(16)	−0.1000(16)	−0.1005(17)	

Table 3. Effects of the inclusion of the two-body currents for the muon capture rate Γ_0 (in $\sec^{-1}$) and angular correlation parameters A_v, A_t, and A_Δ. The PHH wave functions are obtained using the AV18/UIX Hamiltonian model. The column labelled "One-body" lists the contributions associated with the one-body vector and axial charge and current operators. The column labelled "Mesonic" lists the results obtained by including, in addition, the contributions from meson-exchange mechanisms. Finally the column labelled "Δ" lists the results obtained by including also the Δ-excitation contributions. The experimental values of Γ_0 and A_v are taken from ref. [31] and [28], respectively. Here, we have assumed $R_{PS} = 1$ in Eq. (2).

Observable	One-body	Mesonic	Δ	Expt.
Γ_0	1316	1384	1484	1496(4)
A_v	0.5749	0.5511	0.5350	0.63(15)
A_t	−0.3565	−0.3679	−0.3650	
A_Δ	−0.0686	−0.0810	−0.1000	

with the one-body terms of the vector and axial charge and current operators, including relativistic corrections proportional to $1/m^2$. The column labelled "Mesonic" lists the results obtained by including, in addition, the contributions from two-body vector and axial charge and current operators, associated with pion- and vector-meson-exchanges. Finally, the column labelled "Δ" lists the values of the observables obtained by including the contributions arising from Δ excitation.

Among the observables, Γ_0 and A_Δ are the most sensitive to two-body contributions in the weak current. These are in fact crucial for reproducing the experimental capture rate, see Table 3.

An important motivation of the present work is to test the sensitivity of the muon capture observables to the induced pseudo-scalar form factor G_{PS} and, eventually, infer its value from the Γ_0 measurement. Therefore, we have repeated the calculation using AV18/UIX PHH wave functions and several different values of the parameter R_{PS}, defined in Eq. (2). The variation of each observable in terms of R_{PS} is displayed in Fig. 4. The angular correlation parameters, in particular A_t and A_Δ, are more sensitive to changes in R_{PS} than the total capture rate, as first pointed out in ref. [27]. A precise measurement of these polarization observables could therefore be useful to ascertain the extent to which the induced pseudo-scalar form factors deviate from their PCAC values.

By enforcing perfect agreement between the experimental and theoretical values, taken with their uncertainties, for the total capture rate Γ_0, it is possible to obtain an estimate for the range of values allowed for R_{PS}, and we have found

$$R_{PS} = 0.94 \pm 0.06 \ . \tag{10}$$

This 6 % uncertainty is significantly smaller than that found in previous studies [26, 33, 34]. This substantial reduction in uncertainty can be traced back to

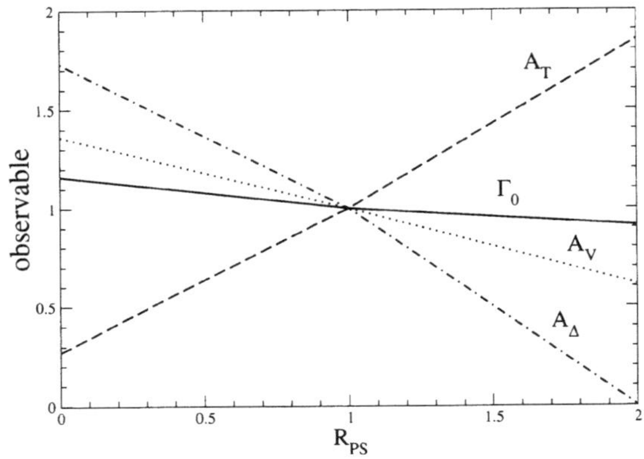

Figure 4. Variation of the capture rate Γ_0 and angular correlation parameters A_v, A_t, and A_Δ with the parameter R_{PS} entering the expression for the induced pseudoscalar coupling G_{PS} given in Eq. (2). The AV18/UIX PHH wave functions are used. For each observable, the ratio between the result obtained with the given value of R_{PS} and that one with the PCAC value $R_{PS} = 1$ is plotted as function of R_{PS}.

the procedure used to constrain the (model-dependent) two-body axial currents discussed in the Introduction.

4 Summary and Outlook

We have reported new calculations of $p-d$ radiative capture observables at energies above the DBT, and of observables in the process $^3\text{He}(\mu^-, \nu_\mu)^3\text{H}$. These calculations have been based on the Argonne v_{18} two-nucleon and Urbana IX three-nucleon interactions. For the muon capture reaction also the older Argonne v_{14} two-nucleon and Tucson-Melbourne three-nucleon interactions have been used. Accurate bound and continuum wave functions have been obtained with the PHH method. The model for the EM and weak transition operators has been taken to consist of one- and two-body components. In recent studies, this theory has been shown to correctly predict the static properties of the three-nucleon systems [32], as well as their associated elastic and transition electromagnetic form factors.

A satisfactory description of measured $p-d$ capture observables at $E_{\text{c.m.}} = 3.33$ MeV has emerged with the exception of the T_{20} and T_{21} tensor analyzing powers. Interestingly, the very same discrepancies were observed also below the DBT, namely at $E_{\text{c.m.}} = 2$ MeV. Similarly, at higher values of $E_{\text{c.m.}}$, differential cross sections and $A_y(d)$ seem to be well reproduced, while the tensor analyzing powers A_{xx}, A_{yy} and A_{zz} at $E_{\text{c.m.}} = 5.83$ MeV present the same discrepancies mentioned above.

In order to clarify the origin of these discrepancies, we plan to extend the calculation of $p - d$ capture observables at even higher energies, and to investigate alternative models for short-range two-body EM currents.

In regard to the muon capture process, we have found that the predicted total rate is in agreement with the experimental value, and has only a weak model dependence: the AV18/UIX and AV14/TM results differ by less than 0.5 %. As discussed above, the weak model dependence comes about because both Hamiltonians reproduce the binding energies, charge and magnetic radii of the three-nucleon systems, and the Gamow-Teller matrix element in tritium β-decay.

It is important to note that, if the contributions associated with two-body terms in the axial current were to be neglected, the predicted capture rate would be 1316 (1318) sec^{-1} with AV18/UIX (AV14/TM), and so two-body mechanisms are crucial for reproducing the experimental value. The present work shows that the procedure adopted for constraining these two-body contributions leads to a consistent description of available experimental data on weak transitions in the three-body systems. It also corroborates the robustness of our recent predictions for the cross sections of the proton weak captures on ^{1}H [15] and ^{3}He [7], which were obtained with the same model for the nuclear weak current.

Finally, it would be interesting to extend our investigation to the ^{3}He(μ^-,ν_μ)nd and ^{3}He(μ^-,ν_μ)nnp processes, both of which have been studied experimentally in ref. [35] and theoretically in ref. [36]. Work along these lines is vigorously being pursued.

Acknowledgement. The work of R.S. is supported by the U.S. Department of Energy contract number DE-AC05-84ER40150 under which the Southeastern Universities Research Association (SURA) operates the Thomas Jefferson National Accelerator Facility.

References

1. A. Kievsky et al.: Phys. Rev. **C58**, 3085 (1998)
2. W. Glöckle et al.: Phys. Rep. **274**, 107 (1996)
3. A. Kievsky et al.: Nucl. Phys. **A607**, 402 (1996)
4. R. Schiavilla et al.: Phys. Rev. **C40**, 2294 (1989)
5. R. Schiavilla and D.O. Riska: Phys. Rev. **C43**, 437 (1991)
6. J. Carlson and R. Schiavilla: Rev. Mod. Phys. **70**, 743 (1998)
7. L.E. Marcucci et al.: Phys. Rev. Lett. **84**, 5959 (2000); Phys. Rev. **C63**, 015801 (2001)
8. L.E. Marcucci et al.: Phys. Rev. **C66**, 054003 (2002)
9. T.R. Hemmert et al.: Phys. Rev. **D51**, 158 (1995)

10. J.D. Walecka: In: *Muon Physics; Weak Interactions*, edited by V.W. Hughes and C.S. Wu, p. 114. New York: Academic Press 1975

11. J.D. Walecka: *Theoretical Nuclear and Subnuclear Physics*, New York: Oxford University Press 1995

12. D.O. Riska: Phys. Scr. **31**, 471 (1985);
A. Buchmann et al.: Nucl. Phys. **A443**, 726 (1985)

13. R. Schiavilla et al.: Phys. Rev. **C45**, 2628 (1992)

14. M. Kirchbach et al.: Nucl. Phys. **A542**, 616 (1992)

15. R. Schiavilla et al.: Phys. Rev. **C58**, 1263 (1998)

16. A. Kievsky et al.: Nucl. Phys. **A551**, 241 (1993); **A577**, 511 (1994)

17. M. Viviani et al.: Phys. Rev. **C54**, 534 (1996)

18. M. Viviani et al.: Phys. Rev. **C61**, 064001 (2000)

19. A. Kievsky et al.: Phys. Rev. **C64**, 024002 (2001)

20. F. Goeckner et al.: Phys. Rev. **C45**, R2536 (1992)

21. B.D. Belt et al.: Phys. Rev. Lett. **20**, 1120 (1970)

22. H. Akiyoshi et al.: Phys. Rev. **C64**, 034001 (2001)

23. R.B. Wiringa et al.: Phys. Rev. **C51**, 38 (1995)

24. B.S. Pudliner et al.: Phys. Rev. Lett. **74**, 4396 (1995)

25. D.F. Measday: Phys. Rep. **354**, 243 (2001)

26. J.G. Congleton and H.W. Fearing: Nucl. Phys. **A552**, 534 (1992)

27. W.-Y.P. Hwang: Phys. Rev. **C17**, 1799 (1978)

28. P.A. Souder et al.: Nucl. Instr. and Meth. in Phys. Res. **A402**, 311 (1998)

29. R.B. Wiringa et al.: Phys. Rev. **C29**, 1207 (1984)

30. S.A. Coon et al.: Nucl. Phys. **A317**, 242 (1979)

31. P. Ackerbauer et al.: Phys. Lett. **B417**, 224 (1998)

32. L.E. Marcucci et al.: Phys. Rev. **C58**, 3069 (1998)

33. J.G. Congleton and E. Truhlík: Phys. Rev. **C53**, 956 (1995)

34. J. Govaerts and J.-L. Lucio-Martinez: Nucl. Phys. **A678**, 110 (2000)

35. S.E. Kuhn et al.: Phys. Rev. **C50**, 1771 (1994)

36. R. Skibinski et al.: Phys. Rev. **C59**, 2384 (1999)

Few-Body Systems Suppl. 15, 99–104 (2003)

Few-Body Systems

Nuclear Medium Effects in Hadron Leptoproduction

N. Bianchi*

INFN - Laboratori Nazionali di Frascati, Via E. Fermi 40, I-00044, Frascati, Italy

Abstract. Recent results from HERMES in the hadron leptoproduction off nuclei are presented. The possible interpretations in terms of medium modifications of the parton fragmentation function and the implications on the parton energy loss and the hadron formation time are discussed.

1 Introduction

The inclusive Deep-Inelastic-Scattering (DIS) of high energy leptons on free nucleons has been used from late 60s to measure the partonic distributions and, by using nuclear targets, to study the medium modification to these distributions. These modifications were found to be large and were explained by invoking medium effects at both hadronic and partonic levels (see [1] for a review). Semi-Inclusive DIS (SIDIS) of high energy leptons on free nucleons can be used to study another fundamental property of the quarks, the fragmentation function. This function can be also modified by the nuclear medium and only recently this effect has been clearly revealed by new and precise data.

2 Fragmentation Functions and Multiplicities

The SIDIS process can be factorized in a parton distribution function q_f, which is function of the Bjorken scaling variable $x = Q^2/(2M\nu)$, a hard scattering cross section calculable in QCD, and a parton fragmentation function.

The fragmentation function $D_f^{\pi}(z)$ represents the probability for a quark of flavor f to fragment into a specific hadron (a pion in this case) with relative energy $z = E_\pi/\nu$ with respect to the initial one of the struck quark. In analogy with the structure functions, the fragmentation functions depend on a scaling variable z and their logarithmic Q^2-evolution is determined by the DGLAP equations. In principle, fragmentation functions are universal and do

* *E-mail address:* nicola.bianchi@lnf.infn.it

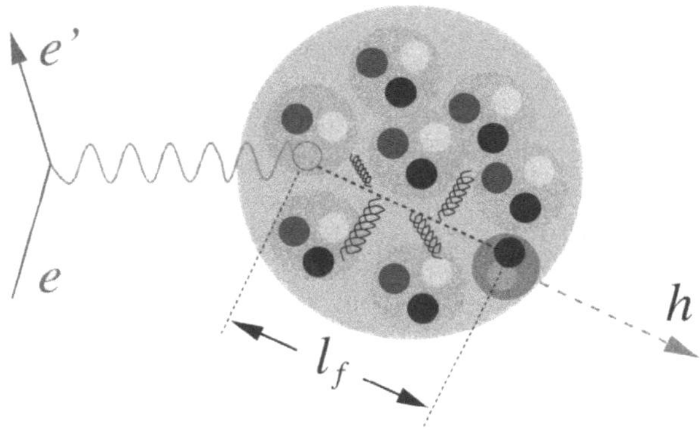

Figure 1. Pictorial view of SIDIS in nuclei. The observed hadron yield is reduced and modified by the interaction of the struck quark and of the formed hadron inside the nucleus.

not depend from the generating process (e^+e^-, $p\bar{p}$, ep, γp, $\gamma\gamma$). Fragmentation functions are not calculable in QCD, but several phenomenological models are used to describe the mechanism of the hadron formation (see [2] for a review).

In the Quark Parton Model (QPM) description of SIDIS, the fragmentation functions are multiplied by the parton distributions and experimentally hadron multiplicities are determined by normalizing the SIDIS yield to the DIS rate:

$$\frac{1}{N_{DIS}} \frac{dN^h(x,z)}{dz} = \frac{\sum_f e_f^2 q_f(x) D_f^h(z)}{\sum_f e_f^2 q_f(x)}$$

It is easy to demonstrate that within the QPM and under the u-quark dominance assumption, the pion multiplicities are almost equivalent to the pion fragmentation functions. In QCD the relation between fragmentation functions and multiplicities is more complicated since both the LO and the NLO terms of the cross section are convoluted with the initial photon and parton distributions and with the final hadron fragmentation functions [3]. Nevertheless, it has been shown [4] that, once integrating over a broad range of x, the SIDIS multiplicities are in good agreement with the NLO-QCD evolution of fragmentation functions determined from e^+e^- measurements at LEP.

3 Hadronization in the Nuclear Medium

The study of the medium modification of fragmentation functions can be performed in SIDIS on nuclear targets. As illustrated in Fig. 1, the nucleus acts as an ensemble of targets which reduce the multiplicity of fast hadrons due to both partonic and hadronic interactions. The partonic and hadronic effects are discriminated according to the time $\tau_h = l_h/c$ of formation of the hadron which may occur inside or outside the nucleus. The hadron formation times, which are basic ingredients also in hadron-nucleus and in heavy-ion reactions, are currently not well determined. In particular, the study of the Quark Gluon Plasma requires the knowledge of the proper time at which the plasma is formed and this time is generally derived from the hadron formation time.

After the first measurements performed at SLAC and CERN [5], the more recent results from HERMES [6, 7] provided a new strong boost of interest in this field. HERMES [8] is an experiment at DESY which is mainly devoted to the study of the spin structure of the nucleon by using a polarized lepton beam and polarized internal gas targets. However, some HERMES runs have been performed with unpolarized nuclear gas targets (D, ^{4}He, N, Ne, Kr) with densities up to $\sim 10^{16}$ nucl$\cdot$ cm^{-2}, allowing the extraction of interesting informations on the hadron multiplicities in the nuclei.

With an incident positron (or electron) beam of 27.5 GeV, the kinematic region explored by HERMES is about $0.004 < x < 0.8$ and $0.2 < Q^2 < 10$ GeV^2. This region is slightly reduced by the additional constrains ($W > 2$ GeV and $Q^2 > 1$ GeV^2) which are needed to ensure the DIS regime. The large acceptance and the complete particle identification make HERMES well suited for several semi-inclusive reactions. To minimize the fraction of hadrons produced in the target fragmentation and to ensure the factorization of the SIDIS process, only fast hadrons with $z > 0.2$ are considered.

The complete particle identification in HERMES, with the installation of the RICH detector in 1998, allows to fully disentangle the information for different hadron types. This is shown in Fig. 2 where the multiplicity ratios for the ^{84}Kr target are presented as function of the virtual photon energy ν (left panel) and of the scaling variable z (right panel). As it is seen, the medium effects for π^+ and π^- are equal. Also medium effects on K^- production are similar to the charged pion case. Quite interesting and unexpected is the difference between K^+ and K^- and especially between p and $\overline{p}$. In particular the effect for the proton is strongly different at low-z, where a contamination from the target fragmentation is more probable. The different results for different hadrons may reveal differences in the modification of q and $\overline{q}$ fragmentation functions, in the hadron formation times and in the different hadronic interaction cross sections.

4 Modification of Fragmentation Functions

The new HERMES results on nuclear multiplicity ratios have been interpreted in terms of modification of fragmentation functions.

In ref. [9] the modification of quark fragmentation functions and their QCD evolution are described in the framework of multiple parton scattering and induced gluon radiation with additional higher-twist terms in the DGLAP equations. As consequence of the partonic energy loss Δz, due to the gluon radiation, the fragmentation function is modified as $\tilde{D}(z) \approx 1/(1-\Delta z) \cdot D\left(z/(1-\Delta z)\right)$. In Fig. 3 these higher twist predictions are compared with the HERMES data. In the framework of this calculation, for a cold and static system like the nucleus, the quark energy loss involved in SIDIS reaction has been determined

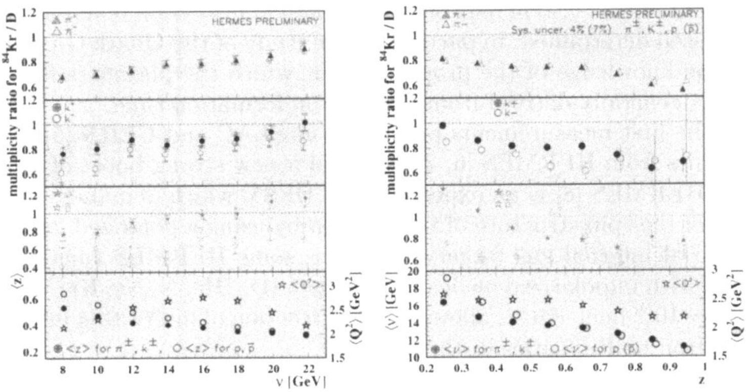

Figure 2. HERMES preliminary multiplicity ratio ^{84}Kr/D for different charged hadrons as a function of the virtual photon energy ν (left panel) and of the scaling variable z(right panel). In the lower panels the relevant average kinematic variables are presented for each ν- or z-bin (from ref. [7]).

and resulted in $dE/dx \approx 0.3$ GeV/fm for the Kr nucleus. This result has been compared with the energy loss ($dE/dx \approx 0.25$ GeV/fm) in a hot, dense and but expanding system like the one involved in heavy ion reaction at RHIC energies. An increase of the gluon density of a factor ~ 15 has been derived for the initial hot stage of Au + Au reaction with respect to the one in a static nucleus.

In ref. [10] the modification of quark fragmentation functions have been considered according to the Q^2-rescaling model, originally developed to interpret the EMC-effect in the nuclear structure functions. In this work the multiplicity in the medium included a rescaling by a factor $\xi_A(Q)$ of both structure $q_f^A(x,Q) = q_f(x, \xi_A(Q)Q)$ and fragmentation functions $D_f^{h|A}(z,Q) = D_f^h(x, \xi_A(Q)Q)$ where the deconfinement scale is assumed to be proportional to the degree of overlap of the nucleons inside the given nucleus.

In ref. [11], the combined effect of the fragmentation function modification due to the nuclear induced gluon radiation and of the hadron interaction inside the nucleus has been calculated. In this work, a pion formation time of ~ 1.2 $(E/\mathrm{GeV})\cdot$ fm/c has been determined in the comparison with HERMES data.

5 Modification of p_t Distribution

The scattering of quark and gluon increases the transverse momentum k_t of the struck quark with respect to its intrinsic value and, as consequence, the transverse momentum p_t of the leading hadron. This effect is shown in the left panel of Fig. 4. The HERMES data for $p_t^2 < 1$ GeV^2 indicate the hadron

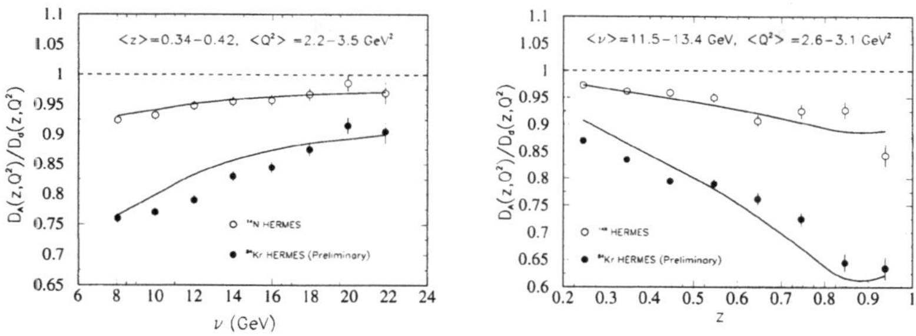

Figure 3. Predicted nuclear modification of quark fragmentation functions compared with HERMES data (from ref. [9])

attenuation due to the fragmentation function modification while the data for $p_t^2 \geq 1$ GeV^2 suggest the onset of a hard component due to incoherent parton scattering. The increase of p_t is larger for the heavier nucleus.

The p_t broadening is known as Cronin effect [12]. This effect has been observed in many hadronic reactions like in pA collisions or in high energy heavy-ion collisions performed at CERN SPS (see right panel of Fig. 4). The latter data has been described by multiple parton scattering within the Glauber formalism. The p_t broadening in SIDIS is mainly dependent on the multi-parton correlation function and it is proportional to the nuclear size. Therefore it will be important to measure it on several nuclei and for different hadron types.

6 Conclusions

The recent and precise HERMES results in SIDIS on N and Kr nuclei have shown a significant nuclear attenuation for fast hadrons which are interpreted as a modification of the parton fragmentation function. For the first time, different medium effects have been observed for different hadron types. In particular while the effect for the pion do not depend on the charge, a sizable difference has been observed between K^+ and K^- and between p and $\overline{p}$. The so called Cronin effect in the p_t broadening of the leading hadron has also been observed in SIDIS.

The SIDIS results provide important informations on the parton energy loss and on the hadron formation time which are basic ingredients for other hadronic reactions and in particular for the heavy-ion interactions at high energies, which are currently studied to provide hints for the Quark Gluon Plasma formation.

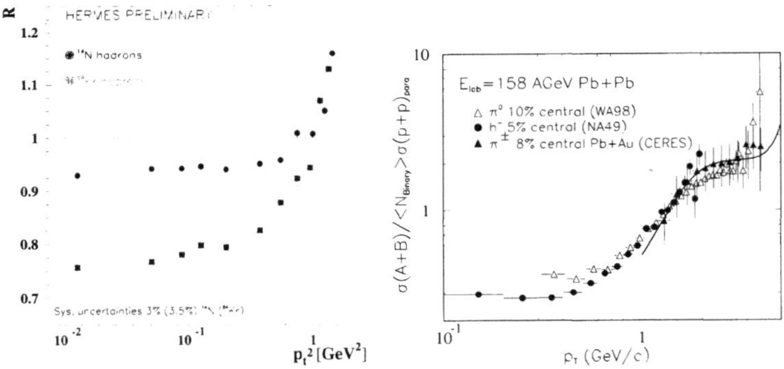

Figure 4. p_t-dependence of nuclear multiplicity ratio in SIDIS (left) and of nuclear modification factor in heavy ion reactions (right). The curve is a microscopic calculation which considers multiple parton scattering

References

1. M. Arneodo: Phys. Rep. **240**, 301 (1994)

2. B.R. Webber: Int.J. Mod. Phys. **A15S1**, 577 (2000)

3. B. Kniehl, G. Kramer, B. Potter: Nucl. Phys. **B582** 514 (2000); J. Binnewies, B. Kniehl, G. Kramer: Z. Phys. **C65** 471 (1995); Phys. Rev. **D52**, 4947 (1995)

4. HERMES Collaboration: A. Airapetian: Eur. Phys. J. **C21**, 599 (2001)

5. L. Osborne et al.: Phys. Rev. Lett. **40**, 1624 (1978); EMC Collaboration: J. Ashman et al. Z. Phys. **C52**, 1 (1991)

6. HERMES Collaboration: A. Airapetian: Eur. Phys. J. **C20**, 479 (2001)

7. V. Muccifora for the HERMES Collaboration hep-ex/0106088 (2001)

8. HERMES Collaboration: K. Ackerstaff: Nucl. Instr. Meth. **A417**, 230 (1998)

9. X. Guo, X.N. Wang: Phys. Rev. Lett. **85**, 3591 (2000); E. Wang, X.N. Wang: Phys. Rev. Lett. **89**, 162301-1 (2002)

10. A. Accardi, V. Muccifora, H.J. Pirner: nucl-th/0211011 (2002)

11. B. Kopeliovich et al.: hep-ph/9511214 (1995)

12. J.W.Cronin: Phys. Rev. **D11**, 3105 (1975); B. Kopeliovich et al.: hep-ph/0201010 (2002)

Few-Body Systems Suppl. 15, 105–118 (2003)

Few-Body Systems

Recent Results and Prospects from the STAR Detector and RHIC

W.W. Jacobs* for the STAR Collaboration†

Indiana University Cyclotron Facility and Department of Physics,
2401 Milo B. Sampson Lane, Bloomington, Indiana 47408 USA

Abstract.
Selected recent results from STAR and highlights from the RHIC complex of detectors, using soft and hard probes in heavy ion Au + Au collisions, are presented. As well, a glimpse to future investigations employing p + Au and nucleon spin structure studies with polarized proton collisions.

1 Introduction

The Relativistic Heavy Ion Collider (RHIC) is a new accelerator having completed its second running period for physics. RHIC provides for collisions of heavy nuclei, of spin-polarized protons, and of protons with nuclei at center-of-mass energies an order of magnitude higher than previously available. These capabilities allow access to a new generation of issues concerning hot dense matter formed in Relativistic Heavy Ion collisions, as well as investigations of parton distributions in cold matter via $p+A$ and spin polarized $p+p$ collisions.

Physics issues addressed at RHIC include: 1) do nucleons effectively dissolve in the hot matter produced in relativistic heavy-ion collisions, leaving behind a region of partonic (colored) matter? 2) do gluon densities in cold nuclear matter "saturate" at small momentum fractions? 3) do gluons carry a major fraction of the proton's spin? These questions span a large breadth of physics but are also related (e.g., soft gluon distributions in cold matter impact the early time evolution of the hot systems ignited by hard scattering processes in heavy ion collisions, etc...). We discuss aspects of this new investigative era below.

* *E-mail address: jacobs@iucf.indiana.edu*
† *Collaboration list may be found at:* http://www.star.bnl.gov, by selecting "publications"

2 RHIC, Detectors and Observables

RHIC comprises two independent cryogenic rings 3.83 km in circumference that can accelerate everything from p to Au. It is the first high-energy accelerator of polarized protons. Luminosity goals at the top energy are: 10^{32} cm^{-2}s^{-1} for $p+p$ at $\sqrt{s} = 500$ GeV and 10^{26} cm^{-2}s^{-1} for Au + Au at $\sqrt{s_{\mathrm{NN}}} = 200$ GeV. Au + Au collisions at $\sqrt{s_{\mathrm{NN}}} = 130$ GeV were run in the year 2000; Au + Au at $\sqrt{s_{\mathrm{NN}}} = 200$ GeV were run and 200 GeV polarized $p+p$ collisions commissioned in 2001.

Two large detectors (STAR and PHENIX) and two small detectors (Phobos and Brahms) are located at beam interaction regions. The detectors generally complement each other and they all involve the ability to measure particle multiplicities, have particle identification for a range of particles and momenta and generally some form of momentum analysis. A description of the detectors, their capabilities and technical details, is published elsewhere [1]. Fig. 1 (left panel) gives a view of the components that make up the STAR detector.

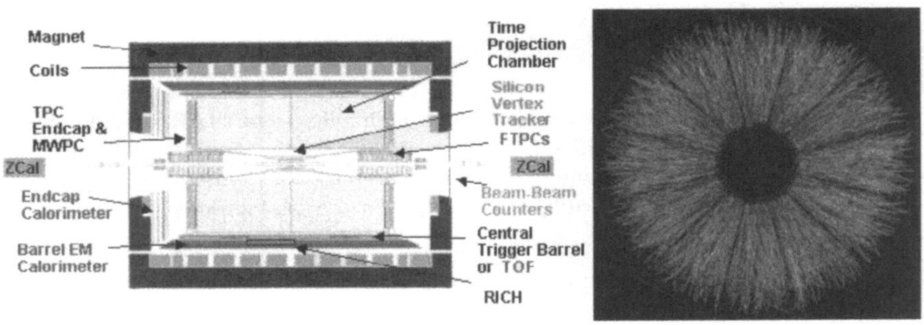

Figure 1. The STAR detector (left) comprises a large solenoid field with tracking TPC's, Si-vertex tracking, RICH, TOF and large EM Calorimetry. An end-on view (right) of the tracks in the STAR central TPC following a central Au + Au collision.

At RHIC, the average momenta of the particles produced in central collisions is well below 2 GeV/c. These soft particles are emitted with $\sim$ constant yield ($dN/d\eta$) over a broad range near pseudorapidity ($\eta = -\ln(\tan(\theta/2)) = 0$, i.e., orthogonal to the colliding beams. At STAR, charged particles are detected using the large time projection chamber (TPC) that surrounds the interaction region. Level 3 trigger on-line TPC tracking (right panel, Fig.1) vividly displays the large number of particles that accompany a central Au + Au collision.

Heavy ion collisions can be characterized in terms of their centrality on an event-by-event basis by relating the number of observed charged particles to the impact parameter. The most central collisions produce the largest number of charged particles (at STAR for the top 5% of the cross section, typically $\sim$ 570 charged particles for $|\eta| < 0.5$ at $\sqrt{s_{\mathrm{NN}}} = 130$ GeV), whereas for peripheral collisions the number is considerably less. A smooth binning provides a continuous scale of centrality. A minimum bias trigger is generated by measurement of neutrons in zero degree calorimeters at each experiment. Constraints on charged particle multiplicity enrich the more central events in a trigger sample.

3 Soft Probes

The space-time evolution of heavy ion collisions involves an initial formation stage, followed by a fireball of hot dense matter which may pass through a phase characterized by a quark-gluon plasma. Expansion/cooling eventually results in hadronization. The point in time when inelastic scattering ceases is called "chemical freeze-out", fixing particle species abundances and making particle ratios useful observables. After further expansion/cooling, elastic scattering stops ("kinematic freeze-out"), and particle transverse momentum distributions and effective temperatures are useful observables in this regime. Below are selections of data and interpretation obtained from these latter phases.

3.1 Particle Ratios

Identified particle spectra from the RHIC detectors for many species are now available, including results from last year's $\sqrt{s_{NN}} = 200$ GeV running [2]. Such spectra are generally plotted versus particle transverse momenta p_T or mass m_T for the different event centrality classes. Their analysis sharpen the answers obtained as compared to simple charged particle multiplicities alone. Of particular interest is the $\overline{p}/p$ ratio. Recent spectra from STAR at $\sqrt{s_{\rm NN}} = 200$ GeV have been fit and extrapolated. For the most central collisions, the preliminary $\overline{p}/p$ ratio is ~ 0.8, implying that at mid-rapidity 4 times as many protons are pair-produced in the collision than are redirected from those brought in by the colliding partners. This is up from 0.72 ± 0.05 (sys) at $\sqrt{s_{\rm NN}} = 130$ GeV from STAR [3] (and similar values at the other RHIC detectors), and a considerable increase over that observed for much lower energies ($\sqrt{s_{\rm NN}} = 17$ GeV) at SPS/CERN. While the mid-rapidity region is not completely baryon free, for the first time at RHIC energies, more baryons are pair produced in the collisions than are brought in by the colliding partners. An analytical description of the ratio $\overline{p}/p$ vs. $\sqrt{s_{\rm NN}}$ shows the curve flattening already at RHIC, indicating that further significant increase will come only at very high energies.

Strange particles are also of interest and measurements of kaons, ϕ, Λ, Ξ, Ω and their anti-particles have been underway at STAR and RHIC. Results at $\sqrt{s_{\rm NN}} = 130$ GeV from STAR indicate that a fraction of the net-baryon number at mid-rapidity is due to an excess of hyperons to anti-hyperons, but the trend is for the anti-hyperon/hyperon ratio to monotonically approach unity with increasing strangeness content. Strange anti-baryon to baryon ratios have long been predicted as a signature to help differentiate between a hadron gas and a deconfined plasma of quarks and gluons, and measurements of $\overline{\Lambda}/\overline{p}$ at lower energies have suggested a large enhancement. The result from STAR [2] after feed-down corrections is close to one ($\overline{\Lambda}/\overline{p} \sim 0.93$), showing a significant increase from elementary collisions, but a lower value and little centrality

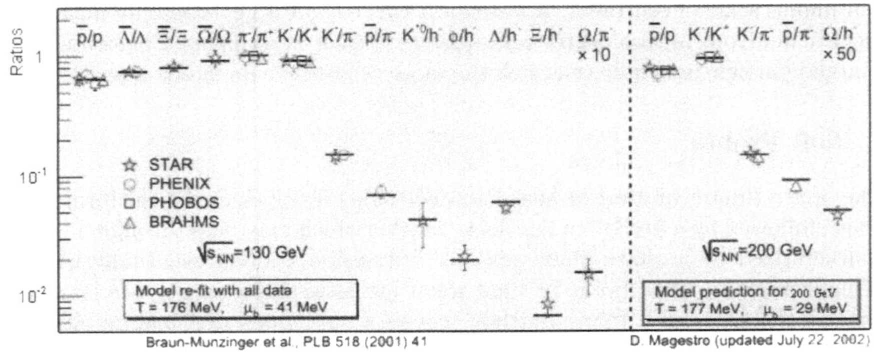

Figure 2. Particle ratios and statistical model calculations. Left: RHIC data ($\sqrt{s_{NN}} =$ 130 GeV) fit with equilibrium temperature $T_{ch} = 176$ MeV, baryon chemical potential $\mu_B = 41$ MeV; right: predictions for $\sqrt{s_{NN}} = 200$ GeV compared to STAR data.

dependence compared to SPS/CERN results. Whether these observations are indicative of collisional aspects or strangeness equilibrium is a matter of continued debate.

Many other particle species have been identified and their spectra characterized. Ratios at $\sqrt{s_{NN}} = 130$ GeV from all the RHIC detectors are summarized in the left panel of Fig. 2 [4, 2]. A thermal model has been fit to the anti-particle/particle ratios employing a partition function with parameters T_{ch}, μ_B, μ_S, μ_{I3}, where T_{ch} is the equilibrium temperature at chemical freeze-out. The baryon chemical potential μ_B is also freely varied in the fits, while the strangeness and isospin potentials are constrained. The horizontal bars in the figure show the results of the fit for which typically $T_{ch} = 175 - 200$ MeV, $\mu_B < 50$ MeV, $\mu_S \sim 0$. In the right panel the model predictions (with $T_{ch} = 177$ MeV, $\mu_B = 29$ MeV) are shown with STAR data at $\sqrt{s_{NN}} = 200$ GeV. Overall, fits/predictions of this simple model work surprisingly well. While simpler systems can show similar behavior, the above is taken by some as evidence for system chemical equilibrium.

It is interesting at this point to identify where we are on the phase diagram, plotted generally as T_{ch} vs. μ_B. At RHIC, for $\mu_B < 50$ MeV we are clearly near the baryon free values of the early universe ($\bar{p}/p \sim 1.0$), and the freeze-out equilibrium temperature ~ 175 MeV puts the system within the bounds of the region predicted by lattice QCD for a hadron gas/deconfined phase transition. Thus we know our location on the phase diagram, but we now need to know other features such as the equation of state (EOS), dynamics, etc. If possible we would like to study the dynamics well into the "deconfined" region.

3.2 Particle Dynamics

Particle yields, when plotted vs. transverse mass (m_T) show a strong slope dependence on particle species. The inverse slopes, and hence the effective temperature deduced, increase with particle mass (e.g., π^-; K^-; $\bar{p}$). A simple explanation is that the source has a component of radial expansion which adds kinetic energy to the thermal energy of the particles, with the more massive particles receiving more kinetic energy. The real situation is a bit more complex.

To understand the properties of the systems formed, a Blast Wave model [5] is used to fit the differential (d^2N/dp_Tdy) spectra, plotted as a function of p_T, to obtain values for the kinetic freeze-out temperature (T_{kin}) and average radial velocity of expansion $\langle\beta_r\rangle$ [2]. The model is inspired from a hydrodynamic point of view and includes both longitudinal (along the colliding beams) and transverse boosts as well as effects from resonance decays. Fits to π, K, p, Λ ... spectra at RHIC typically give $T_{kin} \sim 100$ MeV and $\langle\beta_r\rangle \sim 0.55$ c. As a function of mass, the average p_T for central collisions deduced in these calculated fits increases with roughly the same slope and has similar values for $\sqrt{s_{NN}} = 130$ and 200 GeV implying similar underlying dynamics. The corresponding increase for peripheral collisions is more modest and in agreement with observations in p+p data. Overall, as a function of $\sqrt{s_{NN}}$, T_{kin} shows a nearly constant value for STAR data from $\sqrt{s_{NN}} = 19.6$ to 200 GeV, while $\langle\beta_r\rangle$ is large and increasing. These results, are generally consistent with observations from the other RHIC detectors, and indicate that an explosive transverse expanding (high pressure) system is formed in the most central collisions at RHIC energies.

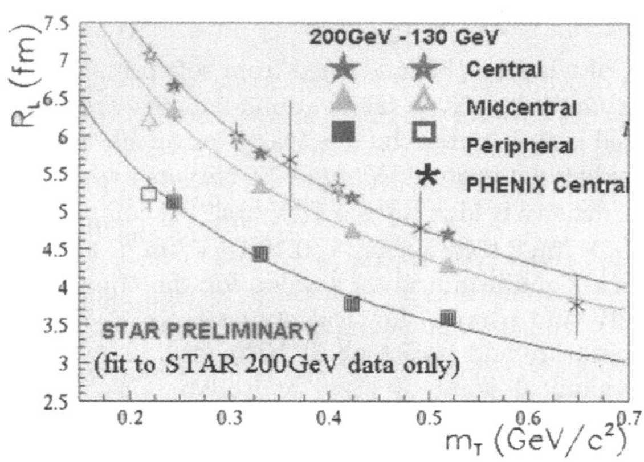

Figure 3. Particle interferometry parameter R_L for 3 different event centralities as a function of transverse mass (e.g., see Lopez-Noriega in ref. [2]). Curves are simple fits used to extract kinematic freeze-out times as described in the text.

Particle interferometry is a powerful tool that can be used to further characterize collision dynamics. For systems formed by relativistic heavy ions, emerging pairs of the same (or in some cases different) particle species can be used to probe the size, shape orientation and time evolution of the source. The momentum difference of the two-particles (e.g., similar to traditional Hanbury-Brown and Twiss interferometry) is decomposed into components q_{long}, q_{out},

q_{side}, with "long" along the beam direction, "out" in the transverse momentum direction of the particle pair and "side" perpendicular to "long" and "out". At STAR the corresponding radial dimensions, R_L, R_O and R_S, deduced at $\sqrt{s_{NN}} = 200$ GeV show a similar behavior with centrality [2]. Due to flow effects they also exhibit an overall steady decrease in magnitude vs. transverse mass. Perhaps the surprising observation is that $R_O/R_S \sim 1$, a result persisting from experiments reported at lower values of $\sqrt{s_{NN}}$, indicating that even at these high collision energies the duration time for particle emission is short. A blast wave calculation gives an emission time of $\sim 1-2$ fm/c.

The above observations appear inconsistent with any relatively long-lived intermediate ("slow burning log") state of the hot matter that is formed. But further information about the system's overall time evolution can also be obtained from these data as shown in Fig. 3. A fit of the R_L parameter (plotted vs. transverse mass) with the simple relation $R_L = \langle t_{fo} \rangle \sqrt{T_{kin}/M_T}$ allows one to extract the average freeze-out time $\langle t_{fo} \rangle$. Using a value $T_{kin} = 110$ MeV (from $\sqrt{s}_{NN} = 130$ GeV data), a fit to STAR $\sqrt{s_{NN}} = 200$ GeV data gives $\langle t_{fo} \rangle_{central} \sim 10$ fm/c and $\langle t_{fo} \rangle_{pherip} \sim 7.6$ fm/c [2]. These fits are quite consistent with R_L vs. m_T data presented by PHENIX and imply a very rapid overall evolution time scale (in addition to the short particle emission duration) for the hot dense systems formed in RHIC central heavy ion collisions.

3.3 Soft Probes Summary

The high points of what has been learned from soft probe studies includes the following [2]: baryon/anti-baryon ratios at mid-rapidity are $\sim$ 0.7-0.8 indicating the matter formed is close to baryon free (early universe), but not quite (e.g., as a function of rapidity, net proton $dN/dy \sim 10$; baryon transport $\Delta y \sim 5.5$). The apparent energy density is high [6] ~ 5 GeV/fm^3 (as compared to lattice phase transition ~ 1 GeV/fm^3; cold matter ~ 0.16 GeV/fm^3). Fitting particle ratios assuming chemical equilibrium gives a value for the equilibrium temperature of $T_{ch} \sim 175$ MeV and baryon chemical potential $\mu_B < 50$ MeV, close to the interesting temperature and lattice phase boundary.

From the dynamical point of view, hydrodynamics works well at RHIC energies, although the message from studies of Bose-Einstein correlations is that the hadronization is rapid; there appears to be no long lived "mixed" phase (e.g., such as one might expect from strong 1st order phase transitions). Many further dynamical fluctuation and correlation analysis are in progress.

Overall the picture is one of a system that appears to be in equilibrium but "explodes" and hadronizes rapidly, suggesting also a high initial pressure. Further studies using soft probes will help elucidate the detailed dynamical nature of the systems formed at RHIC.

4 Hard Probes

Experiments at RHIC have reached a new domain for hard scattering, and hard probes are just what is required to further study the initial conditions and properties of hot matter formed in these heavy ion collisions. Hard scattering of nucleon constituents in the initial formation and high density phase of $A + A$ collisions will result in large transverse momentum recoiling partons. Bjorken first predicted [7] that hard scattered partons would suffer a large energy loss (via gluon radiation) when traversing a hot dense colored medium. Recent calculations for various plasma conditions indicate a linear dependence of the energy loss on gluon density; at anticipated densities the energy loss is $\sim$ one order of magnitude greater than that deduced for cold nuclear matter (e.g., $dE/dx \sim 0.5$ GeV/fm for a 10 GeV quark in cold matter [8]). Hence, measurement of hadronized fragments following hard collisions may be a sensitive probe of the properties (e.g., gluon density) of hot matter formed early in heavy ion collisions.

4.1 High p_T Hadron Suppression

Hadron momentum spectra are measured at all four RHIC experiments, typically out to $p_T \sim 10$ GeV/c (e.g., at PHENIX and STAR). At STAR, power law fits reveal systematic differences in shape as a function of centrality: for the peripheral bins, they resemble those obtained in $p+p$ collisions, while those for central bins exhibit a steeper slope. One way [9, 10] to quantify the difference observed in these spectra is through the "nuclear modification factor", which relates heavy ion collision yields to a NN reference:

$$R_{AA}(p_T) = \frac{d^2N^{AA}/dp_T d\eta}{T_{AA} d^2\sigma^{NN}/dp_T d\eta} \quad (1)$$

where N^{AA} is the measured yield in $A + A$ collisions as a function of p_T and pseudorapidity, and σ^{NN} is the cross section observed for $p + p$. The quantity $T_{AA} = \langle N_{bin} \rangle / \sigma_{inel}^{NN}$ is the average number of binary collisions for the centrality bin, with the equivalent number of binary collisions $\langle N_{bin} \rangle$, usually calculated in a Glauber model. At low momenta, R_{AA} is less than unity, reflecting a scaling with the (large) number of $A+A$ collision participants, whereas hard scattering scales as $\langle N_{bin} \rangle$ (e.g., $R_{AA} = 1$).

Figure 4 shows results for inclusive charged hadrons from STAR [9, 2]. In the left panel, values of the nuclear modification factor are plotted as a function of transverse momentum p_T for different centrality bins (0-5% being the most central). Horizontal lines indicating expected limits of participant scaling (lower) and binary scaling (upper) are shown for each spectrum. In all cases the data indicate an increase in R_{AA} with p_T, starting from a value consistent with participant scaling, and steadily approaching unity (binary scaling) up to about 2 GeV/c. For the most peripheral bins the ratio holds up at larger values of p_T, but with increasing centrality, R_{AA} falls off, and for the most central bin shows significant deviation from binary scaling in the region $p_T > 2$ GeV/c. These comparisons rely on an accurate extrapolation of nucleon-nucleon spectra to RHIC energies and detector acceptances. However, the right-hand panel in Fig. 4 shows a comparison largely immune to any such difficulties yet containing similar information. The ratio of the measured yields from STAR for two choices

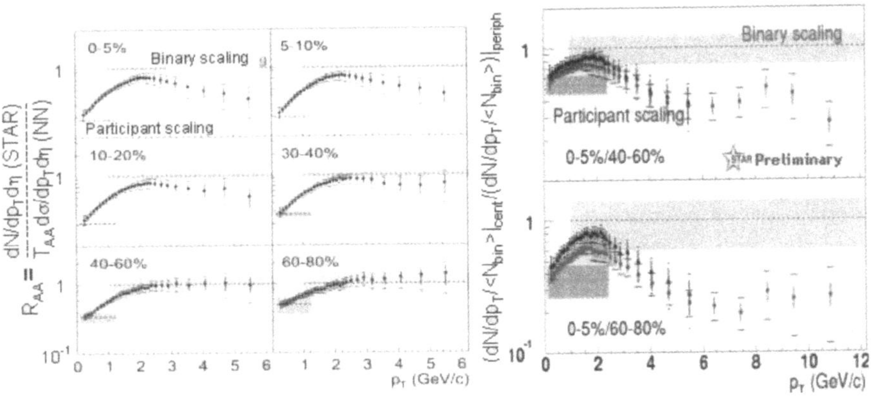

Figure 4. Left: nuclear modification factor vs. p_T and event centrality from STAR at $\sqrt{s}_{NN} = 130$ GeV. Participant and binary scaling limits are indicated. Right: ratio of central to peripheral yields for $\sqrt{s}_{NN} = 130$ and 200 GeV [9, 2].

of central vs. peripheral event classes, normalized to the appropriate $\langle N_{bin} \rangle$, respectively, are shown as a function of p_T. Data from both $\sqrt{s}_{NN} = 130$ and 200 data sets are combined. The message here is similar, with the ratio starting out near participant scaling, peaking close to unity near $p_T \sim 2$ GeV/c, then showing a clear and strong suppression at higher transverse momenta (data from $\sqrt{s}_{NN} = 200$ extend out to $p_T \sim 12$ GeV/c).

Similar results have been obtained at PHENIX [10]. The nuclear modification factor was calculated using $p + p$ pion production yields measured at PHENIX, avoiding a number of normalization issues. The most recent PHENIX data have been compared to various calculations (e.g., see Fig. 2 of Peitzmann in Ref. [2]). The calculations that qualitatively match the magnitude of the data include some form of enhanced parton energy loss (e.g., increased gluon density).

The above and other RHIC data provide clear evidence for suppression of high p_T (leading particle) yields in Au + Au central collisions. The data suggest substantial "in-medium" interactions [9, 10], extending out to very high p_T.

4.2 Jets in Au + Au *Collisions*

Jets, reflecting hard scattering in hadronic collision processes where the momentum of the recoiling parton is carried by the subsequent fragmentation products, have long since been established at high transverse momentum (p_T) in nucleon-nucleon collisions at high energy [11]. Today, identifying jet events in such collisions is relatively straightforward. However, in Au + Au collisions at RHIC, cross sections for other processes are large, there are many soft particles produced and a high number of tracks in the more central collisions. How to see jets? Well, jets as characterized in $p+p$ have a "jet core" $\Delta\phi \times \Delta\eta \sim 0.5 \times 0.5$, so one could statistically look for such correlations of high p_T hadron pairs in Au + Au. In heavy ion collisions, however, an added complication is that there are high p_T hadrons correlated with the reaction plane (e.g., elliptic flow) that are also correlated with each other. But the latter correlations are long range so that their effects may be separated in phase space.

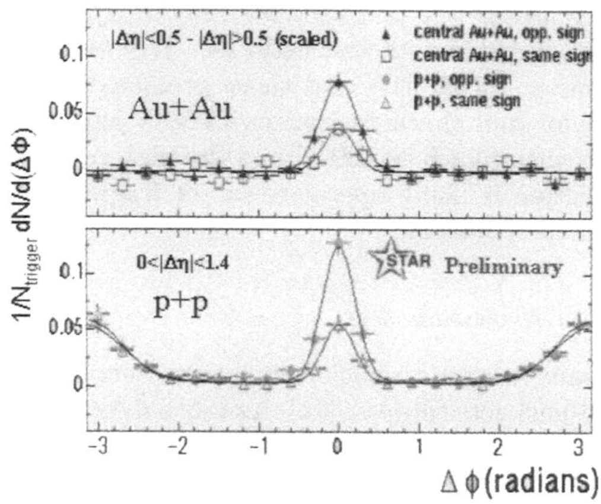

Figure 5. Jet structures in 2-particle correlation data from STAR for central Au + Au (top panel) and *pp* (bottom panel) collisions [11, 12]. For Au + Au, a scaled contribution (region $\Delta\eta > 0.5$) has been subtracted to better exhibit the correlation at $|\Delta\phi| = 0$.

To search for evidence of jets in Au + Au, we search for azimuthal statistical correlations of pairs of high p_T hadrons: 1) choose a leading hadron with $p_T > p_{T(trig)}$ as a "trigger" particle, 2) look for correlations with "associated" particles in the range $p_{T(assoc)} < p_T < p_{T(trig)}$ vs. azimuthal difference $\Delta\phi$ between the pair of particles, 3) to accommodate multiple jets in Au + Au, use a correlation function of the form:

$$C_2(\Delta\phi) \equiv \frac{1}{N_{\text{trigger}}} \frac{1}{\epsilon} \int d\Delta\eta N(\Delta\phi, \Delta\eta), \tag{2}$$

where N_{trigger} is the number of tracks satisfying the trigger requirement and ϵ is a track reconstruction efficiency for the associated particles.

High p_T pair correlation data from STAR for Au + Au central collisions (0-10%) at $\sqrt{s_{NN}} = 200$ GeV are shown in the top panel of Fig. 5 and compared to minimum bias $p+p$ data in the bottom panel. For Au + Au, the small pseudorapidity (η) difference azimuthal correlation ($|\Delta\eta| < 0.5$) as displayed has subtracted from it the correlation at large pseudorapidity difference ($|\Delta\eta| > 0.5$), scaled so as to match in the region $0.75 < |\Delta\phi| < 2.24$ radians. The resultant spectrum ($4 < p_{T(trig)} < 6$ GeV/c; $2 < p_{T(assoc)} < p_T < p_{T(trig)}$)

clearly shows a peak near $\Delta\phi = 0$, indicating jet-like correlations in $\Delta\phi$ and $\Delta\eta$ for Au + Au. This enhancement is qualitatively similar to that observed near $\Delta\phi = 0$ for $p + p$ in the lower panel.

Further evidence that jets are observed in Au + Au collisions come from dynamical charge correlations ("leading" and "next-to-leading" tracks tend to have opposite sign) that are known to be a feature of jet fragmentation. The correlations in Fig. 5 are plotted separately for opposite and same sign tracks. Opposite sign correlations are preferred for the Au + Au case as they are for $p + p$. Numerical comparison of the ratio $(+-)/(++ \& --)$ for Au + Au and $p + p$ collisions at STAR [11] and those predicted by HIJING indicate a common jet behavior and characterization. Clearly jets are being observed in Au + Au collisions with STAR at RHIC and the high p_T suppression discussed in the previous section is likely operative on jet fragments formed from hard scattering early in the collision.

4.3 Suppression of Away-Side Jets

Hard scattering leads to back-to-back correlations between high p_T hadrons, or correlated back-to-back jets (di-jets) [11]. Azimuthal correlation data from Au+Au at STAR ($\sqrt{s} = 200$ GeV) are shown in Fig. 6 for a range of pseudorapidity difference $|\Delta\eta| < 1.4$. Data for peripheral events (60-80% centrality) are shown at the left, whereas data from central collisions (0-5%) are shown on the right. The Au + Au data have a sizable elliptic flow contribution (discussed further below). In order to compare Au + Au back-to-back high p_T hadron production rates to p+p, a superposition (histogram curve) of measured p+p triggered rates with a combinatoric elliptic flow background is used. Effectively the correlation is:

$$C_2^{\mathrm{Au+Au}} \equiv C_2^{\mathrm{p+p}} + A[1 + \nu_2^2 \cos(2\Delta\phi)], \tag{3}$$

where ν_2 is the elliptic flow parameter (measured separately) and A is a normalization factor fitting the $\cos(2\phi)$ behavior (curves in figure) to the non-jet region $0.75 < |\Delta\phi| < 2.24$.

For the peripheral Au + Au data in the left panel of Fig. 6 we see that the away side jet correlation behavior (e.g., near $|\Delta\phi| \sim 3$) is reproduced fairly well by the *pp* data plus flow histogram curve. On the other hand, for central events, the Au + Au data show a strong suppression of the away side jet correlation. This result can be quantified [11] by forming the ratio of 2-particle correlation strengths "(AA-flow)/*pp*", from which it is seen that for two different choices of trigger threshold (4 GeV/c, 6 GeV/c) the ratio monotonically falls with increasing centrality, indicating a steadily increasing suppression of away side jets in Au + Au collisions [11].

Fixed target experiments at Fermilab indicate that the shape of the back-to-back di-hadron azimuthal distribution is sensitive to the intrinsic parton k_T. Whether such effects could cause the observed deficit in back-to-back azimuthal correlations has been checked by varying the k_T parameter to unrealistic values

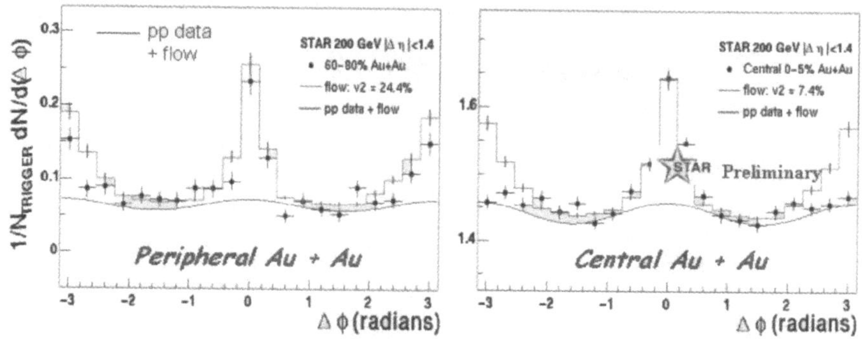

Figure 6. 2-particle correlation data for $\sqrt{s}_{\rm NN} = 200$ GeV Au + Au collisions from STAR. Data in the left (right) panels are for 60-80% (0-5%) centralities. The $\cos(2\phi)$ curve is the Au+Au elliptic flow contribution added to *pp* data giving the "histogram" (expected) curve. The away side jet $|\Delta\phi| \sim 3$ is strongly suppressed in central events.

in standard hard scattering calculations. Such drastic changes resulted in less than a 20% reduction in the predicted correlation strength.

Suppression of away side jets, as seen in the STAR Au+Au data at $\sqrt{s}_{\rm NN} =$ 200 GeV is a robust result and consistent with a picture in which the observed hadrons are fragments of hard scattered partons at the periphery of the hot dense medium formed by the interaction, and these partons, or their hadronized fragments, are strongly absorbed in the nuclear medium.

4.4 Elliptic Flow

Elliptic flow is described by the 2nd harmonic component of a Fourier decomposition of particle momentum distributions, made with respect to a reaction plane $(x - z)$ determined event-by-event. It arises because the collision overlap region is generally anisotropic in coordinate space; strong and collective effects among constituents of the hot matter formed in the overlap region generate a pressure gradient that transforms the spatial anisotropy into the observed momentum anisotropy.

Data are presented from STAR [12] at $\sqrt{s}_{\rm NN} = 130$ GeV for the parameter ν_2 ($\equiv \langle\cos 2\phi\rangle$, with $\phi \equiv {\rm atan}(p_x/p_y)$) as a function of $p_{\rm T}$ in the left-hand panel of Fig. 7. In agreement with hydrodynamic models, elliptic flow is significant and increases with: 1) decreasing centrality, and 2) increasing $p_{\rm T}$ up to about ~ 1 GeV/c. It begins to deviate from pure hydrodynamic models above 1 GeV/c (see hydro curve in right panel) and "saturates" in the region $\sim 3 - 6$ GeV/c. Preliminary STAR data at $\sqrt{s}_{\rm NN} = 200$ GeV show the saturation extends out beyond $\sim$ 10 GeV/c. Because of their hard scattering origin, these high $p_{\rm T}$ hadrons are not expected to participate fully in collective flow. Curves in the right-hand panel of the figure show calculations for a model encompassing both

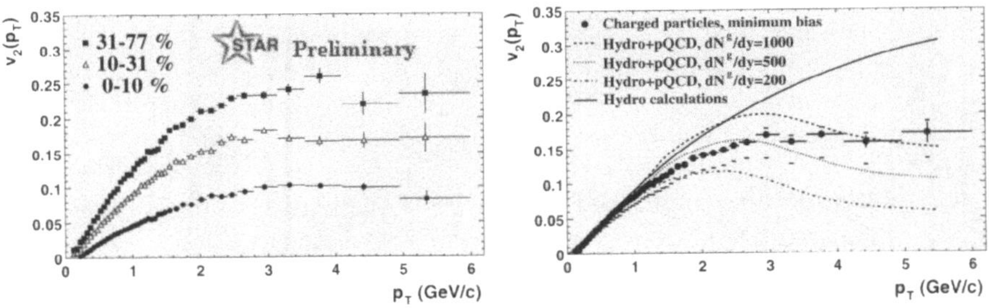

Figure 7. Elliptic flow ν_2 vs. transverse momentum. Left: results for several different event centralities at $\sqrt{s}_{\mathrm{NN}} = 130$ GeV; right: minimum bias data is compared with pure hydro and a model calculation consisting of "soft" and "hard" components [13].

soft (hydro) and hard (with dissipative coupling) components [13], the latter arising from emission followed by substantial parton energy loss in traversing a spatially anisotropic system. A strong dissipative coupling (here, increased gluon density) is required to describe the high $\mathrm{p_T}$ behavior of ν_2.

4.5 Hard Probes Summary

Prominent hard probe results at RHIC are: 1) hadrons at $p_T \gtrsim 3$ GeV/c are likely jet fragments, 2) high $\mathrm{p_T}$ hadron yields are suppressed in central Au+Au (but near-side pair correlations still clearly reveal jet production), 3) away-side jets are strongly (slightly) suppressed in central (peripheral) Au+Au collisions, and 4) elliptic flow saturates for high $\mathrm{p_T}$ hadrons.

A possible explanation for the above is that high $\mathrm{p_T}$ hadrons arise preferentially from partons scattered near the surface of the hot dense interaction region, favoring high $\mathrm{p_T}$ elliptic flow. The partner partons lose much of their energy in traversing the partonic matter created in the more central collisions, resulting in a softening of the high $\mathrm{p_T}$ spectrum and a suppression of away-side jets. This scenario, however, is by no means proven. Future measurements must check (among other things): 1) the role of cold nuclear matter effects via $p(d)$ + Au collisions, 2) the role of hadron energy loss in hadronic matter via extension to higher $\mathrm{p_T}$ (e.g., longer hadronization time) or via fragmentation γ studies, and 3) quantitative agreement with theoretical models, e.g., via γ-tagged jets.

5 Gluons and Spin

We will touch on parton structure questions briefly: Next-to-leading order perturbative QCD analysis of $\mathrm{F}_2(x, Q^2)$ and $\mathrm{g}_1(x, Q^2)$ structure functions indicate that the density of soft gluons appears to be still growing at the lowest x-values [14]. Does the gluon density saturate? If so where? The second question is related to the proton spin puzzle and the fact that the quarks (u, d, s + antiquark alignment) account for $\lesssim 30\%$ of the proton spin. Little is known about the gluon polarization and if and how it may contribute [15].

Experimentally, both of these questions can be addressed via Compton scattering of quarks and gluons ($q + g \rightarrow q(jet) + \gamma$) at RHIC. Gluon density, and in particular its modification in the nucleus, is investigated via measured cross sections in $p + A$ collisions down to low x gluons. To study possible gluon spin contributions, two-spin helicity asymmetries in $q + g$ Compton scattering (measured at STAR with completely reconstructed partonic collision kinematics) will be used to extract the gluon spin structure function $\Delta G(x)$. Expected sensitivity [15] will allow selection among currently poorly constrained gluon spin structure function distributions and determination of the integral to ± 0.5.

These experiments generally involve very asymmetric hard partonic collisions, for which an Endcap ElectroMagnetic Calorimeter (EEMC) at STAR [1] is an essential detector component. The first Endcap installations are presently taking place; we look forward to exciting results in p + A and spin physics!

6 Summary

The current physics program at RHIC in relativistic heavy ions and the near future spin program are vibrant. For the former, several observations from soft and hard probes taken together indicate formation of a hot dense medium, expanding rapidly under considerable pressure with hadronization occurring over a short time scale; its further characterization is a primary goal at RHIC. Suppression of the away-side jets as seen at STAR seems a particularly important observation at this point in the continuing phase of discovery.

Acknowledgement.

We wish to thank the RHIC Operations Group and the RHIC Computing Facility at Brookhaven National Laboratory, and the National Energy Research Scientific Computing Center at Lawrence Berkeley National Laboratory for their support. This work was supported by the Division of Nuclear Physics and the Division of High Energy Physics of the Office of Science of the U.S. Department of Energy, the United States National Science Foundation, the Bundesministerium fuer Bildung und Forschung of Germany, the Institut National de la Physique Nucleaire et de la Physique des Particules of France, the United Kingdom Engineering and Physical Sciences Research Council, Fundação de Amparo a Pesquisa do Estado de São Paulo, Brazil, the Russian Ministry of Science and Technology, the Ministry of Education of China, and the National Natural Science Foundation of China.

References

1. See http://www.bnl.gov.rhic for RHIC overview; a dedicated volume on Detectors and the Accelerator at RHIC is in press, Nucl. Inst. Meth. (2003)
2. Proc. of *Quark Matter 2002*, Nantes, 2002, in press and refs. therein.

3. Adler, C., et al.: Phys. Rev. Lett. **87**, 262302 (2001), refs. therein; note: an updated $\overline{p}/p$ ratio is quoted here, see nucl-ex/0211021

4. Braun-Munzinger, P., et al.: Phys. Lett. **B518**, 41 (2001)

5. Schnedermann, E., et al.: Phys. Rev. **C48**, 2462 (1993)

6. Adcox, K., et al.: Phys. Rev. Lett. **87**, 052301 (2001)

7. Bjorken, J.D.: In: *Energy Loss of Energetic Partons in Quark-Gluon Plasma*, FERMILAB-Pub-82/59-THY (1982)

8. Wang, E., Wang, X.: Phys. Rev. Lett. **89**, 162301 (2002), refs. therein

9. Adler, C., et al.: Phys. Rev. Lett. **89**, 202301 (2002), refs. therein

10. Adcox, K., et al.: Phys. Rev. Lett. **88**, 022301 (2002)

11. Adler, C., et al.: nucl-ex/0210033, Phys. Rev. Lett. in press, refs. therein

12. Adler, C., et al.: nucl-ex/0206006, Phys. Rev. Lett. in press refs. therein

13. Gyulassy, M., et al.: Phys. Rev. Lett. **86**, 2537 (2001)

14. De Roeck, A.: In: *Gluon Distributions at HERA*, Proceedings EPIC99 Workshop, Bloomington, 4/99, eds. Bland et al., p. 225, Singapore: World Scientific 2000

15. Jacobs, W.W. for the STAR collaboration: In: *Spin Physics Program at STAR*, AIP Conference Proceedings, **549**, p. 689, 2000, refs. therein

Few-Body Systems Suppl. 15, 119–122 (2003)

Few-Body Systems

An Estimate of the Percolation Parameter in Heavy Ion Collisions

J. Dias de Deus*, Yu.M. Shabelski†,**

CENTRA, Instituto Superior Técnico, 1049-001 Lisboa, Portugal

Abstract. From existing hadron and heavy ion collisions data on $\bar{p}/p$ central production we estimate the value of the percolation parameter at RHIC energies.

1 Introduction

It is well-know that in high energy hadron-nucleus collision there exists inelastic screening [1, 2] experimentally confirmed, especially for the case of hadron-deuteron interactions. The same inelastic screening has to exist as well in high energy heavy ion collision. This effect is very small for integrated cross sections (because many of them are determined by geometry), but it is very important [3] for the calculations of secondary multiplicities and inclusive densities. Similar results are obtained [4, 5] in the framework of string fusion [6], or percolation [7] models, where string fusion/percolation effects directly correspond [8] to pomeron interactions and are responsible for the suppression of particle production.

The effect of percolation for central heavy ion collision is, in good approximation determined by the reduction factor [9]

$$F(\eta) = \sqrt{\frac{1 - \exp^{-\eta}}{\eta}}, \tag{1}$$

η being the transverse density parameter,

$$\eta = \frac{r_s^2 N_s}{R^2}, \tag{2}$$

* *E-mail address:* jdd@fisica.ist.utl.pt
† *E-mail address:* shabelsk@thd.pnpi.spb.ru
** *Alternative address:* Petersburg Nuclear Physics Institute, Gatchina, St. Petersburg, Russia

where r_s is the string transverse radius (phenomenological estimates [10] give $r_s \sim 0.2 - 0.3$fm), R^2 the square of nuclear overlapping which for central collisions is equal to nuclear radius squared and N_s is the number of produced strings. At $\eta \to 0, F(\eta) \to 1$ (no percolation/no inelastic screening) and at $\eta \to \infty, F(\eta) \to 1\sqrt{\eta}$ (maximal screening).

2 An Estimate of the Percolation Parameter η

In the present paper we will give an estimate for the percolation parameter η from experimental data, not all of them connected to heavy ion physics.

Let us, in fact, start from γp collisions at $W \sim 200$ GeV (HERA). In the laboratory frame the asymmetry between comparatively slow p and $\bar{p}$ was observed to be [11]

$$A_B = 2\frac{N_p - N_{\bar{p}}}{N_p + N_{\bar{p}}} = 8.0 \pm 1.0 \pm 2.5\%. \tag{3}$$

That corresponds to the yield ratio

$$R^{\gamma p} = N_{\bar{p}}/N_p = 0.92 \pm 0.03. \tag{4}$$

However, the HERA kinematics is an asymmetrical one and the particles which are rather slow in the laboratory frame are rather fast in the center of mass frame. To account for this we can use the Quark-Gluon String Model (QGSM) [12, 13] with string junction diffusion, (see details in [14]). This correction is not numerically large: the model estimation for the center of mass photon-proton frame gives $R^{\gamma p} = 0.86 \pm 0.02$.

In γp collisions we have the baryon number flux from one proton. In the case of pp interactions this flux should be twice as large, thus corresponding to

$$0.72 < R^{pp} < 0.76 \tag{5}$$

in the pp center of mass frame.

The ratio $R^{\mathrm{Au+Au}}$ in central region of Au + Au collisions was measured at the same energy 200 GeV per nucleon at RHIC. The values are $0.74 \pm 0.02 \pm 0.03$ [16] and 0.75 ± 0.04, [17], i.e. practically the same as we obtain for pp collisions. This is in principle unexpected, as the sea contribution, which is $\bar{p}p$ symmetrical, is much more important in heavy ion collisions. The string fusion/screening argument gives a fair explanation of what happens.

Now let us note that in the QGSM as well as in the Dual String Model (DSM) [4, 15] the multiplicity of every secondary h produced in pp collision can be written as

$$< n_h^{pp} >= V_h + S_h, \tag{6}$$

where V_h and S_h are the contributions of valence and sea quarks, respectively.

In the case of heavy ion collisions the corresponding equation can be written as

$$< n_h^{AA} >= N_{\text{pair}} V_h + (< \nu >_{AA} - N_{\text{pair}}) F(\eta) S_h, \tag{7}$$

where N_{pair} is the number of pair of nucleon-nucleon interactions and $< \nu >_{AA}$ is the total number of binary interactions. $F(\eta)$ accounts for the sea strings percolation effect, a negligible effect for pp collisions.

In order to have $R^{pp}/R^{AA} \approx 1$ at RHIC energies, as experimentally observed, the ratio between valence quark and sea quark contributions in Eqs. (6) and (7) should be approximately the same. So we have

$$\frac{(< \nu >_{AA} - N_{\text{pair}}) F(\eta)}{N_{\text{pair}}} \simeq 1. \tag{8}$$

Using the experimental estimates of $< \nu >_{AA}$ and N_{pair} from [18] for high-p_T hadron production (these values are in agreement with standard Glauber-like estimates[19]) we obtain for the most central collisions the value $F_{(\eta)} = 1/4.8$, giving

$$\eta = 23. \tag{9}$$

Due to experimental errors and some model calculations we estimate the error involved in the last value to be of the order 1.5. This result agrees with an estimate of $\eta \simeq 16$ (see [15]), obtained through the use of pp central charged particle densities at $\sqrt{s} \simeq 200$ GeV.

3 Conclusions

The value obtained for η shows that we are well inside the percolation (perhaps the quark-gluon plasma) region. Note that if $R^{pp} > R^{AA}$, as it would happen without the center of mass correction bringing $R^{\gamma p}$ down from 0.92 to 0.86, then the η value should have been even larger.

The contributions of valence and sea strings in the QGSM for pp collisions are of the same order. For Au + Au central interactions the sea contribution is enhanced by the factor

$$\frac{< \nu >_{AA} - N_{\text{pair}}}{N_{\text{pair}}}, \tag{10}$$

which is, without percolations, about 4.8. That should increase the inclusive density of secondary multiplicities. The suppression factor $F(\eta)$ about 1/5 for sea contribution decreases the total rapidity density in central region about 2.5 or 3 times, in good agreement with the previous calculations of refs. [3, 5, 20].

Acknowledgement. We are grateful to N. Armesto for discussions.

References

1. V.N. Gribov: Sov. Phys. JETP **29**, 483 (1969); Sov. Phys. JETP **30**, 709 (1969)

2. O.V. Kancheli, S.G. Matinyan: Yad.Fiz. **11**, 130 (1970)

3. A. Capella, A. Kaidalov, J. Tran Thanh Van: Heavy Ion Physics 9, 169, (1999)

4. J. Dias de Deus, R. Ugoccioni: Phys.Lett. B**494**, 53 (2000)

5. N. Armesto, C. Pajares, D.Sousa: hep-ph/0104269

6. N. S. Amelin, M. A. Braun, C. Pajares: Phys. Lett. B**306**, 312 (1993) ; Z.Phys. C**63** 507 (1994)

7. M. A. Braun, C. Pajares: Phys.Rev. Lett. **85**, 4864 (2000)

8. N. Armesto, E. G. Ferreiro, C. Pajares, Yu. M. Shabelski: Z.Phys. C**73**, 309 (1997)

9. M. A. Braun, C. Pajares: Eur.Phys.J. C**16**, 2019 (2000)

10. N. Armesto, M. A. Braun, E. G. Ferreiro, C. Pajares: Phys. Rev. Lett. **77**, 3736 (1996)

11. C. Adloff et al.: H1 Coll. Submitted to the 29th Int. Conf. on High Energy Physics ICHEP98, Vancouver, July 1998

12. A.B. Kaidalov, K.A. Ter-Martirosyan: Yad.Fiz. **39**, 1545 (1984); Yad.Fiz. **40**, 211 (1984)

13. A.B. Kaidalov, Yu. M. Shabelski, K.A. Ter-Martirosyan: Yad.Fiz. **43**, 1282 (1986)

14. G.H. Arakelyan, A. Capella, A. Kaidalov, Yu. M. Shabelski: hep-ph/0103337

15. J. Dias de Deus, R. Ugoccioni: Phys. Lett. B**491**, 253 (2000)

16. PHOBOS Coll.: nucl-ex/0206012

17. BRAMS Coll.: nucl-ex/0207006

18. A. Armesto, C. Pajares: Int. J. Mod. Phys. A**15**, 2019 (2000)

19. PHENIX Coll.: Nucl-ex/0207009

20. J. Dias de Deus, Yu. M. Shabelski: hep-ph/010713

Few-Body Systems Suppl. 15, 123–132 (2003)

Few-Body Systems

Pionic Decays of Hadrons and the Coupling of Pions to Quarks

D.O. Riska*,†

Helsinki Institute of Physics, POB 64, 00014 University of Helsinki, Finland

Abstract. The pion decay rates of excited heavy-light flavor mesons states can provide direct information on the coupling of pions to light-flavor constituent quarks because pions only couple to their light flavor constituent. The chiral quark model describes the empirical widths of excited D-mesons fairly satisfactorily. In this model the pion coupling is determined by the axial coupling of the light flavor quarks. The reaction $D_s^* \to D_s\pi^0$ provides information on the η-meson coupling to strange constituent quarks, as it is mediated by $\pi^0 - \eta$ mixing. The two pion decay rates of heavy quarkonia implies pion exchange between heavy quarks and antiquarks, which is phenomenologically significant.

1 Introduction

Peter Sauer has long insisted on the need for a coherent systematic treatment of the interacting system of nucleons, $\Delta(1232)$ resonances and pions. The nature of the $\Delta(1232)$ resonance is indeed a classic problem in nuclear physics. While Chew and Low [1] showed that iterations of the pion-nucleon scattering amplitude can generate the $\Delta(1232)$ resonance dynamically, this resonance appears as a spin-isospin 3/2 partner of the nucleon in the quark model, and thus has an intrinsic qqq structure as the nucleon. The most realistic description of the $\Delta(1232)$ resonance is likely to a compromise between these extremes, as a qqq state with a substantial pion cloud [2].

This may in fact be inferred directly from the under-prediction of the decay width of the $\Delta(1232)$ by the static quark model. In the quark model, the value for the $\pi N\Delta$ transition coupling $f_{\pi N\Delta}$ is related to the πN coupling as:

$$f_{\pi N\Delta} = 2\sqrt{2} f_{\pi NN} \simeq 1.84. \tag{1}$$

This value is $\sim 16\%$ smaller than the value $f_{\pi N\Delta} = 2.2 \pm 0.4$, that is obtained from the empirical decay width of the Δ resonance.

* *Alternative address:* Department of Physical Sciences, University of Helsinki
† *E-mail address:* riska@pcu.helsinki.fi

Decay	Γ_{exp}	$\text{Ratio}_{\text{exp}}$	qqq_{ratio}
$\Delta \to N\pi$	120	12	9
$\Sigma^* \to \Sigma\pi$	39	3.9	4
$\Xi^* \to \Xi\pi$	10	1	1

Table 1. Pion decay widths [19] of the lowest decuplet states (in MeV) and their ratios.

This under-prediction of the width of the $\Delta(1232)$ resonance indicates the importance of the pion cloud around the $\Delta(1232)$. As pions do not couple to strange quarks, the ratios of the widths for the transitions $\Delta \to N\pi$, $\Sigma^* \to \Sigma\pi$ and $\Xi^* \to \Xi\pi$ should be as 9:4:1. Empirically the ratios of these widths is, however, 12:4:1 (Table 1). That suggests that while the Σ^* and Ξ^* may be fairly clean three-quark states (sqq and ssq respectively, where q denotes a light-flavor quark), the Δ has a considerable 5 quark ($qqq(q\bar{q})$) (or "pion cloud") component. This argument applies not only if the pion coupling is to a single quark, but also if the pion couples through a seagull type interaction mechanism to two quarks. As the static quark model only takes into account the qqq component of the baryons, it is only natural that it under-predicts the decay rates of baryon resonances with large seaquark admixtures.

2 The Baryon Spectrum

The seaquarks may play a significant role in the appearance of low lying positive parity states in the baryon spectrum. In the case of the nucleon and the Δ resonance the lowest excitations (the $N(1440)$ and the $\Delta(1600)$) have positive parity. If nucleons are formed of 3 constituent quarks in a monotonically rising confining potential well, these states should lie above the lowest negative parity states. Only by means of a "hyperfine" interaction between the quarks, which is strong enough to cause multiplet splittings, that are of the same size as the spacings of the major shells in the confining well, is it possible to describe the empirical spectral ordering.

The importance of the seaquarks is revealed by a comparison of the pion decay widths of the low lying positive parity states in the spectrum (Table 2).

If these resonances were pure 3 quark states the ratio of the pion decay

Decay	Γ_{exp}	$\text{Ratio}_{\text{exp}}$	qqq_{ratio}
$N(1440) \to N\pi, \Delta\pi$	350	17	9
$\Delta(1600) \to N\pi, \Delta\pi, N(1440)\pi$	350	17	9
$\Lambda(1600) \to \Sigma\pi$	~ 80	$\sim 4?$	4
$\Sigma(1660) \to \Lambda\pi, \Sigma\pi$	~ 80	~ 4	4

Table 2. Pion decay widths of the lowest positive parity states (in MeV) [19] and their (approximate) ratios.

widths of the non-strange positive parity states to the strange ones should be 9:4. Empirically it is more like 17:4, which suggests that the both the $N(1440)$ and the $\Delta(1600)$ contain substantial $qqq(q\bar{q})$ admixtures. This also suggest a reason for why the in the quark model the $N - N(1400)$ transition coupling is only about a third of the value required to explain the empirical decay width for $N(1440) \to N\pi$ [3].

3 Lattice Calculation Results

Numerical lattice calculations of the baryon spectrum should in principle be able to determine the structure of the low lying positive parity resonances. Hitherto the orbitally excited states have only been studied in the quenched approximation, in which only 3 valence quark configurations are considered. For the nucleon and its lowest negative parity excitation ($N(1535)$) the calculated energy values appear to converge onto the empirical values as the quark mass is reduced [4, 5].

For the range of quark masses hitherto accessible, the published quenched lattice calculations do not, however, show any convergence toward the positive parity states. Linear extrapolation of the calculated values in refs. [4, 5] leads to masses above 2 GeV, in gross disagreement with experiment. Given the argument above, that these states have larger amplitudes for $qqq(q\bar{q})$ than for qqq, this failure is perhaps not all that surprising, as the 5-quark configurations are left out by definition in the quenched approximation.

Two very recent lattice calculations of the low lying baryon resonance spectrum, which employ Bayesian statistics to extract the lowest states, show that it may still be possible to obtain realistic spectra in the quenched approximation [6, 7]. The first of these shows a sudden decline in the predicted masses of the positive parity resonances, when the quark mass falls below values, that correspond to pion mass values of about 600 MeV. This calculation led to realistic energies for both the $N(1440)$, the $\Delta(1600)$ and the $\Sigma(1660)$ positive parity resonances.

4 Quark Interactions

To describe the empirical baryon spectrum, with correct ordering of the states, in the 3-quark constituent quark model a flavor dependent hyperfine interaction is required. A schematic flavor and spin dependent interaction, which gives a qualitative description of the spectrum is in ref. [8].

$$V_{FS} = -C_\chi \sum_{i<j} \lambda_F^i \cdot \lambda_F^j \, \boldsymbol{\sigma}^i \cdot \boldsymbol{\sigma}^j, \tag{2}$$

where the λ_F matrices are generators of the flavor $SU(3)$ group and C_χ is a constant.

This interaction contributes the following energy shifts to the lowest states in the nucleon spectrum (with inclusion of the contribution from the harmonic confining interaction)

$$\begin{aligned} N(939): &\quad -14C_\chi, \\ N(1440): &\quad -14C_\chi + 2\hbar\omega, \\ N(1535): &\quad -2C_\chi + \hbar\omega. \end{aligned} \tag{3}$$

With $C_\chi \simeq -30$ MeV and $\omega = 220$ MeV the empirical ordering is obtained. The $N(1440)$ state falls 440 MeV above the nucleon (exp: ~ 430-500 MeV) and 80 MeV below the $N(1535)$ (exp. ~ 95-165 MeV).

The contributions to the corresponding states in the spectrum of the Δ are

$$\begin{aligned} \Delta(1232): &\quad -4C_\chi, \\ \Delta(1600): &\quad -4C_\chi + 2\hbar\omega, \\ \Delta(1700): &\quad +4C_\chi + \hbar\omega. \end{aligned} \tag{4}$$

This spectrum in also correctly ordered with the same parameter values. The $\Delta(1600)$ falls 440 MeV above the $\Delta(1232)$ (exp: ~ 370 MeV) and 20 MeV below the $\Delta(1700)$ (exp: ~ 100 MeV).

This satisfactory ordering carries over to the spectrum of the Λ hyperon, for which the lowest states obtain the contributions

$$\begin{aligned} \Lambda(1115): &\quad -14C_\chi, \\ \Lambda(1405): &\quad -8C_\chi - \hbar\omega, \\ \Lambda(1600): &\quad -14C_\chi + 2\hbar\omega. \end{aligned} \tag{5}$$

With the parameter values above the Λ(1600 falls 440 MeV above the Λ^0 (exp: 485 MeV) and 40 MeV above the Λ(1405) (exp: 195 MeV).

The flavor dependence of the interaction achieves the desired situation, in which the lowest excited states in the spectra of the nucleon and the Δ have positive parity, while in the spectrum of the Λ the flavor singlet negative parity state Λ(1405) is the lowest excited state [8].

5 The Chiral Quark Model

The chiral quark model extends the chiral coupling of pions to hadrons to the constituent quarks, which form the hadrons. The pion coupling to a constituent quark is, accordingly,

$$\mathcal{L}_{\pi qq} = \mathrm{i}\frac{g_A^q}{2f_\pi}\bar{\psi}_q\gamma_5\gamma_\mu\partial_\mu\boldsymbol{\pi}\cdot\boldsymbol{\tau}\psi_q. \tag{6}$$

Here ψ_q is the isospinor for the constituent quark and $\boldsymbol{\pi}$ the isovector pion field, f_π the pion decay constant ($f_\pi \simeq 93$ MeV), and g_A^q the axial coupling constant of the light flavor constituent quark. The value of g_A^q is of the order 0.87 [9] (typical values fall in the range 0.75-1.0).

The coupling strength in (1) may be re-expressed in terms of a π-quark pseudovector coupling $f_{\pi qq}$ constant by setting $f_{\pi qq}/m_\pi = g_A^q/2f_\pi$. For $g_A^q =$ 0.87 it follows that $f_{\pi qq} = 0.65$, which compares well to the static quark model value $f_{\pi qq} = 3/5 f_{\pi NN} \simeq 0.6$.

The Lagrangian (6) defines the "chiral quark" model, which gives an effective description of the key dynamics of nucleon structure in the baryon range, defined as the range between the chiral restoration scale $1/4\pi f_\pi \sim 0.2$ fm and the confinement scale $\Lambda_{QCD}^{-1} \sim 1$ fm [10].

The pion-quark coupling implies pion exchange interactions between quarks. In fact the one and two-pion exchange interactions combine to produce an interaction that approximately has the form of Eq. (2). The tensor components of the pion and two-pion exchange interactions largely cancel against each other, while the spin-spin components of these interactions combine to a strong attractive interaction [11]. Higher order pion exchange interactions should be suppressed by the cut-off of the pion-quark coupling at the chiral restoration scale.

The simplest observables that can provide information on the strength of the pion-quark coupling (6) are the pion decay rates of the heavy-light mesons. This is because the pions only couple to the light flavor constituent quarks in these quark-antiquark systems, but not to the heavy flavor quarks.

6 The Spectra of Heavy-Light Mesons

The simplest model for charm mesons are that they are $q\bar{c}$ (or $\bar{q}c$) systems, where q is a u or d quark and c a charm quark. Because of the small mass and high velocity of the confined light flavor quark, these systems call for a relativistic description. A convenient covariant approach is to describe the confined $q\bar{c}$ system by the Blankenbecler-Sugar equation [12].

This equation may be written in the form of Schrödinger equation

$$\left(\frac{\boldsymbol{p}^2}{2\mu} + V(r)\right)\psi(\boldsymbol{r}) = \epsilon\psi(\boldsymbol{r}), \qquad (7)$$

where μ is the relative mass of the light and heavy quarks. The energy of the two-quark system is obtained from the eigenvalue ϵ as

$$E = \sqrt{M^2 + m^2 + 4\mu\epsilon + 2\sqrt{M^2m^2 + 2\mu(M^2 + m^2)\epsilon + 4\mu^2\epsilon^2}}. \qquad (8)$$

In the large mass (non-relativistic) limit this expression reduces to $E = M + m + \epsilon$, which is the result for the Schrödinger equation.

A satisfactory description of the spectra of the D and D_s mesons may be achieved by assuming that the interaction between the heavy and light quarks is formed by the sum of a scalar linear confining interaction and a one-gluon exchange interaction. This interaction model describes heavy quarkonia well, including their small spin-orbit splittings. The small spin-orbit splittings are the result of a near cancellation between the spin-orbit components of the scalar confining interaction and the gluon exchange interaction.

To maintain this near cancellation and the correct spin-orbit splitting in heavy-light mesons, the momentum dependence of the running color coupling α_s has to be taken into account [13], as the static gluon exchange interaction would otherwise cause inversion of the spin-orbit splittings of the $L = 1$ charm mesons [14], in conflict with experiment.

7 The $D^* \to D\pi$ Decay Rates

In the single quark approximation based on the quark-pion coupling (1) the pionic decays of the D^* to $D\pi$ are mediated by the spatial part of the axial current of the quark. With the standard values of g_A^q the single quark coupling leads to decay rates that are smaller than the present empirical values (Table 3).

Table 3. Calculated pionic widths (in MeV) for the D^* for $g_A^q = 0.87$. The (relativistic) single quark approximation is denoted SQA, and the result with inclusion of the two-quark contribution is denoted FULL. The complete result is also given for $g_A^q = 1$ [18]. The experimental results are from [19, 20].

Decay	SQA	FULL	$g_A^q = 1$	Expt
$D^{*\pm} \to D\pi^0$	0.026	0.029	0.038	<0.04
$D^{*\pm} \to D^0\pi^\pm$	0.056	0.064	0.084	<0.09
$D^{*0} \to D^0\pi^0$	0.036	0.041	0.54	<1.3
$D^{*\pm} \to D^\pm\pi^0 + D^0\pi^\pm$	0.082	0.093	0.12	$0.096 \pm 0.04 \pm 0.022$

The axial current of the $D-$mesons also obtains two-quark contributions that are associated with the effective interaction between the constituent quarks. The most obvious of these are the "seagull" or "pair" current contributions, which arise in the elimination of explicit negative energy components from the wave functions. While the analogous two-quark contributions to the axial current of nucleon pairs are known to be insignificantly small [21], these contributions are not negligible in the case of quark pairs, because of the small constituent mass of the light flavor quarks. As an example, the axial two-quark pair current that is associated with the scalar confining interaction leads to an enhancement of about 30 % of the calculated pion decay rates, as also shown in Table 3. It is apparent, that only with inclusion of this exchange current contribution, the chiral quark model gives a satisfactory description of the pion decay rates of the charmed vector meson D^*.

8 Pion Decay of the $D_1(2420)$ and $D_2^*(2460)$

The total widths of two of the $D-$meson resonances with $L = 1$ have been measured:

$$\begin{aligned} \Gamma(D_1(2420)) &= 18.9^{+4.6}_{-3.5}\,\text{MeV}, \\ \Gamma(D_2^*(2460)) &= 25^{+8}_{-7} \quad \text{MeV}. \end{aligned} \tag{9}$$

These widths are mainly due single π decay to the D and D^*, although $\pi\pi$ decay also contributes significantly to the widths [22].

For these $\Delta L = 1$ transitions the axial charge component of the coupling (1) also gives a contribution to the decay rate in addition to the axial current. In the single quark approximation the axial charge operator leads to a single pion decay width of the $D_1(2420)$ meson, which is of the order of 60 MeV, and therefore exceeds the empirical width by about a factor 4 [18].

The axial charge operator of the light flavor quark is, however, strongly suppressed by the confining interaction, if that has the form of a scalar interaction. The simplest way to see how this suppression arises is to note, that scalar interaction potentials are additive with the mass operators of the constituents. This gives rise to a shift of the constituent mass in the axial charge operator $-g_A^q \boldsymbol{\sigma} \cdot \boldsymbol{P}/m$, so that

$$m \rightarrow m + cr - b, \tag{10}$$

if the confining interaction is $cr - b$, where c is the string tension and b a constant. If $c = 1120$ MeV/fm and $b = 320$ MeV as in [13], this mass shift reduces the matrix element of the axial charge operator by about 45 %.

An alternate way of deriving the suppression of the axial charge operator from the scalar confining interaction, is to derive it as a two-quark operator, by elimination of the negative energy components from the wavefunction in the same way as employed for the axial current operator above. This latter method is more general, as it may be applied to the gluon exchange hyperfine interaction as well. The vector coupled gluon exchange contribution also causes a suppression of the matrix element of the axial charge operator of the light flavor constituent quark in $D-$mesons, which is similar to, if somewhat weaker than, the suppression caused by the confining interaction.

Once the suppression of the axial charge operator has been taken into account, reasonable values for the single pion decay widths of the excited positive parity charm mesons are obtained (Table 4).

The branching ratios for single pion decay also agree with the empirical values:

$$\begin{aligned} \frac{\Gamma(D_2^* \to D\pi)}{\Gamma(D_2^* \to D^*\pi)} &= 2.1 \quad \text{Expt}: \quad 2.3 \pm 0.6 \\ \frac{\Gamma(D_2^* \to D\pi + D^*\pi)}{\Gamma(D_1 \to D^*\pi)} &= 1.5 \quad \text{Expt}: \quad 1.5 \end{aligned} \tag{11}$$

The two-pion decay widths of the $L = 1$ charm mesons may be calculated from the Born terms that are associated with the two different time ordered iterations of the chiral coupling in Eq. (1).

With this model two-pion decay was found to contribute ~ 2 MeV to the decay width of the $D_2^*(2460)$ and ~ 1 MeV to the decay width of the $D_1(2420)$ [22] (Table 4).

Table 4. π and $\pi\pi$ decay widths (in MeV) of the $D_1(2420)$ and $D_2^*(2460)$ mesons for $g_A^q = 1$. The complete result is also given for $g_A^q = 0.87$ [18, 22]. The experimental results are from [19, 20].

Decay	π	$\pi + \pi\pi$	$g_A^q = 0.87$	Expt
$D_2^*(2460)$	15.76	18.8	13.7	25^{+8}_{-7}
$D_1(2420)$	13.6	14.9	11.1	$18.9^{+4.6}_{-3.5}$

In the case of the two-pion exchange amplitude, there are also exchange current contributions that are associated with the scalar confining and gluon exchange hyperfine interactions which reduce the value of the transition amplitude, which is obtained in the single quark approximation by about 25 %.

The net results for the decay widths of the $L = 1$ charm mesons are closer to the empirical values if $g_A^q = 1$ than if $g_A^q = 0.87$. This result is obviously dependent on the Hamiltonian model for the charm mesons. The degree of this model dependence is revealed by the fact that in the work of ref.[15], which employed a similar, but static, interaction model and the Gross quasipotential equation, the authors obtained good results with $g_A^q = 0.75$. The qualitatively significant result is nevertheless, that the chiral quark model is able to describe the pionic decays of the charm meson resonances.

9 Pion Decay of the D_s^*

The isospin forbidden π^0 decay of strange charm vector meson $D_s^* \to D_s\pi^0$ is interesting, because it indicates the presence of pion coupling to both the strange and the charm quark, which is mediated by the η meson through $\pi^0\eta$ mixing [16]. If the $\pi^0\eta$ mixing angle is known, this decay can provide information on the strength of the η-meson coupling to both the strange and the charm quarks. Direct evidence for the latter coupling is provided by the empirically known decay branching ratio for $\psi' \to J/\psi\,\eta$.

As the total decay width of the D_s^* has not yet been determined experimentally, the phenomenological analysis has to be based on the known branching ratios for $D_s^* \to D_s\pi^0$ and $D_s^* \to D_s\gamma$. Thus a reliable model for the electromagnetic decay of heavy-light mesons is required.

A phenomenological analysis of $D_s^* \to D_s\pi^0$ along this line shows that the η coupling to the strange quark is likely to be only about 60% of the value from $SU(3)$ flavor extended chiral quark model value ($f_{\eta ss} = -m_\eta g_A^q/\sqrt{3}f_\eta$) [17]. If the quark model relation for baryon couplings can be relied upon, this would suggest that the η couplings to nucleons should be small. With the first order chiral perturbation theory value for the $\pi^0\eta$ mixing angle ~ 0.01 [23] this analysis suggests that $f_{\eta NN} \sim 0.8$ [17]. If the $\pi\eta$ mixing angle is larger, e.g. 0.026 as suggested in ref. [24], the value for $f_{\eta NN}$ would be proportionally smaller.

10 Pion Coupling to Charm Quarks

Single pions couple to charm quarks only through the $\pi^0\eta$ mixing. Two pions do however couple to both charm and beauty quarks as revealed by the empirical $\pi\pi$ decay branches of excited heavy quarkonia. Once two pions couple to single charm quarks, there is inevitably a pion exchange contribution to the two-pion decay modes of the ψ' and Υ'. This has to be taken into account in a phenomenological determination of the strength of the coupling of the two-pion system to charm and beauty quarks [25].

11 Conclusion

The chiral quark model is able to provide a qualitatively satisfactory description of the pionic decays of the charm meson resonances, with standard values for the parameters in the model. Thus g_A^q should fall in the range 0.75-1. The coupling of pions to quarks implies pion and multipion exchange between quarks. This interaction has the flavor dependence, which is required for a description of the empirical baryon spectrum.

The role of the two-quark contributions to both the axial current and charge operators of the D-meson systems is significant. The two-quark contributions to the axial charge causes a strong suppression of the net matrix element of the axial charge operator, which is necessary for realistic values for the decay width of the $D_1(2420)$ meson.

Acknowledgement. Research supported in part by the Academy of Finland under grant 54038.

References

1. G. E. Chew, F. E. Low: Phys. Rev. **106**, 1570 (1977)
2. T. Sato, T.S.H. Lee: Phys. Rev. **C63**, 055201 (2001)
3. D. O. Riska, G. E. Brown: Nucl. Phys. **A679**, 577 (2001)
4. S. Sasaki et al.: Phys. Rev. **D65**, 074503 (2002)
5. W. Melnitchouk et al.: eprint hep-lat/0202022
6. F. X. Lee et al.: eprint hep-lat/0208070
7. S. Sasaki, K. Sasaki, T. Hatsuda, M. Asakawa: hep-lat/0209059
8. L. Ya. Glozman, D. O. Riska: Phys. Rep. **B366**, 305 (1996)
9. S. Weinberg: Phys. Rev. Lett. **67**, 3473 (1991)
10. A. Manohar, H. Georgi, Nucl. Phys. B234, 189 (1984)
11. D. O. Riska, G. E. Brown: Nucl. Phys. **A653**, 251 (1999)
12. R. Blankenbecler, R. Sugar: Phys. Rev. **142**, 1051 (1966)
13. T.A. Lähde, C.J. Nyfält, D.O. Riska: Nucl. Phys. **A674**, 141 (2000)
14. H. Schnitzer: Phys. Lett. **B76**, 461 (1978)
15. J. L. Goity, W. Roberts: Phys. Rev. **D64**, 094007 (2001)
16. P. Cho, M. Wise: Phys. Rev. **D49**, 6228 (1994)
17. T. A. Lähde, D. O. Riska: Nucl. Phys. **A710**, 99 (2002)
18. K. O. E. Henriksson et al.: Nucl. Phys. **A686**, 355 (2001)
19. D. E: Groom et al.: Eur. Phys. J. **C15**, 1 (2001)
20. S. Ahmed et al.: Phys. Rev. Lett. **87**, 251801 (2001)
21. K. Tsushima, D. O. Riska: Nucl. Phys. **A549**, 313 (1992)
22. T. A. Lähde, D. O. Riška: Nucl. Phys. **A693**, 755 (2001)
23. J. Gasser, H. Leutwyler: Nucl. Phys. **B250**, 465 (1985)
24. W. B. Tippens et al.: Phys. Rev. **D63**, 052001 (2001)
25. T. A. Lähde, D. O. Riska: Nucl. Phys. **A707**, 425 (2002)

Few-Body Systems Suppl. 15, 133–138 (2003)

Few-Body Systems

Search for the Quark Shell Structure Using the Non-Topological Soliton Model

N. Sawado*, N. Shiiki, S. Oryu

Department of Physics, Tokyo University of Science, 2641, Yamazaki, Noda, Chiba 278-8510, Japan

Abstract. We investigate higher baryon-number states in the Chiral Quark Soliton Model using the rational map ansatz for the background chiral fields. The soliton solutions are obtained self-consistently. We show that the baryon number density has point-like symmetries and the corresponding valence quark spectrum of the lowest energy exhibits approximate four-fold degenerate. Our results indicate the possibility of existence of the shell-like structure in the multi-baryonic system.

The Chiral Quark Soliton Model (CQSM) was developed in 1980's as an effective theory of QCD interpolating between the Constituent Quark Model and the Skyrme Model [1, 2]. In the large N_c limit, these models are identical [3].

The CQSM is derived from the instanton liquid model of the QCD vacuum and incorporates the non-perturbative feature of low-energy QCD, spontaneous chiral symmetry breaking. The vacuum functional is defined by

$$\mathcal{Z} = \int \mathcal{D}\pi\mathcal{D}\psi\mathcal{D}\psi^\dagger \exp\left[i\int d^4x\,\bar{\psi}\left(i\partial\!\!\!/ - mU^{\gamma_5}\right)\psi\right], \tag{1}$$

where the SU(2) matrix

$$U^{\gamma_5} = \frac{1+\gamma_5}{2}U + \frac{1-\gamma_5}{2}U^\dagger \quad \text{with} \quad U = \exp\left(i\boldsymbol{\tau}\cdot\boldsymbol{\pi}/f_\pi\right)$$

describes chiral fields, ψ are quark fields, and m is the dynamical quark mass. f_π is the pion decay constant, and experimentally $f_\pi \sim 93$ MeV. Since our concern are the tree-level pions and one-loop quarks according to the Hartree mean field approach, the kinetic term of the pion fields which gives a contribution to higher loops can be neglected. Due to the interaction between the valence

* *E-mail address:* sawado@ph.noda.tus.ac.jp

quarks and the Dirac sea, soliton solutions appear as bound states of quarks in the background of self-consistent mean chiral field. N_c valence quarks fill each bound state to form a baryon. The baryon number is thus identified with the number of bound states filled by the valence quarks [4]. For $B = 1$ and 2, the spherically symmetric soliton [5, 6] and the axially symmetric soliton [7] were found respectively. Upon quantization, the intermediate states of nucleon and deuteron between the Constituent Quark Model and Skyrme Model were obtained.

The vacuum functional in Eq. (1) can be integrated over the quark fields to obtain the sea quark energy [8]

$$E_{\text{sea}}[U] = \frac{1}{4\sqrt{\pi}} N_c \int_{1/\Lambda^2}^{\infty} \frac{d\tau}{\tau^{3/2}} \left(\sum_{\nu} \mathrm{e}^{-\tau\epsilon_\nu^2} - \sum_{\nu} \mathrm{e}^{-\tau\epsilon_\nu^{(0)2}} \right) . \tag{2}$$

where Λ is the cut-off parameter in the proper-time regularization [9] and $\epsilon_\nu, \epsilon_\nu^{(0)}$ are the eigenvalues of the Dirac equation for a single quark with the vacuum sector defined by

$$\left(\frac{\boldsymbol{\alpha}\cdot\boldsymbol{\nabla}}{\mathrm{i}} + \beta m U^{\gamma_5} \right) |\nu\rangle = \epsilon_\nu |\nu\rangle \ , \ \left(\frac{\boldsymbol{\alpha}\cdot\boldsymbol{\nabla}}{\mathrm{i}} + \beta m \right) |\nu\rangle^{(0)} = \epsilon_\nu^{(0)} |\nu\rangle^{(0)} \ . \tag{3}$$

In the Hartree picture, the baryon states are the quarks occupying all negative Dirac sea and valence levels. Hence, if we define the total soliton energy E_{total}, the valence quark energy should be added;

$$E_{\text{total}}[U] = N_c \sum_i E_{\text{val}}^{(i)}[U] + E_{\text{sea}}[U] \ . \tag{4}$$

where E_{val}^i is the valence quark contribution to the ith baryon.

For higher winding number solitons, the solution is expected to have point-like symmetries from the study of the Skyrme model [10]. Therefore, we shall impose the same symmetries on the chiral fields using the rational map ansatz. According to the ansatz, the chiral field can be expressed as [11]

$$U(r, z) = \exp\left(\mathrm{i} f(r) \boldsymbol{n}_R \cdot \boldsymbol{\tau}\right) \ , \tag{5}$$

where

$$\boldsymbol{n}_R = \frac{1}{1 + |R(z)|^2} \left(2\mathrm{Re}[R(z)], 2\mathrm{Im}[R(z)], 1 - |R(z)|^2 \right) \ ,$$

and $R(z)$ is a rational map.

Rational maps are maps from $CP(1)$ to $CP(1)$ (equivalently, from S^2 to S^2) classified by winding number. It was shown in [11] that $B = N$ skyrmions can be well-approximated by rational maps with winding number N. The rational map with winding number N possesses $(2N + 1)$ complex parameters whose values can be determined by imposing the symmetry of the skyrmion. Thus,

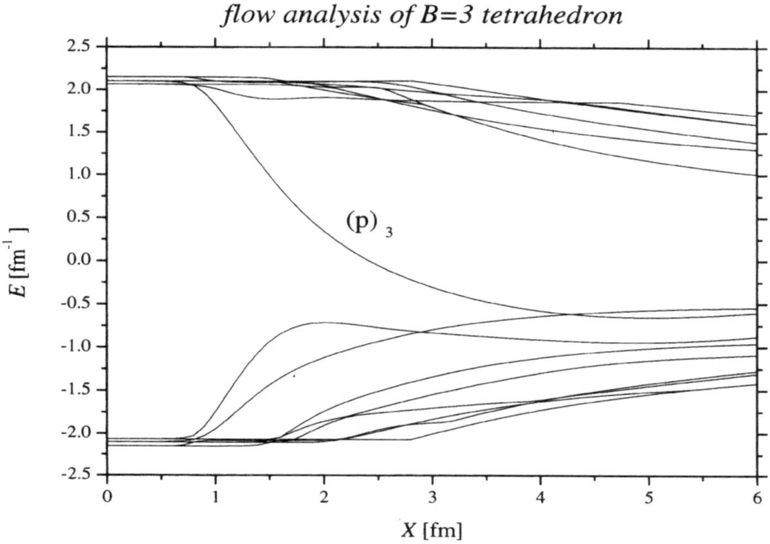

Figure 1. The spectrum of the quark orbits for the $B = 3$ soliton is illustrated as a function of the "soliton size" X. The trial profile function is defined by $f(r) = -\pi + \pi r/X$ for $X \le r$ and $f(r) = 0$ for $X > r$. The bound states are triply degenerate.

the rational maps for $B = 2 \sim 7$ and $B = 17$ take the form

$$
\begin{aligned}
&R(z) = z^2 \ (B=2)\ , \ R(z) = \frac{\sqrt{3}\mathrm{i}z^2 - 1}{z(z^2 - \sqrt{3}\mathrm{i})} \ (B=3)\ , \\
&R(z) = \frac{z^4 + 2\sqrt{3}\mathrm{i}z^2 + 1}{z^4 - 2\sqrt{3}\mathrm{i}z^2 + 1} \ (B=4)\ , \\
&R(z) = \frac{z(z^4 + 3.94z^2 + 3.07)}{3.07z^4 - 3.94z^2 + 1} \ (B=5)\ , \ R(z) = \frac{z(z^4 - 5)}{-5z^4 + 1} \ (B=5^*)\ , \\
&R(z) = \frac{z^4 + 0.16\mathrm{i}}{z^2(0.16z^4\mathrm{i} - 1)} \ (B=6)\ , \\
&R(z) = \frac{7/\sqrt{5}z^6 - 7z^4 - 7/\sqrt{5}z^2 - 1}{z(z^6 + 7/\sqrt{5}z^4 + 7z^2 - 7/\sqrt{5})} \ (B=7)\ , \\
&R(z) = \frac{17z^{15} - 187z^{10} + 119z^5 - 1}{z^2(z^{15} + 119z^{10} + 187z^5 + 17)} \ (B=17)\ ,
\end{aligned} \tag{6}
$$

where the complex coordinate z on $CP(1)$ is identified with the polar coordinates on S^2 by $z = \tan(\theta/2)\mathrm{e}^{\mathrm{i}\varphi}$ via stereographic projection. Substituting (6) into (5), one obtains the complete form of the chiral fields with appropriate symmetry and winding number. Since the chiral fields in (5) are parameter-

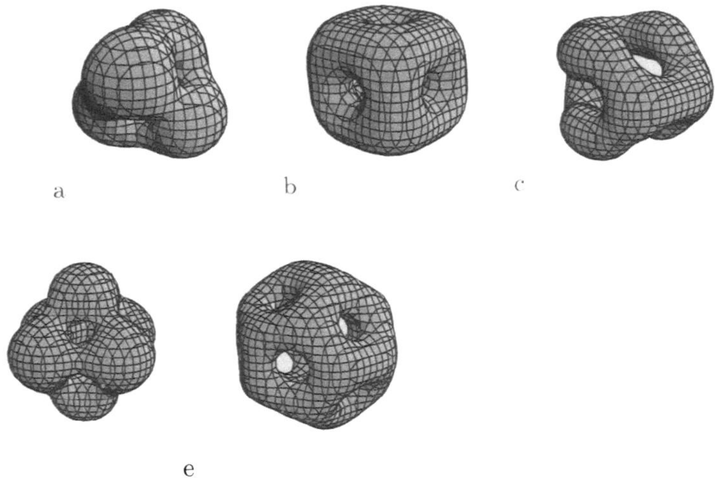

Figure 2. Surface of the baryon-number density: (a) $B=3$ tetrahedron, (b) $B=4$ cube, (c) $B=5$ with D_{2d} symmetry, (d) $B=5$ octahedron, (e) $B=7$ dodecahedron.

ized by polar coordinates, one can apply the numerical technique developed for $B=1$ to find higher soliton solutions [6].

Demanding that the total energy in (4) be stationary with respect to variation of the profile function $f(r)$,

$$\frac{\delta}{\delta f(r)} E_{\text{total}} = 0 \ ,$$

yields the field equation

$$S(r)\sin f(r) = P(r)\cos f(r)\,, \tag{7}$$

where

$$S(r) = N_c \sum_{\nu} \big(n_\nu \theta(\epsilon_\nu) + \text{sign}(\epsilon_\nu)\mathcal{N}(\epsilon_\nu)\big)\langle \nu|\gamma^0 \delta(|x|-r)|\nu\rangle \tag{8}$$

$$P(r) = N_c \sum_{\nu} \big(n_\nu \theta(\epsilon_\nu) + \text{sign}(\epsilon_\nu)\mathcal{N}(\epsilon_\nu)\big)\langle \nu|\mathrm{i}\gamma^0\gamma^5 \boldsymbol{n}_R\cdot\boldsymbol{\tau}\delta(|x|-r)|\nu\rangle \ . \tag{9}$$

The procedure to obtain the self-consistent solution of Eq. (7) is to 1) solve the eigenequation in (3) under an assumed initial profile function $f_0(r)$, 2) use the resultant eigenfunctions and eigenvalues to calculate $S(r)$ and $P(r)$, 3) solve Eq. (7) to obtain a new profile function, 4) repeat 1) – 3) until self-consistency is obtained.

Figure 1 shows the spectra of the quark orbits for $B = 3 \sim 7$ as a function of the soliton-size parameter X. In $B=3$, the P orbit diving into the negative energy region is triply degenerate. As discussed in [4], the baryon number of the

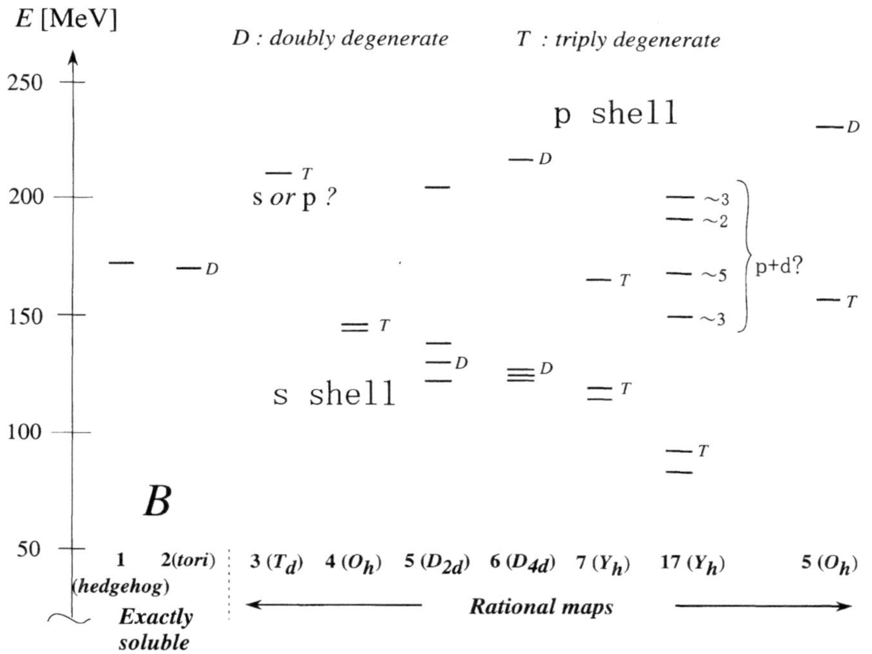

Figure 3. Valence quark spectra of $B = 1 \sim 7$ and $B = 17$.

soliton equals the number of diving levels occupied by N_c valence quarks. Thus, putting $N_c = 3$ valence quarks in each of the degenerate levels, one obtains the $B = 3$ soliton solution. Similarly, for B higher than 3, the same number of orbits as B dive into the negative energy region.

Figure 2 shows the corresponding baryon density. This result is consistent with the result for skyrmions obtained by Braaten *et.al* [10] and Houghton et al. [11].

In Fig. 3, the valence quark spectra are shown for $B = 1 \sim 7$ and $B = 17$. The approximate four-fold degeneracy can be observed for the lowest level. However, in the CQSM, a quark has only two quantum numbers: colour and grand spin defined by $\boldsymbol{K} = \boldsymbol{L} + \frac{1}{2}\boldsymbol{\sigma} + \frac{1}{2}\boldsymbol{\tau}$, which means the lowest level ($K = 0$) should show no degeneracy. At present, all we can say is that if one employs point-like symmetries for the background chiral fields, one obtains a larger degeneracy of single quark states than in the spherically symmetric case. Consequently, the number of states can be reduced and the large shell gaps emerge. These results indicate the possibility of the existence of the lowest energy shell filled up with four quarks, and this may also be the reason why ^{4}He is stable compared with other nuclei.

Table 1. Spectra of the valence quarks, vacuum sea, and their sum (in MeV). $\bar{M}$ is defined as $\bar{M} \equiv M/BM_{B=1}$, which means the energy per baryon. The results for $\bar{M}$ in Skyrme model calculations [11] are also quoted.

B	Valence							Sea	Total	$\bar{M}$ (Skyrme)
1	173							674	1192	1.00 (1.00)
3	210	210	210					1554	3444	0.96 (0.96)
4	153	156	156	156				2544	4407	0.92 (0.92)
5	123	131	131	139	210			3265	5467	0.92 (0.93)
5*	157	157	157	232	232			2874	5680	0.95 (1.00)
6	123	126	129	129	216	216		3184	6632	0.93 (0.92)
7	115	120	120	120	166	166	166	4554	7478	0.90 (0.90)

Acknowledgement. We are grateful to N.S. Manton for suggesting this subject to us and the possibility of a shell structure for quarks.

References

1. D.I. Diakonov, V.Yu. Petrov: Nucl. Phys. **B306**, 457 (1986)
2. R. Alkofer, H. Reinhardt, H. Weigel: Phys. Rep. **265**, 139 (1996); C.V. Christov, A. Blotz, H.-C.Kim, P. Pobylitsa, T. Watabe, Th. Meissner, E. Ruiz Arriola, K. Goeke: Prog. Part. Nucl. Phys. **37**, 91 (1996)
3. A. Manohar: Nucl. Phys. **B248**, 19 (1984)
4. S. Kahana, G. Ripka: Nucl. Phys. **A429**, 462 (1984)
5. H. Reinhardt, R. Wünsch: Phys. Lett. **B215**, 577 (1988)
6. M. Wakamatsu, H. Yoshiki: Nucl. Phys. **A524**, 561 (1991)
7. N. Sawado, S. Oryu: Phys. Rev. **C58**, R3046 (1998); N. Sawado, Phys. Rev. **C61**, 65206 (2000)
8. Th. Meissner, E. Ruiz Arriola, F. Grümmer, H. Mavromatis, K. Goeke: Phys. Lett. **B214**, 312 (1988)
9. J. Schwinger: Phys. Rev. **82**, 664 (1951)
10. E. Braaten, S. Townsend, L. Carson: Phys. Lett. **B235**, 147 (1990)
11. C.J. Houghton, N.S. Manton, P.M. Sutcliffe: Nucl. Phys. **B510**, 507 (1998)

Few-Body Systems Suppl. 15, 139–150 (2003)

Few-Body Systems

Baryons as Relativistic Three-Quark Systems

W. Plessas

Institut für Theoretische Physik, Universität Graz, Universitätsplatz 5, A-8010 Graz, Austria

Abstract. We discuss some recent developments in the description of baryons as three-quark systems within relativistic constituent quark models. In particular we address the issues of excitation spectra, electroweak structure, and mesonic resonance decays. The necessities of implementing simultaneously the symmetries of low-energy quantum chromodynamics and of relativistic invariance are emphasized.

1 Introductory Remarks

The consideration of few-quark systems has become a very exciting topic in few-body physics. On the one hand it represents a big challenge for modern few-body methods to solve few-quark problems, since the corresponding bound and continuum states have to be solved in a consistent relativistic framework. Thus, in order to treat hadron phenomena on a microscopic level the formalism of calculation has to be essentially relativistic. On the other hand the issues of quark dynamics are very important. In principle, the microscopic description of hadronic bound states and reactions amounts to solving quantum chromodynamics (QCD) at low energies. Evidently, this is not strictly possible nowadays, and one must resort to effective theories and/or models. For any of such approaches, however, it is required to implement the essential properties of low-energy QCD, which are still not known in a satisfactory manner. In working with few-quark problems it is an intriguing task to find out the forces binding the constituents of hadrons and finally governing their behaviours. Evidently it is of eminent interest from many aspects of nuclear and subnuclear physics to be able to describe and to understand the properties (and reactions) of hadrons in a microscopic manner on the basis of the fundamental theory of strong (and also electroweak) interaction.

A very powerful tool for low- and intermediate-energy hadron physics consists in employing constituent quark models (CQMs). They have already witnessed a long and reasonably successful development. Starting out from the naive nonrelativistic quark model of hadrons in the 60's one has constantly improved the quark-quark dynamics and one has learned how to apply a relativistic formalism[1]. At present we have several refined CQMs available, which at least prove the approach to be promising. Of course, they advocate various types of dynamics and use different calculational methods, what leads to partly dissimilar results. However, this just makes up the flavour of present research in this field. By considering more and more hadron processes within the concept of CQMs one will gradually find out which is the most promising variant of the existing models.

In hadronic physics one would like to have an effective theory/model of QCD that is able to deal with all phenomena in a certain energy domain (at low and intermediate energies, say). It is not sufficient to consider just a limited aspect, for instance, only spectroscopy. This is especially true also for CQMs. Therefore, any realistic CQM should be subject to as many tests as possible. A model that has proven to be successful in describing the hadron spectra (bound state and resonance energies) should then be applied to hadronic reactions in order to check its global usability (notably also the wave function behaviour).

In this context, we shall here mainly discuss results of the relativistic CQM whose dynamics is based on Goldstone-boson exchange (GBE) vis-à-vis corresponding results from other types of relativistic CQMs. In particular, we shall consider the GBE CQM by the Graz group [1, 2] and compare its results to the ones by a typical representative of one-gluon-exchange (OGE) CQMs, namely the relativistic variant of the Bhaduri-Cohler-Nogami (BCN) CQM [3] as parametrized in ref. [4], and by the CQM relying on instanton-induced (II) forces [5]. We shall address the aspects of light and strange baryon spectroscopy, nucleon electroweak structure, electric radii and magnetic moments of octet and decuplet baryon ground states, and pionic decays of nucleon and Δ resonances.

2 Baryon Spectroscopy

The baryon spectra as provided by CQMs have been discussed from many viewpoints, especially also because of the big interest from the experimental side. While the description was rather qualitative in the early stages of CQMs, it is clear that considerable improvements have been achieved in the past years. Finally one has met the demand of describing all the light- and strange baryon spectra within one and the same framework. Due to the very nature of these spectra it is immediately evident that one needs flavor-dependent forces. As a result, the OGE CQM being flavor-independent encounters difficulties in describing simultaneously the N and Λ spectra, in addition to the notorious shortcomings in the level orderings of positive- and negative-parity states in the N (and tentatively Δ) spectrum. The reasons have already been discussed in much detail in the literature (see, e.g., ref. [2]). Here we only recall the typical behaviour through a comparison of the OGE, II, and GBE CQMs for

[1] By 'relativistic' we mean that a model should at least meet the requirements of Poincaré invariance. Thus it need not be field-theoretic, with infinitely many degrees of freedom, but can also be based on a covariant mass operator or Hamiltonian, defined on a Hilbert space with a finite number of degrees of freedom.

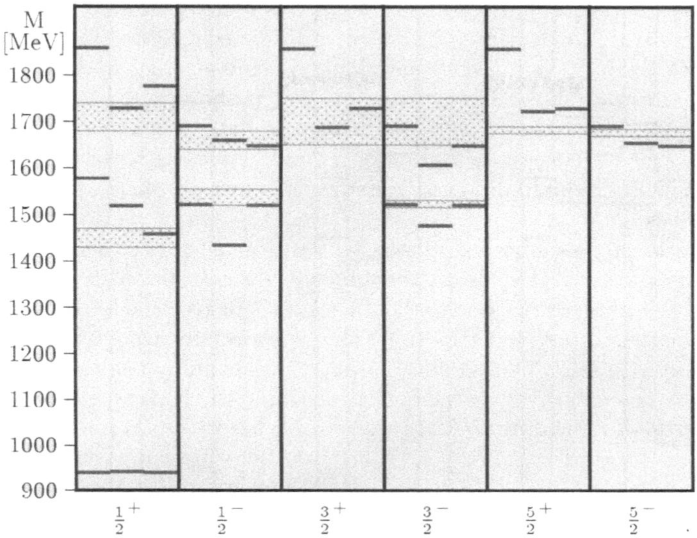

Figure 1. Nucleon excitation spectra of three different types of relativistic CQMs. In each column the left horizontal lines represent the results of the BCN OGE CQM [4], the middle ones of the II CQM (Version A) [5], and the right ones of the GBE CQM [1]. The shadowed boxes give the experimental data with their uncertainties after the latest compilation of the PDG [6].

the N and Λ spectra in Figs. 1 and 2 (the complete spectra can be found in the original refs. [4], [5], and [1], respectively). This comparison actually shows the (most striking) current problems still prevailing in the various types of CQMs. Notably, all of them encounter severe shortcomings in the Λ spectrum, especially with regard to the $\Lambda(1405)$, the first excitation above the ground state. Furthermore the OGE and the II CQMs are left with the wrong level orderings of the first positive- and negative-parity excitations above the nucleon ground state. Only the GBE CQM can reproduce both the $N^*(1440)$ $\frac{1}{2}^+$ and the $N^*(1535)$ $\frac{1}{2}^-$ states in the right places, in accordance with experiment. Except for the open problem with $\Lambda(1405)$, it thus represents the only CQM that is able to describe in a unified framework all light and strange baryon spectra in reasonable agreement with experiment.

With regard to the problem of the correct level ordering of positive- and negative-parity states the following remark is in order. It has repeatedly been claimed above all by Valcarce et al. that a proper description, especially of the N and Δ spectra, can also be obtained within a hybrid (i.e. one-gluon exchange plus meson exchange) CQM keeping a predominant contribution from the OGE interaction. In the first paper of the Salamanca-Valencia group [7], where such a claim was originally made, they used a variant of their CQM

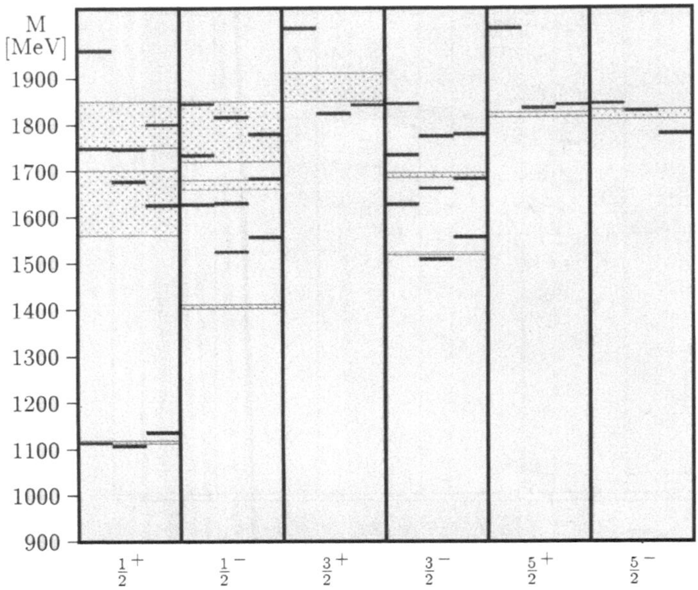

Figure 2. Λ excitation spectra. Same caption as in Fig. 1.

with a value $\alpha_s = 0.485$ of the effective quark-gluon coupling and a cut-off parameter $r_0 = 0.0367$ fm in the OGE part of the hyperfine interaction. Due to a nonconverged variational calculation of the nucleon spectrum they obtained a wrong level structure, with the positive-parity $N^*(1440)$ resonance practically degenerate with the negative-parity $N^*(1535)$ state. Upon recalculating this model in a converged variational calculation and also in a rigorous Faddeev approach, the Graz group established the right N and Δ spectra and found them in disagreement with the original paper [7] and thus also at variance with phenomenology [2]. The problem was reiterated in ref. [8], where also some technical details about treating the δ-function smearing of the color-magnetic interaction in three-quark calculations were elucidated. Still, in ref. [9] a similar claim as before was made by the Salamanca-Valencia group, then with a different parametrization of their CQM, namely with the parameters $\alpha_s = 0.65$ and $r_0 = 0.8$ fm in the OGE Part. As it turned out in another recalculation by the Graz group, the results reported in ref. [9] were again not converged. Similar errors were repeated in a subsequent paper by Garcilazo et al. [10]. There, some further modified versions of the Salamanca-Valencia CQM were considered reporting results for the N and Δ spectra calculated in the framework of Faddeev equations. One claimed a reasonable reproduction of the ground-state and resonance levels, in particular of the Roper $N^*(1440)$ state, while maintaining a considerable contribution from the OGE part in the Q-Q hyperfine interaction. Upon checking these results the Graz group found another time that they were not satisfactorily accurate. This was established by applying on the one hand the stochastic variational method (SVM) [11] and on the other hand a modified type of Faddeev approach as in ref. [12]; the latter was especially designed to treat infinitely rising confining potentials along with Faddeev equations and leads to improved convergence. Only in another subsequent paper [13] Garcilazo et al. produced practically correct N and Δ spectra by including into their Faddeev calculation enough angular momentum states. As is clearly seen from the figures given in that work, a proper ordering of the positive- and negative-parity levels is not achieved with neither one of the different versions of the Salamanca-Valencia hybrid CQM. On the contrary, the negative-parity $N^*(1535)$ falls clearly below the positive-parity $N^*(1440)$ Roper resonance, leaving the N spectrum (and in addition also the Δ spectrum) at variance with experiment.

Now, there has also been a contribution to this conference by A. Valcarce. Therein it is stated again that 'a quite reasonable description of the baryon spectrum' is obtained with the Salamanca-Valencia CQM. By inspecting fig. 5 of this contribution [14] (which shows basically correct results but due to missing higher states rather incomplete excitation spectra) one cannot find this claim confirmed by any means. The negative-parity $\frac{1}{2}^-$ resonances in both the N and Δ spectra lie far below the experimental values and notably also below the respective first positive-parity excitations above the ground states. In addition, the spectrum shown in fig. 5 of ref. [14] does not belong to the parameter set quoted in table 1 of the same paper. Rather it belongs to the parameter set given in table 3 of ref. [13], which produces even inferior spectra with regard to the relative positions of positive- and negative-parity excitations (cf. figs. 1 and 2 in ref. [13]); in addition the first $\frac{3}{2}^+$ and $\frac{5}{2}^+$ N^* states (not shown neither in ref. [13] nor by Valcarce in the contribution to this conference) both lie more than 100 MeV below the experimental levels! Thus the argumentation in Valcarce's contribution is badly misleading in connection with the performance of the Salamanca-Valencia CQM in baryon spectroscopy. The truth is that correct N and Δ spectra have so far not been achieved in neither one of the presented parametrizations of this particular CQM. In principle, within a pure OGE CQM or a hybrid model that relies on a predominant contribution from OGE in the hyperfine interaction it is not possible to describe specifically the first positive- and negative-parity excitations in the N and Δ spectra, the N-Δ splitting, and the Λ spectrum correctly at the same time. This fact can be readily derived from the spin-isospin symmetry of the color-magnetic interaction [2, 15, 16].

Obviously a correct quantitative description of all light and strange baryon spectra in a unified relativistic framework is of importance for applications of the CQM wave functions in dynamical processes involving ground states and resonances. Certain constraints are, of course, imposed by the spectra on the spin-flavor symmetry of the Q-Q interaction, and the various CQMs may not yet be complete in these respects[2]. However, one must also bear in mind that

[2]See, e.g., the attempts of extending the GBE CQM to include further force components [17].

the description of resonances in terms of $\{QQQ\}$ bound states (without explicit couplings to decay channels) might be totally inadequate (see Sec. 4 below).

3 Electroweak Nucleon/Baryon Structure

The particular relativistic CQMs considered in the previous section have already been tested for nucleon form factors and, partly, also for electric radii and magnetic moments of the light and strange baryon ground states [18, 19, 20, 21]. For the cases of the OGE and GBE CQMs one obtained manifestly covariant results along the point-form spectator approximation (PFSA) whereas the predictions of the II CQM were obtained also with one-body currents but in a Bethe-Salpeter approach [22]. In fig. 3 we present a comparison of the results for the elastic proton and neutron electroweak form factors. The electric radii and magnetic moments of the nucleons, and likewise of all other measured octet and decuplet ground states, are quoted in tables 1 and 2.

Table 1. Predictions for electric radii of the proton, neutron, and the Σ^- baryon as obtained in PFSA from the GBE CQM [1] in comparison to the case with the confinement potential only and results from the BCN OGE CQM [4].

Baryon	r_E^2 [fm^2] GBE CQM	CONF	OGE CQM	Experiment [6]
p	0.8176	0.7595	0.8859	0.757 ± 0.014
n	−0.1332	−0.0088	−0.1958	-0.1161 ± 0.0022
Σ^-	0.4946	0.5329	0.4440	0.61 ± 0.15

In general, the relativistic results appear quite reasonable in comparison to experiments, regarding all observables considered here. This is the more remarkable since in all instances only one-body operators were employed for the electromagnetic and axial currents. Of course, one must use realistic nucleon wave functions, notably with mixed-symmetry spatial components included, in order to describe especially also the neutron and similar ground states[3]. Beyond that the specific dynamical ingredients in the wave functions are of secondary importance while relativistic (boost) effects turn out to be very important. This is true for the results in point-form quantum mechanics as well as for the ones from the Bethe-Salpeter approach. Especially the GBE and OGE CQMs produce quite similar results in PFSA. But also the predictions of the II CQM are qualitatively similar even though they were calculated in a completely different relativistic formulation. Thus, it has become quite evident that kinematical relativistic effects in any case play a big role, and every nonrelativistic theory for electroweak form factors must be considered as inadequate [18, 20, 22].

[3] With a spatially symmetric wave function, such as the one from a confinement potential alone, one would fail in reproducing, in particular, the neutron electric form factor, cf. the dash-dotted line in left middle pane of fig. 3, which produces essentially no structure.

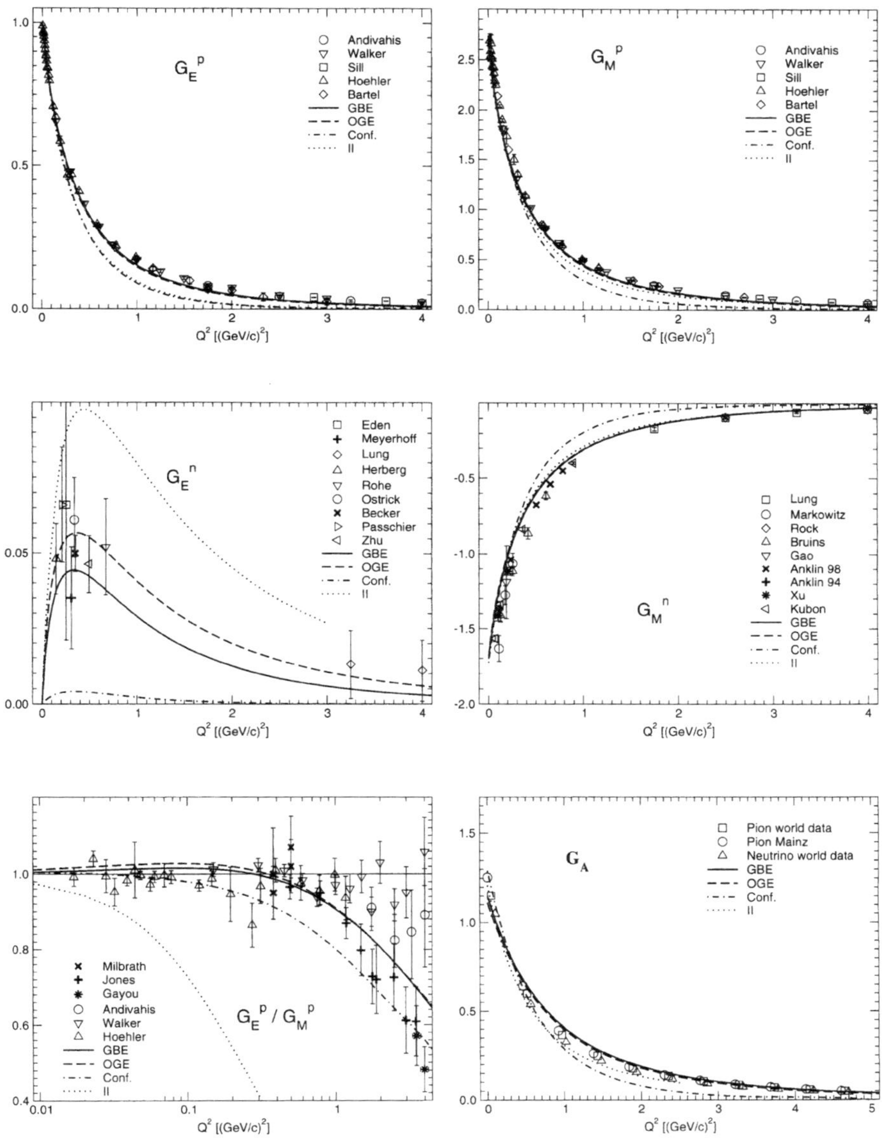

Figure 3. Predictions of different CQMs for the nucleon electromagnetic and axial form factors. The solid and dashed lines represent the PFSA results for the GBE CQM and the BCN OGE CQM, respectively; the dash-dotted lines refer to the case with confinement only. The dotted lines show the results of the II CQM within the Bethe-Salpeter approach after ref. [22].

Table 2. Predictions for magnetic moments of the octet baryons as well as Δ^+, Δ^{++}, and Ω^- as obtained in PFSA from the GBE CQM [1] in comparison to the case with the confinement potential only and results from the BCN OGE CQM [4].

Baryon	μ [n.m.] GBE CQM	CONF	OGE CQM	Experiment [6], [23]
p	2.6980	2.6478	2.7257	2.793
n	-1.7004	-1.7287	-1.6951	-1.913
Λ	-0.5894	-0.5802	-0.5883	-0.613
Σ^0	0.7001	0.6865	0.6571	-
Σ^+	2.3372	2.2294	2.2046	2.458
Σ^-	-0.9371	-0.8564	-0.8905	-1.160
Ξ^0	-1.2744	-1.2927	-1.2717	-1.250
Ξ^-	-0.6663	-0.5490	-0.5657	-0.6507
Δ^+	2.0827	2.0880	2.0708	$2.7 \pm 1.5 \pm 1.3$
Δ^{++}	4.1654	4.1761	4.1417	$3.7 - 7.5$
Ω^-	-1.5907	-1.5939	-1.5762	-2.0200

While the final insight has certainly not yet been gained, in view of the conceptual incompleteness of the calculations managed so far, the description of the nucleon (ground state) properties within CQMs nevertheless appears promising. Still, a number of further investigations, e.g., employing two- and three-body currents within a relativistic framework, will be necessary in order to further manifest the theoretical understanding of the nucleon structure at low momentum transfers.

4 Strong Decays of Baryon Resonances

Once having considered dynamical reactions involving the ground state(s), it is natural to proceed to the baryon resonances in order to further test the CQMs. Despite a number of investigations available in the literature, unfortunately, not many evidences that can really be considered reliable have so far been gained. In view of the experiences made with nucleon form factors and the enormous relativistic effects observed there, one must take any nonrelativistic study of resonances with considerable doubt. In this connection it is certainly not surprising that, notably, the mesonic baryon resonance decays are by far not yet understood. Theoretical predictions obtained hitherto with genuine CQM wave functions (i.e. without introducing additional phenomenological ingredients beyond the CQMs) compare quite unfavourably with existing experiments (see, e.g., ref. [4]).

Following the studies of nucleon form factors addressed in the previous section, the Graz group has recently performed calculations of (pionic) decay widths of N and Δ resonances with CQMs in relativistic point-form quantum

Table 3. Predictions for pionic decay widths by the GBE CQM [1] along the EEM in PFSA in comparison to experiment, an analogous calculation with the BCN OGE CQM [4], and results from a nonrelativistic EEM approach.

Decays	Experiment [6]	Rel. PFSA CQM		Nonrel. EEM GBE CQM	
		GBE	OGE	dir	dir+rec
$N^{\star}_{1440} \to \pi N_{939}$	$(227 \pm 18)^{+70}_{-59}$	30.3	37.1	4.85	6.16
$N^{\star}_{1520} \to \pi N_{939}$	$(66 \pm 6)^{+9}_{-5}$	16.9	16.2	22.0	38.3
$N^{\star}_{1535} \to \pi N_{939}$	$(67 \pm 15)^{+55}_{-17}$	93.2	122.8	24.3	574.3
$N^{\star}_{1650} \to \pi N_{939}$	$(109 \pm 26)^{+36}_{-3}$	28.8	38.3	11.3	160.3
$N^{\star}_{1675} \to \pi N_{939}$	$(68 \pm 8)^{+14}_{-4}$	5.98	6.20	7.65	15.1
$N^{\star}_{1700} \to \pi N_{939}$	$(10 \pm 5)^{+3}_{-3}$	0.91	1.19	1.43	2.87
$N^{\star}_{1710} \to \pi N_{939}$	$(15 \pm 5)^{+30}_{-5}$	4.06	2.28	23.4	5.95
$\Delta_{1232} \to \pi N_{939}$	$(119 \pm 1)^{+5}_{-5}$	33.7	32.1	59.1	81.2
$\Delta_{1600} \to \pi N_{939}$	$(61 \pm 26)^{+26}_{-10}$	0.116	0.503	74.2	55.7
$\Delta_{1620} \to \pi N_{939}$	$(38 \pm 8)^{+8}_{-6}$	10.4	14.6	4.82	74.8
$\Delta_{1700} \to \pi N_{939}$	$(45 \pm 15)^{+20}_{-10}$	2.92	3.10	7.12	14.4

mechanics. A first report thereof was given in ref. [24]. In table 3 we provide a comparison of some results for pionic decay widths of N and Δ resonances from the GBE and BCN OGE CQMs. So far one has only employed a decay operator according to a relativistic generalization of the elementary emission model (EEM), i.e. a one-body operator. Relativistic boost effects, however, have been included exactly in the calculation of matrix elements. By comparison to a standard nonrelativistic calculation one has again found that relativistic effects are of considerable importance, with their magnitudes exceeding by far the influences of dynamical ingredients in the wave functions. Still, at this instance, one has not yet arrived at a satisfactory explanation of the decay widths in accordance with existing experimental data. Here, the problems appear much more intricate than anticipated. Due to the involvement of resonance wave functions in evaluating the transition matrix elements one might tentatively need a more elaborate description of the resonance states with finite rather than zero widths. This means that one should include the couplings to the decay channels in the resonance wave functions. In a microscopic approach, ideally, one should provide additional configurations beyond three-quark states explicitly or at least take care of their roles in an effective description. Similarly one must consider further refinements in the decay operator and/or alternative mechanisms for the meson emission.

5 Conclusions

From the evidences highlighted above, a number of important conclusions can be drawn on the performance and adequacy of CQMs (as effective models of QCD for low-energy hadronic physics). These insights will be important for the future development of the CQM approach and, more generally, for the physics of low- and intermediate-energy hadron phenomena.

Regarding spectroscopy a further improvement of CQMs is most desirable. Certainly, with the GBE CQM (and to a lesser quality also with the II CQM) a description of all light and strange baryons is at hand within a single model[4]. It would be satisfying to extend the existing models in order to reach an even more comprehensive unified framework for the treatment of baryons (including also the heavy flavors) and mesons. From the applications considered here, one might be content with the behaviour of ground states (as $\{QQQ\}$ bound states) but the treatment of resonances must certainly be made more realistic. In this respect it is also not clear how far the concept of CQMs can be pushed up in excitation energy, even under the inclusion of further, say mesonic, degrees of freedom. The assumption of the constituent quark itself, as a quasiparticle of low-energy QCD, could soon break down in higher excitations of hadrons, namely, when chiral symmetry gets restored.

From dealing with hadron reactions in CQMs (in our case lepton scattering on nucleons and strong decays of baryon resonances) it has become fairly evident that a relativistic formulation has to be employed. In this light, any nonrelativistic treatment must be considered as inadequate from the very beginning. Still, it appears viable to work with a finite number of degrees of freedom instead of a genuine field-theoretic framework. This paves the way for approaches following Poincaré-invariant quantum mechanics. While being well founded axiomatically, it seems they also provide a proper tool for fulfilling Lorentz covariance and allow at the same time the application of refined few-body techniques to solve a variety of few-quark problems without having to resort to drastic approximations. Over the past decades few-body physics has been especially successful in accurate descriptions of nuclear phenomena. For that purpose a nonrelativistic quantum theory was mostly sufficient. In subnuclear physics, when dealing with a microscopic description of hadrons as few-quark systems, this is no longer the case, and we have to employ relativistic approaches. Their solution usually requires much more technical efforts, and it represents a challenge to apply them for all kind of hadron phenomena. We may well be confident that the corresponding studies will reward us with new exciting flavours in the traditional field of few-body physics.

[4] We recall that, e.g., for the Capstick-Isgur OGE CQM [25] two separate parametrizations are necessary, one for positive-parity and one for negative-parity states. There is not a unique parameter set allowing for a simultaneous reproduction of all states in a uniform manner.

Acknowledgement. I want to express my sincere gratitude to Teresa Peña and her team for organizing the 2002 Lisbon Workshop and for dedicating it to Peter Sauer. Indeed, it is justified to acknowledge Peter's contributions to few-body physics over the years by a topical workshop. I have been knowing him since the early 70's and I have always appreciated him as a critical colleague and an enthusiastic as well as helpful collaborator. The scientific links between Hanover and Graz have produced a number of essential results and also several friendly personal relationships. I feel uncomfortable when thinking about a future where all this could be lost due to the retirement of Peter and a possible disruption of the Hanover few-body-physics group.
The results discussed in this work have mostly originated from combined efforts of the few-body physics group in Graz, partly in collaboration with the Pavia group (Sigfrido Boffi and Marco Radici) and with Bill Klink from the University of Iowa. It is a pleasure to acknowledge this fruitful collaboration and here above all the individual contributions by Robert Wagenbrunn, Katharina Berger, and Thomas Melde. This work was supported by the Austrian Science Fund, project no. P14806-TPH.

References

1. Glozman, L.Ya., Plessas, W., Varga, K., and Wagenbrunn, R.F.: Phys. Rev. D **58**, 094030 (1998).

2. Glozman, L.Ya., Papp, Z., Plessas, W., Varga, K., and Wagenbrunn, R.F.: Phys. Rev. C **57**, 3406 (1998).

3. Bhaduri, R.K., Cohler, L.E., and Nogami, Y.: Nuovo Cim. **A65**, 376 (1981).

4. Theußl, L., Wagenbrunn, R.F., Desplanques, B., and Plessas, W.: Eur. Phys. J. **A12**, 91 (2001).

5. Löring, U., Metsch, B.C., and Petry, H.R.: Eur. Phys. J. **A10**, 395 (2001); ibid. **A10**, 447 (2001).

6. Hagiwara, K. et al.: Phys. Rev. D **66**, 010001 (2002).

7. Valcarce, A., González, P., Fernández, F., and Vento, V.: Phys. Lett. **B367**, 35 (1996).

8. Glozman, L.Ya., Papp, Z., Plessas, W., Varga, K., and Wagenbrunn, R.F.: Phys. Rev. C **61**, 019804 (1999).

9. Valcarce, A., González, P., Fernández, F.: Few-Body Systems Suppl. **10**, 395 (1988).

10. Garcilazo, G., Valcarce, A., and Fernández, F.: Phys. Rev. C **63** 035207 (2001).

11. Varga, K. and Suzuki, Y.: Phys. Rev. C **52**, 2885 (1995); Suzuki, Y. and Varga K.: *Stochastic Variational Approach to Quantum-Mechanical Few-Body Problems*. Berlin: Springer 1998.

12. Papp, Z., Krassnigg, A., and Plessas, W.: Phys. Rev. C **62**, 044004 (2000).

13. Garcilazo, G., Valcarce, A., and Fernández, F.: Phys. Rev. C **64** 058201 (2001).

14. Valcarce, A.: Contribution to these Proceedings.

15. Glozman, L.Ya. and Riska, D.O.: Phys. Rep. **268**, 263 (1996).

16. Glozman, L.Ya., in: *Perturbative and Nonperturbative Aspects of Quantum Field Theory* (Lecture Notes in Physics, vol. 479). Eds.: Latal, H. and Schweiger, W. Berlin: Springer 1997; p. 363.

17. Glantschnig, K., Kainhofer, R., Plessas, W., Sengl, B., and Wagenbrunn, R.F.: Few-Body Systems Suppl. **14**, 55 (2003).

18. Wagenbrunn, R.F., Boffi, S., Klink, W., Plessas, W., and Radici, M.: Phys. Lett. **B511**, 33 (2001); Glozman, L.Ya., Radici, M., Wagenbrunn, R.F., Boffi, S., Klink, W., and Plessas, W.: ibid. **B516**, 183 (2001); Boffi, S., Glozman, L.Ya., Klink, W., Plessas, W., Radici, M., and Wagenbrunn, R.F.: Eur. Phys. J. **A14**, 17 (2002).

19. Wagenbrunn, R.F., Boffi, S., Glozman, L.Ya., Klink, W., Plessas, W., and Radici, M.: Few-Body Systems Suppl. **14**, 411 (2003). arXiv: hep-ph/0212190.

20. Berger, K., Plessas, W., and Wagenbrunn, R.F.: Few-Body Systems Suppl. **14**, 53 (2003).

21. Berger, K., Plessas, W., and Wagenbrunn, R.F.: Prog. Part. Nucl. Phys., in print.

22. Merten, D., Löring, U., Kretzschmar, K., Metsch, B., and Petry, H.R.: Eur. Phys. J. **A14**, 477 (2002).

23. Kotulla, M.: Acta Phys. Polon. B 33 (2002) 957.

24. Melde, T., Plessas, W., and Wagenbrunn, R.F.: Few-Body Systems Suppl. **14**, 37 (2003). arXiv: hep-ph/0211356.

25. Capstick, S. and Isgur, N.: Phys. Rev. D **34**, 2809 (1986).

Few-Body Systems Suppl. 15, 151–163 (2003)

Few-Body Systems

Relativistic Theory of Few-Body Systems

Franz Gross*

Theory Group, MS12H2, Jefferson Laboratory, Newport News, VA 23606. USA

Abstract. Very significant advances have been made in the relativistic theory of few-body systems since I visited Peter Sauer and his group in Hannover in 1983. This talk provides an opportunity to review the progress in this field since then. Different methods for the relativistic calculation of few-nucleon systems are briefly described. As an example, seven relativistic calculations of the deuteron elastic structure functions, A, B, and T_{20}, are compared. The covariant SPECTATOR© theory, among the more successful and complete of these methods, is described in more detail.

1 Prologue

In 1983, Peter Sauer invited me to Hannover to lecture to his students on my SPECTATOR© approach to the relativistic theory of few-body systems.[1] This was an exciting time, and I am very glad to have this opportunity today to thank Peter and his students for their interest, and for the many good questions and interesting discussions we had during and after those lectures. Peter and I never wrote a paper together, but my collaboration with Alfred Stadler and Teresa Peña eventually grew from this beginning. The calculation of the relativistic three nucleon bound state done with Alfred Stadler is based in part on notes I originally prepared for the Hannover lectures.

In preparing this talk I decided to summarize progress made since 1983, as if I were resuming discussion with Peter again on a Monday morning in 1983 (after an unusually long "19 year weekend"). Many of the results discussed here are covered in more detail in the recent review I wrote with Ron Gilman [1].

---

* *E-mail address:* gross@jlab.org

[1] At this conference some speakers used the term "spectator" to refer to an approximation in point-form quantum mechanics. When I pointed out that this usage would increase the confusion already present in this field, they jokingly suggested that I should copyright the term, and my notation is a response to this suggestion.

Table 1. Relativistic calculations referred to in Figs. 1 and 2.

Greens function dynamics	
VODG	Manifestly covariant SPECTATOR© theory [2]
PWD	"Equal time" calculation based on the Mandelzweig-Wallace equation [3]
Hamiltonian dynamics	
SPR	instant-form; no v/c expansion [4]
ARW	instant-form; with v/c expansion [5]
CK	front-form; dynamical light-front [6]
LPS	front-form; fixed light-front [7]
AKP	point-form [8]

2 Relativistic Approaches Reviewed and Compared

Since 1983 several alternative relativistic approaches have been developed. Attempts to classify and compare them is a continuing challenge.

These relativistic approaches fall into two major "schools." Both of these schools, and several alternatives within each, have recently been used to calculate the deuteron structure functions and form factors relativistically. These seven calculations (referred to as the "c7" below) are summarized in Table 1.

2.1 Greens Function Dynamics

In applications to the deuteron and NN scattering below the pion production threshold, methods based on Greens function dynamics

- start from a covariant effective field theory that takes nucleons and mesons to be the effective degrees of freedom,
- use this effective theory to define the covariant generalized meson-exchange ladder sum (the sum of all ladder and crossed ladder diagrams, either omitting vertex corrections and self energies entirely, or treating them phenomonologically using form factors),
- *define* a free NN progagator, G_x, and use it to reorganize the generalized ladder sum into an infinite series of "irreducible" kernels (potentials) $V^{(i)}$ (involving the exchange of i mesons), and products of these kernels, such as $V^{(i)}G_xV^{(j)}$, $V^{(i)}G_xV^{(j)}G_xV^{(k)}$, etc.,
- evaluate the ladder (ie. one boson exchange) amplitude using the integral equation $M_x^{\rm OBE} = V^{(1)} + V^{(1)}G_xM_x^{\rm OBE}$, and
- evaluate the contributions form multi-boson exchange kernels ($V^{(i)}$ with $j \geq 2$), perturbatively.

For $j \geq 2$, it is usually either assumed that $V^{(j)} \ll V^{(1)}$, or that $\sum_j V^{(j)} \simeq V'^{(1)}$ so the last step need not be done. There are several different choices for the free Greens function, G_x, and hence my use of the term "Greens function dynamics."

This approach is manifestly covariant. This means that the action of *all* of the 10 generators of the Poincaré group on matrix elements can be fully specified in terms of the kinematics, so that the action of finite Poincaré transformations can be calculated in simple, closed form. Momentum, energy, and angular momentum are strictly conserved, and finite boosts are given by a simple operation on the matrix elements, with no further calculations needed. [In practice, approximations are sometimes made that spoil this covariance.] A further advantage is that the effective field theory used to construct the generalized ladder sum can also be used to construct consistent, manifestly covariant electromagnetic currents. A very general method of constructing these currents consistently was published in 1987 [9], opening up the application of these methods to electromagnetic interactions. Unfortunately, the construction is not unique, so additional phenomenology is needed to fix the currents.

The disadvantages of the Greens function approach are that

- the Greens function, G_x, will include the propagation of negative energy (or antiparticle) states, requiring that the Hilbert space of quantum mechanical states be expanded, and that we learn to deal with the mathematical and physical interpretation of these states,

- the kernels and the propagator can include unphysical singularities that must be removed or avoided, complicating either the theoretical development or the numerical calculations, and

- the NN scattering amplitudes, which now include contributions from virtual negative energy states, should be fit directly to the NN data, thus increaseing the "investment" required to develop and apply these methods.

2.2 Hamiltonian Dynamics

In elementary quantum mechanics, states are defined at $t = 0$ and their evolution away from $t = 0$ is given by the Schrödinger equation

$$\mathrm{i}\frac{\partial}{\partial t}\Phi(\mathbf{r}, t)\bigg|_{t=0} = H\, \Phi(\mathbf{r}, 0)\,. \tag{1}$$

In 1949, Dirac [10] pointed out that it is possible to generalize the initial hypersurface on which the states are defined, leading also to a generalization of the Hamiltonian. Methods using Hamiltonian dynamics are so named because they use one of the three forms of generalized dynamics identified by Dirac [11]. These are[2]

[2]Please beware of some misprints in the discussion of Hamiltonian dynamics in ref. [1].

- *Instant form*: the states are defined on the $t = 0$ hyperplane, and the Hamiltonian is the usual generator of time translations, H. The generators of space translations (the momentum operators P^i) and rotations (the angular momentum operators J^i) form a subalgebra of the Poincaré group that leaves the $t = 0$ hypersurface invariant, so they may be defined without regard to the interactions living in H. The generators of the boosts, K^i, do not leave the hyperplane invariant, so must also include the interactions, and this is the disadvantage of this method.

- *Front form*: the states are defined initially on the light-front $t_+ \equiv t + z = 0$, and the generalized Hamiltonian is $H_- \equiv H - P^z$. The subgroup leaving this light-front invariant is composed of the *seven* generators H_+, K^z, J^z, $\mathbf{E}_\perp = \{K^x + J^y, K^y - J^x\}$, and $\mathbf{P}_\perp = \{P^x, P^y\}$. The dynamical interactions are in the three generators H_-, and $\mathbf{J}_\perp \simeq \{J^x, J^y\}$. This method is very popular for high energy interactions where the dynamics evolves along the light cone $t = z$ and it is a great advantage to have the boost K^z among the kinematic generators. Its disadvantage is that the conservation of angular momentum becomes a dynamical issue.

- *Point form*: Here the states are defined on the forward hyperboloid: $t > 0$ and $t^2 - \mathbf{r}^2 = a^2$. The homogeneous Lorentz group, generated by the boost and angular momentum operators, K^i and J^i, leaves this surface invariant, and the dynamics are in H and the momentum operators P^i. This method automatically conserves angular momentum and gives boosts independent of the dynamics, but its disadvantage is that the conservation of momentum becomes a dynamical issue.

All of these methods use quantum mechanical states that span the positive energy spectrum only, leading to the standard interpretation of quantum mechanics. In each case it is possible to transform the generalized Schrödinger equation into an equivalent instant form, and hence use the excellent phenomenological potentials that have recently been fit to low energy NN scattering.

The disadvantages of Hamiltonian dynamics depend in part on the specific form chosen. In the instant form it is difficult to boost the states, and the front and point forms require special care to insure that either angular or linear momentum is conserved. Perhaps the most serious disadvantage of Hamiltonian dynamics is that they provide litte or no guidance on how to construct consistent, conserved, covariant electromagnetic currents. The construction of currents is phenomenolgical and sometimes ad-hoc.

2.3 Predictions for the Deuteron Form Factors

The predictions of the c7 listed in Table 1 are shown in Fig. 1. I conclude the following:

- Only the VODG and SPR calculations provide a reasonable description of *all three* of these observables. This is probably due to the fact that these

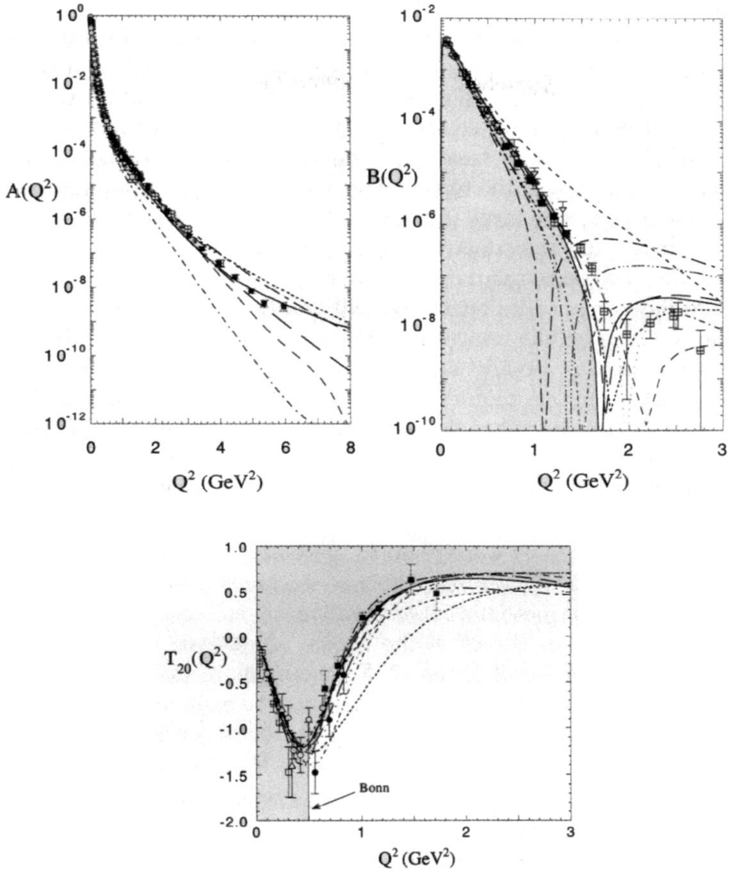

Figure 1. Predictions for the structure functions A, B, and T_{20}. The models, in order of the Q^2 of their mimima in B, are: CK (long dot-dashed line), PWD (dashed double-dotted line), AKP (short dot-dashed line), VODG full calculation (solid line), VODG in RIA (long dashed line), LPS (dotted line), a quark model calculation not discussed here (widely spaced dotted line), SPR (medium dashed line), and ARW (short dashed line). None of the curves shown include the $\rho\pi\gamma$ exchange current. (This figure is taken from ref. [1].)

are the only calculations valid to all orders in $(v/c)^2$ that also include complete currents consistent with the dynamics.

- The magnetic structure function, B, is by far the most sensitive to the calculations, and provides a stringent test of the theory.

To get some insight into the differences between these calculations, it is useful to think of each as built from the same three (roughly defined and overlapping) ingredients: (i) the "leading" nonrelativistic S and D-state deuteron wave functions, (ii) the single-nucleon electromagnetic current, and (iii) all the rest, including $I = 0$ interaction currents and relativistic modifications (which are model dependent and not small) to both wave functions and currents.

All of the c7 use modern "realistic" deuteron wave functions, so the differences between them is not due to the choice of the leading, nonrelativistic part of the wave function. The large differences arise instead from the treatment of relativistic effects, and the construction of the current.

Some of the c7 include relativistic effects to all orders in $(v/c)^2$, while others only include the lowest order terms. By expanding in powers of $(v/c)^2$ it would be possible to compare the relativistic effects obtained from all of the c7. To a limited extent this has already been done; in the 1970's and 1980's Friar and others found that the one pion exchange corrections for the instant-form and SPECTATOR© theory agreed to lowest order (for a review see ref. [12]). The comparison was very informative – it showed that the "pair term" corrections (for example) differed in the two cases, and agreement was only obtained after *all* of the corrections were added together. This leads to the expectation that a similar comparison of the c7 would lead to agreement *only for those calculations that include all possible corrections*. I believe that such a benchmark calculation comparing the different methods could significantly increase our understanding.

Another difference in the c7 is the choice of current. Even the one body current is not the same for all of the c7. For example, in the point-form it turns out [8] that the single nucleon form factors must be evaluated at a momentum transfer Q^2 *larger* than that transmitted to the deuteron as a whole (possible because momentum is not conserved). Because the nucleon form factors decrease rapidly in Q^2, this leads to a strong suppression of the prediction for A, as shown in Fig. 1. This could be corrected by adding an interaction current, but the point-form approach itself does not tell us what this should be. In the absence of an underlying field theory to guide the physics, the currents are purely phenomenological.

The currents in the SPECTATOR© theory are well constrained, but even here they cannot be uniquely determined. I will conclude this discussion by showing how we can exploit this flexibility, and adjust the single nucleon SPECTATOR© current to give an excellent fit to all of the elastic electron-deuteron data.

Following ref. [9], current conservation requires that the single-nucleon current satisfy the Ward-Takashi identity

$$q_\mu\, j_N^\mu(p', p) = S^{-1}(p) - S^{-1}(p') \tag{2}$$

where $S(p)$ is the propagator of a nucleon with four-momentum p. The NN model used in the VODG calculation [13] uses the dressed nucleon propagator

$$S(p) = \frac{h^2(p)}{m - \not{p}} = \frac{h^2(p)}{\Lambda_-(p)} \tag{3}$$

with $h(p)$ a phenomenological scalar function of p^2. The *simplest* solution of Eq. (2) gives the following one nucleon current

$$j_N^\mu(p',p) = F_0 \left\{ F_1(Q^2)\gamma^\mu + F_2(Q^2)\frac{i\sigma^{\mu\nu} q_\nu}{2m} \right\} + G_0 F_3(Q^2) \Lambda_-(p')\gamma^\mu \Lambda_-(p) \quad (4)$$

where terms proportional to q^μ have been dropped (they are required by the identity (2) but vanish when contracted into the conserved electron current),

$$\begin{aligned} F_0 &= \frac{h(p)(m^2 - p'^2)}{h(p')(p^2 - p'^2)} - \frac{h(p')(m^2 - p^2)}{h(p)(p^2 - p'^2)} \\ G_0 &= \left[\frac{h(p')}{h(p)} - \frac{h(p)}{h(p')} \right] \frac{4m^2}{p^2 - p'^2}, \end{aligned} \quad (5)$$

and $F_3(Q^2)$ is *completely undetermined* except for the requirement that $F_3(0) = 1$. The VODG calculation shown in Fig. 1 uses the "standard" dipole form for F_3

$$F_3(Q^2) = \left(\frac{\Lambda^2}{\Lambda^2 - Q^2} \right)^2 \quad (6)$$

with $\Lambda^2 = 0.71$ GeV^2. Clearly other choices of F_3 are possible.

Another source of uncertainty is the famous $\rho\pi\gamma$ exchange current. This current is separately gauge invariant and strongly dependent on the $\rho\pi\gamma$ form factor, $f_{\rho\pi\gamma}(Q^2)$ [14]. The value of $f_{\rho\pi\gamma}(0)$ is constrained by meson radiative decays, and its contribution to the deuteron form factors at small Q^2 is negligible. Its importance at high Q^2 depends strongly on the assumed dependence of $f_{\rho\pi\gamma}(Q^2)$, which is unknown but can be estimated from quark models. One of the best calculations of this form factor is that of the Rome group [15]. To clarify the comparison between models, the versions of the c7 shown in Fig. 1 did not included this current, and it could be added to any of them.

Taking the VODG calculation in RIA approximation (discussed in ref. [2]) as our "standard," I now consider the effect of (i) altering the Q^2 dependence of F_3, or (ii) adding the $\rho\pi\gamma$ exchange current. The "standard" calculation is the long dashed lines shown in Fig. 1 and the dotted lines reproduced in Fig. 2. The effect of changing F_3 to a tripole form

$$F_3(Q^2) = \left(\frac{\Lambda^2}{\Lambda^2 - Q^2} \right)^3 \quad (7)$$

with $\Lambda^2 = 5$ GeV^2 (and continuing to keep $f_{\rho\pi\gamma} = 0$) are shown as solid lines. The effect of adding a $\rho\pi\gamma$ exchange current using a dipole form factor with $\Lambda^2 = 1.5$ GeV^2 (and keeping the standard dipole form for F_3) are shown as the dashed curves. [The F_3 and $\rho\pi\gamma$ form factors themselves are compared to the Rome form factor and the standard dipole (6) in the lower left panel.] To see the effects on A and B more clearly, the middle two panels show the ratios of A/A_0 and B/B_0, where A_0 and B_0 are the standard calculation.

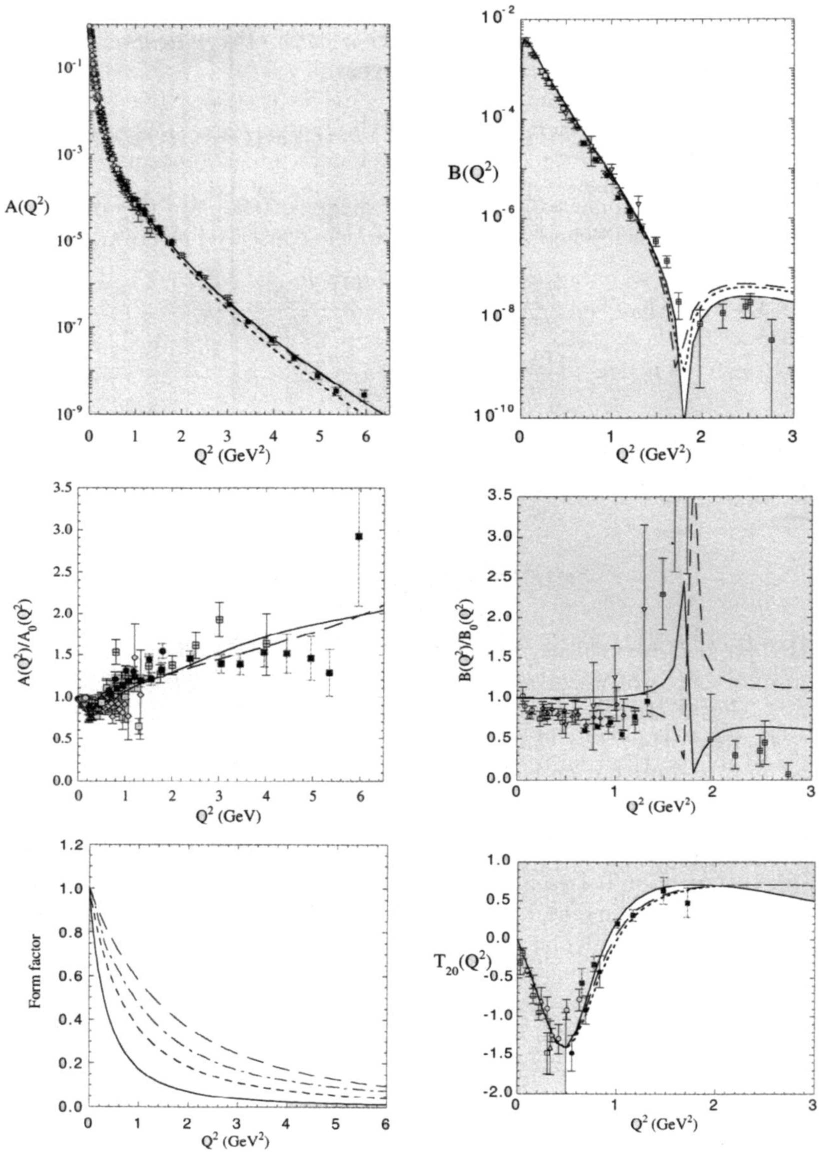

Figure 2. Upper panels (A and B) and lower right panel ($\tilde{T}_{20}$) compare data to three theoretical models based on VODG: (i) "standard" case referred to in the text (dotted line), (ii) model with the tripole F_3 (solid line), and (iii) model with the dipole $f_{\rho\pi\gamma}$ (dashed line). The center two panels show the data and models (ii) and (iii) divided by model (i). The lower left panel shows form factors: standard dipole with $\Lambda^2 = 0.71$ (solid line), dipole with $\Lambda^2 = 1.5$ (short dashed line), Rome $f_{\rho\pi\gamma}$ (dot-dashed line) [15], and the tripole with $\Lambda^2 = 5$ (long dashed line). (This figure is taken from ref. [1].)

We see that a reasonable adjustment of F_3 at high Q^2 can give an excellent description of all of the elastic deuteron observables. This is perfectly permissable within the theory, since the form factor F_3, while it *must* be present, cannot be determined by on-shell data, and must be treated phenomenologically. From this point of view the deuteron data has now determined the unknown form factor F_3, and the isoscalar single nucleon current is now fixed. It remains to be seen whether the same F_3 will give excellent results for other electron scattering observables.

What are we to say about the $\rho\pi\gamma$ exchange current? My own belief is that this exchange current is being overestimated, even using the Rome form factor. While this current is certanly present, it is probably either negligible, or cancelled by the many other short range currents neglected in these calculations.

3 Progress with the SPECTATOR© Theory

In the remainder of this talk I return to the SPECTATOR© theory and report on developments since 1983, and on my expectations for the future.

3.1 *Definition of the Theory*

The SPECTATOR© propagator for two nucleons with four momenta p_1 and p_2 is

$$G_S = 2\pi\mathrm{i}\,\delta_+(m^2 - p_2^2)\, S(p_1) \sum_\lambda u(\mathbf{p}_2,\lambda)\bar{u}(\mathbf{p}_2,\lambda) \tag{8}$$

where the δ_+ function restricts p_{20} to its positive energy mass-shell, so that $p_2 = \{E(p_2), \mathbf{p}_2\}$, and $u(\mathbf{p}_2,\lambda)$ is the free Dirac spinor for a nucleon with three-momentum $\mathbf{p}_2$ and helicity λ, normalized to $\bar{u}(\mathbf{p},\lambda)u(\mathbf{p},\lambda') = 2m\delta_{\lambda\lambda'}$. This propagator insures that the integration over the internal energy will place particle 2 on its positive energy mass-shell, giving the SPECTATOR© equation. It was already known before 1983 that this equation is manifestly covariant, has the right one-body limit, a smooth nonrelativistic limit, satisfies the cluster property, and produces a gauge invariant one-photon-exchange interaction [16]. The manifest covariance means that momentum and angular momentum are trivially conserved, and that the boosts are known exactly, to all orders in $(v/c)^2$.

3.2 *Progress since 1983*

Progress since 1983 has been substantial. Theoretically, we have seen (i) the development of a general method for constructing consistent, conserved currents [9] (already discussed), and (ii) the recent demonstration that all vertex corrections and momentum dependent self energies *cancel* in massive scalar QED [17]. This last result means that the generalized ladder sum gives the full result for the interaction of scalar bosons with a massive photon. Combining this with the already known fact that the generalized ladder sum is well approximated by the ladder approximation to the SPECTATOR© equation, and we have the first demonstration that solutions of the SPECTATOR© equation are a good approximation to the *exact* solutions of massive scalar QED. It is not known if this result also holds for other effective theories.

Many applications have been developed since 1983. These include:

- Mesons as $q\bar{q}$ bound states, including covariant treatment of scalar confinement consistent with chiral symmetry breaking [18, 19].
- Unitary model of $\pi N \to \pi N$ and $\gamma N \to \pi N$ processes up to 600 MeV including Δ, Roper, and D_{13} resonances [20].
- NN scattering and bound state solutions to 350 MeV lab kinetic energy with $\chi^2 \sim 2$ and 13 fitting parameters [13].
- Exact numerical solution of the relativistic three body equations that give a good binding energy for ^{3}H without relativistic three body forces [21].
- Good description of the deuteron form factors [2] reviewed above.
- General classification of inelastic scattering observables [22] and first calculation of inelastic scattering in relativistic impulse approximation [23].
- Derivation of the pA multiple scattering series [24] and applications to proton nucleus scattering [25].

When I visited Peter Sauer in 1983, the goal was to do a relativistic three-body calculation. Peter and I never completed this, but my contacts with Peter lead eventually to my collaboration with Alfred Stadler, and to the desired calculation.

The major result of this relativistic calculation of the three body binding energy [21] is shown in Fig. 3. We obtain good agreement with the binding energy only after introducing an off-shell scalar meson coupling

$$g_s \Lambda(p', p) = g_s \left\{ 1 - \frac{\nu_s}{2m} \left[\Lambda_-(p') + \Lambda_-(p) \right] \right\} \tag{9}$$

with the off-shell coupling parameters ν_σ and ν_δ proportional to the ν shown in Fig. 3. The exciting conclusion is that the same ν that fits the experimental binding energy also gives the best fit to the two-body data.

3.3 The NEW SPECTATOR© Theory

The SPECTATOR© theory raises a number of questions:

- In the n-body problem, why should $n-1$ particles be on shell, and which ones should they be?
- Can all spurious singularities be removed from the theory?

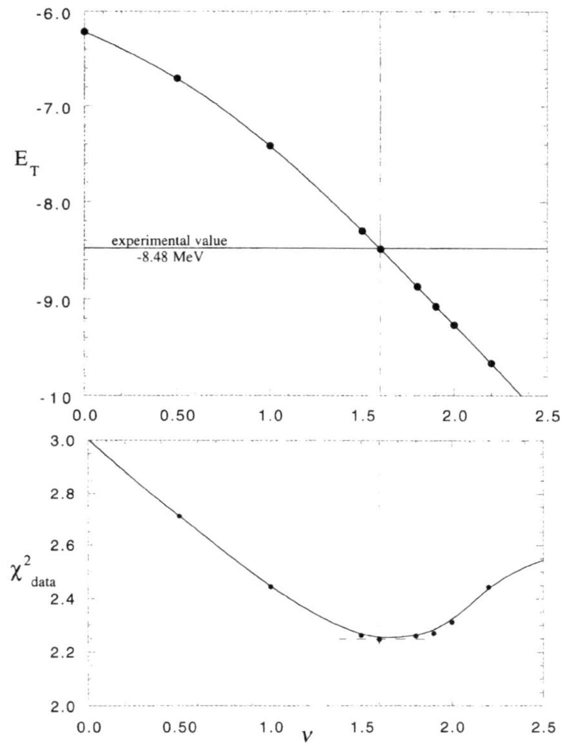

Figure 3. Upper panel shows the three body binding energy as a function of the off-shell coupling parameter ν discussed in the text. The lower panel shows the variation in χ^2 of the fit to the two-body data (up to lab energies of 350 MeV).

- What are the Feynman rules for the general case?
- Can the theory be formulated as a quantum mechanics, with a well defined Hilbert space of states?

Many of these questions were raised initially by Peter Sauer and his students during my visit in 1983, and I have answered some of them to my satisfaction. I am currently preparing a new formulation of the SPECTATOR© theory that I believe will provide more complete answers. Like a good salesman of computer games, I want to advertize my product in advance of its "public" release.

Figure 4 shows the real part of Bethe-Salpeter propagator, iG_{BS}, as a function of the relative energy $p_0 = \frac{1}{2}(p_{10} - p_{20})$ and the square of the relative three-momentum, p^2, evaluated in the c.m. system (to aid in seeing the structure the ϵ's in the $-i\epsilon$ prescriptions were given a finite value). This figure shows clearly that the propagator is dominated by the two positive energy mass-shell poles. Using Cauchy's residue theorem to evaluate the integral over p_0 gives the result that the integration is dominated by the on-shell contribution from

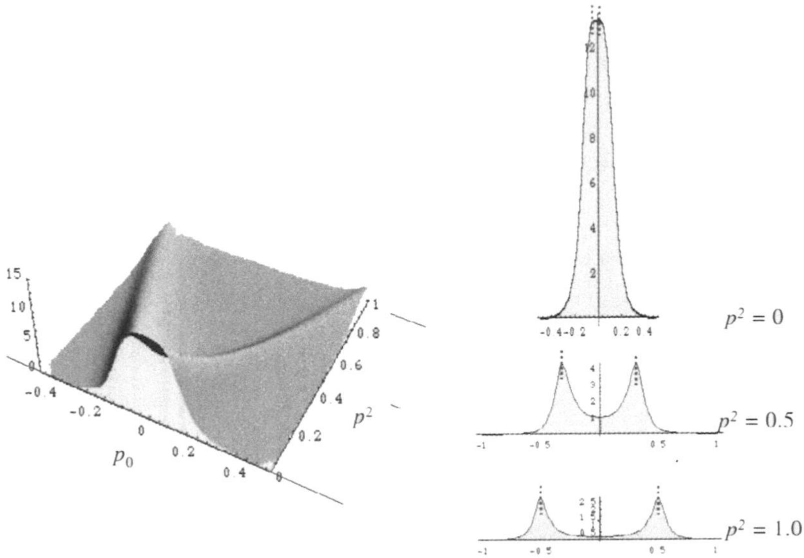

Figure 4. Mass-shell peaks in the free propagator for two particles with equal mass m a total rest mass of $M/m = 1.8$. The figures on the right show slices through the surface at various constant values of p^2. (All quantities are in units of m.)

one particle, but when interpreted as an ordinary integral along the real axis the same result requires the contribution of *both* peaks (each giving one-half of the residue result). Looked at from this point of view, the wave function is the sum of two terms, one with particle 1 on-shell and one with 2 on-shell. These two terms have support in separate regions of phase space, so adding them together leaves some things unchanged, but continuing in this direction leads to a reformulation of the SPECTATOR© theory that gives nice answers to the questions above. Watch for the release of this theory in the near future!

Acknowledgement. I thank Peter Sauer for stimulating so many developments in Few Body physics, Alfred Stadler for his contributions to the relativistic three-body problem, and for conversations leading to the new SPECTATOR© theory, and John Tjon, Wally Van Orden, and Teresa Peña for many useful discussions and long and fruitful collaborations. This work was supported in part by DOE grant No. DE-FG02-97ER41032, held at William and Mary. The Southeastern Universities Research Association (SURA) operates the Thomas Jefferson National Accelerator Facility under DOE contract DE-AC05-84ER40150.

References

1. R. Gilman, F. Gross: J. Phys. **G 28**, R37 (2002)
2. J.W. Van Orden, N.K. Devine, F. Gross: Phys. Rev. Lett. **75**, 4369 (1995)
3. D.R. Phillips, S.J. Wallace, N.K. Devine: Phys. Rev. **C 58**, 2261 (1998)
4. R. Schiavilla, V.R. Pandharipande: Phys. Rev. **C 65**, 064009 (2002)
5. H. Arenhövel, F. Ritz, T. Wilbois: Phys. Rev. **C 61**, 034002 (2000)
6. J. Carbonell, V.A. Karmanov: Eur. Phys. J **A 6**, 9 (1999)
7. F.M. Lev, E. Pace, G. Salmé: Phys. Rev. **C 62**, 064004 (2000)
8. T.W. Allen, W.H. Klink, W.N. Polyzou: Phys. Rev. **C 63**, 034002 (2001)
9. F. Gross, D.O. Riska: Phys. Rev. **C 36**, 1928 (1987)
10. P.A.M. Dirac: Rev. Mod. Phys. **21**, 392 (1949)
11. B.D. Keister, W.N. Polyzou: Adv. Nucl. Phys. **20**, 225 (1991)
12. F. Gross: In *Modern Topics in Electron Scattering*, B. Frois, I. Sick (eds.) p. 219 (1991)
13. F. Gross, J.W. Van Orden, K. Holinde: Phys. Rev. **C 45**, 2094 (1992)
14. H. Ito, F. Gross: Phys. Rev. Lett. **71**, 2555 (1993)
15. F. Cardarelli, I.L. Grach, I. Narodetsky, G. Salme, S. Simula: Phys. Lett. **B 359**, 1 (1995)
16. F. Gross: Phys. Rev. **186**, 1448 (1969); Phys. Rev. **D 10**, 223 (1974); Phys. Rev. **C 26**, 2203 (1982)
17. Ç. Şavklı, F. Gross, J. Tjon: Phys. Lett. **B 531**, 161 (2002)
18. F. Gross, J. Milana: Phys. Rev. **D 43**, 2401 (1991); **45**, 969 (1992); **50**, 3332 (1994)
19. Ç. Şavklı, F. Gross: Phys. Rev. **C 63**, 035208 (2001)
20. F. Gross, Y. Surya: Phys. Rev. **C 47**, 703 (1993); Y. Surya, F. Gross: Phys. Rev. **C 53**, 2422 (1996)
21. A. Stadler, F. Gross: Phys. Rev. Lett. **78**, 26 (1997)
22. V. Dmitrasinovic, F. Gross: Phys. Rev. **C 40**, 2479 (1989); **43**, 1495(E) (1991)
23. J. Adam, Jr., et al.: Phys. Rev. **C 66**, 044003 (2002)
24. F. Gross, K.M. Maung: Phys. Rev. **C 42**, 1681 (1990)
25. K.M. Maung: Phys. Rev. **C 43**, 1378 (1991)

Few-Body Systems Suppl. 15, 165–170 (2003)

Few-Body Systems

Longitudinal Response Functions for Quasielastic Electron Scattering in Relativistic Nonlinear Models

J.C. Caillon, J. Labarsouque

Centre d'Etudes Nucléaires de Bordeaux Gradignan, Université Bordeaux I, IN2P3, le Haut Vigneau, BP 120, 33175 Gradignan Cedex, France

Abstract. The longitudinal response functions for quasielastic electron scattering on ^{56}Fe have been calculated in relativistic nonlinear models taking into account RPA correlations. For these calculations, a covariant, consistent, calculation of the nuclear matter linear response has been performed. The effect of the non-linear terms on the longitudinal response has been discussed.

These last years, relativistic non-linear models have gained more and more success in the description of nuclear matter and ground state properties of finite nuclei. In these models, the Lagrangian includes usual linear nucleon-meson couplings (like in the Walecka model [1]) as well as couplings between the mesonic fields. The nucleons interact self-consistently through meson exchanges in mean field approximation. Among many models, the models NL1 [2] contain only σ self-coupling terms in the Lagrangian while TM1 [3] (for medium and heavy nuclei) include in addition an ω self-coupling term in order to reduce the resulting strong scalar and vector potentials obtained with NL1. More recently, a parameter set G1 [4], coming from an effective field theory which allows all scalar-vector coupling terms up to fourth order in the Lagrangian, has been proposed. By adjusting only few parameters, all of these models lead to results which compare rather well with the ground state properties of finite nuclei.

In the two last decades, the separation of the longitudinal and transverse response functions in quasielastic electron scattering has been performed for various nuclei. It appeared that the experimental values of the longitudinal response functions seemed to exhibit a quenching in comparison to calculations taking into account all possible non-relativistic NN correlations. Recently, two new separations using the "world data" on inclusive quasi-elastic scattering have been performed. One by Jourdan [5] who concludes that the "quenching" is absent and one by Morgenstern and Meziani [6] whose conclusion is the existence of a "quenching" in medium and heavy nuclei. The difference observed between these two separations has been attributed mainly to the Coulomb correction used. Since some uncertainties remain in the experimental data (especially at large energy transfer), a new determination of the longitudinal response functions obtained with data coming from new accelerators, like for example the Thomas Jefferson National Accelerator Facility, is necessary. Such a proposal has been approved recently at Jefferson Lab [7].

In this work, we have determined the longitudinal response functions for ^{56}Fe at a three-momentum transfer of 570 MeV/c taking into account RPA correlations in the relativistic nonlinear models. We have used the local density approximation such that the response of a nucleus is the sum of the responses of its volume elements treated as nuclear matter with local Fermi momenta $k_{\mathrm{F}p}(r)$, $k_{\mathrm{F}n}(r)$ and local effective nucleon mass $M_N^*(r)$. Thus, the relativistic RPA longitudinal response function for finite nuclei (with Z protons and N neutrons) is given by

$$\begin{aligned} R_{\mathrm{L}}(\omega, \|\mathbf{q}\|) &= 4\pi \int r^2 \left(\frac{-1}{\pi}\right)\Big[\frac{Z}{A}\mathrm{Im}\Pi_{\mathrm{L}}(k_{\mathrm{F}p}(r), M_N^*(r); \omega, \|\mathbf{q}\|) \\ &\quad + \frac{N}{A}\mathrm{Im}\Pi_{\mathrm{L}}(k_{\mathrm{F}n}(r), M_N^*(r); \omega, \|\mathbf{q}\|)\Big] dr, \end{aligned} \tag{1}$$

with

$$\Pi_{\mathrm{L}} = \Pi_{\mathrm{is}}^{00\ \mathrm{NM}} + \Pi_{\mathrm{iv}}^{00\ \mathrm{NM}}, \tag{2}$$

where $\Pi_{\mathrm{is}}^{\mu\nu\ \mathrm{NM}}$ and $\Pi_{\mathrm{iv}}^{\mu\nu\ \mathrm{NM}}$ are the isoscalar and isovector parts of the polarization tensor in nuclear matter. Due to the separation of the current operator in Dirac and Pauli parts, each isoscalar and isovector contribution to the polarization tensor can be written as a sum of four contributions: a purely vector piece (VV), a purely tensor piece (TT) and two interference terms (VT and TV), such that:

$$\Pi_{\mathrm{is}}^{\mu\nu\ \mathrm{NM}} = F_{1\mathrm{is}}^2 \widetilde{\Pi}_{\mathrm{VVis}}^{\mu\nu} + F_{1\mathrm{is}} F_{2\mathrm{is}} (\widetilde{\Pi}_{\mathrm{VTis}}^{\mu\nu} + \widetilde{\Pi}_{\mathrm{TVis}}^{\mu\nu}) + F_{2\mathrm{is}}^2 \widetilde{\Pi}_{\mathrm{TTis}}^{\mu\nu}, \tag{3}$$

$$\Pi_{\mathrm{iv}}^{\mu\nu\ \mathrm{NM}} = F_{1\mathrm{iv}}^2 \widetilde{\Pi}_{\mathrm{VViv}}^{\mu\nu} + F_{1\mathrm{iv}} F_{2\mathrm{iv}} (\widetilde{\Pi}_{\mathrm{VTiv}}^{\mu\nu} + \widetilde{\Pi}_{\mathrm{TViv}}^{\mu\nu}) + F_{2\mathrm{iv}}^2 \widetilde{\Pi}_{\mathrm{TTiv}}^{\mu\nu}. \tag{4}$$

Using Dyson equation, each part of the polarization tensor has been written as a function of the meson propagators. The diagrammatic representation of the meson propagators is shown in Fig. 1a where the polarization operators include particle-hole excitations as well as contributions coming from nonlinear terms in the Lagrangian. The full lines in Fig. 1 represent the in-medium nucleon propagator in which the nucleon self-energy includes linear as well as nonlinear terms (taken into account through the nucleon effective mass). The diagrammatic representation of $\widetilde{\Pi}_{\mathrm{ab\ is,iv}}^{\mu\nu}$ is also shown in Fig. 1b.

We show in Figs. 2-4 the longitudinal response functions calculated taking into account relativistic RPA correlations for ^{56}Fe as a function of energy loss at a three-momentum transfer of 570 MeV/c. Our results will be compared to the response functions obtained using the "world data" on inclusive quasi-elastic electron scattering by Jourdan [5] and by Morgenstern and Meziani [6].

We have studied the influence of both relativistic RPA correlations and nonlinear terms on the longitudinal response. In Fig. 2, the short dashed curve is

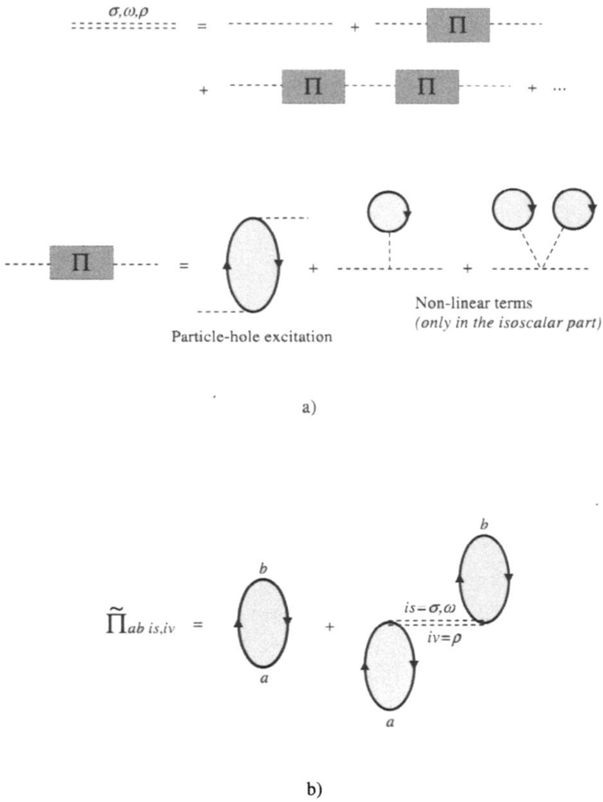

Figure 1. The diagrammatic representation of the meson propagators in non linear models is displayed in part a). Part b) shows the diagrammatic representation of the polarization tensor of nuclear matter.

obtained with a simple relativistic Fermi gas model without any adjusted parameter. The solid curve represents a calculation taking into account relativistic RPA correlations in model G1 while for the long dashed curve the relativistic RPA correlations have been put equal to zero (the long dashed curve represents thus the result of a relativistic Fermi gas calculation where the density-dependent nucleon effective mass obtained in the G1 model has been used). As it can be seen on Fig. 2, this curve significantly differs from the solid one obtained with the full G1 model. Thus, although a large part of the response is given by the use of the appropriate nucleon effective mass, we observe that the effect of relativistic RPA correlations is important (17% around the peak energy). Moreover, we can note that the important difference between the relativistic Fermi gas result (short dashed curve) and the one corresponding to

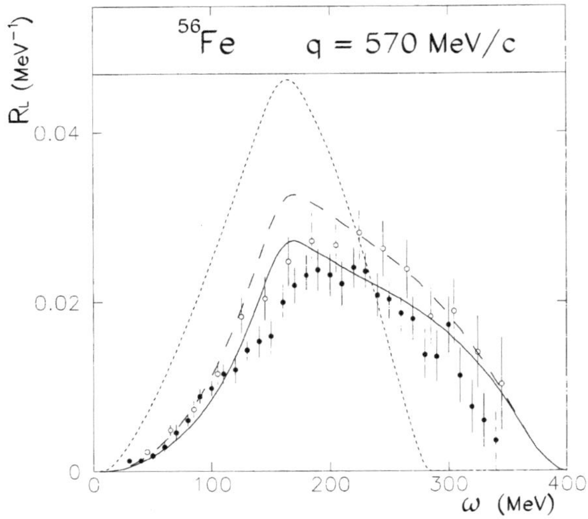

Figure 2. Longitudinal response functions for ^{56}Fe as a function of energy transfer ω for a three-momentum transfer of 570 MeV/c in various approximations (see text). The experimental data are taken from ref. [5] (open circles) and ref. [6] (filled circles).

the relativistic Fermi gas model but with the G1 nucleon effective mass (long dashed curve) results from the effect of the nucleon self-energy and thus also from correlations in the medium.

To study precisely the influence of non-linear terms in a realistic way, we have to compare models obtained with and without non-linear terms but built from a fit to the same observables. To do that, we have used various models coming from the work of Furnstahl, Serot and Tang [4]. In Fig. 3, the short dashed curve has been obtained with the W1 model where only linear terms are included in the Lagrangian (similar to the Walecka model) while other curves take into account non-linear effects. The long dashed curve has been obtained with the C1 model in which only third order self-coupling terms are taken into account while the solid curve represents the full G1 model including fourth order non linear terms. We can see that the longitudinal responses obtained with the C1 and G1 models significantly differ from that given by the simple W1 model, showing that the influence of non-linear terms is not small.

In Fig. 4 we show the longitudinal response functions calculated with the non linear models NL1, TM1 and G1. We have found very similar results for these three models in suitable agreement with the data. Note that a relatively good agreement with the data is also obtained for other nuclei, like for example,

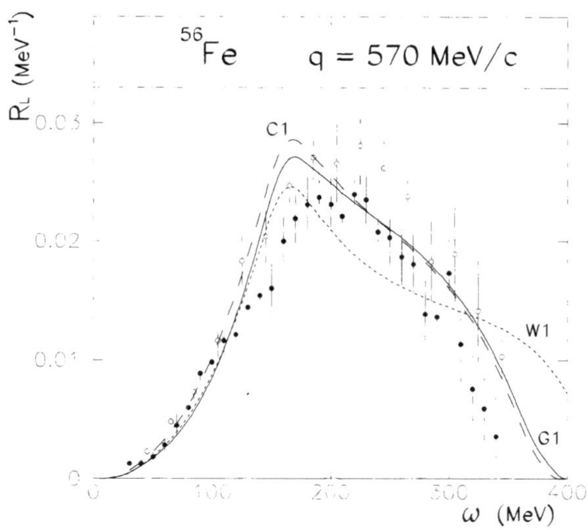

Figure 3. Longitudinal response functions for as a function of energy transfer ω for a three-momentum transfer of 570 MeV/c in various models obtained by Furnstahl, Serot and Tang (see text). The experimental data are taken from ref. [5] (open circles) and ref. [6] (filled circles).

for calcium and carbon.

In this work, we have determined the relativistic RPA longitudinal response functions for quasielastic electron scattering using non-linear models. We have obtained the following results. First, although a large part of the response is given by the use of the appropriate density dependent nucleon effective mass, we have found a significant effect of relativistic RPA correlations. Second, since the non-linear terms lead also to a change of the coupling constants of the linear terms, we have seen that their influence is not small. This study was motivated partly by the fact that the different non-linear models give very different results for the energies of the isoscalar giant monopole resonances [8]. These energies, directly connected to the compressibility of nuclear matter, vary from one model to another since the incompressibility parameter strongly depends on the model considered. This is not the case here for the longitudinal response functions where we have found very similar results for all the models considered, in relatively good agreement with the data. A more precise comparison of our results with the data will not be possible before that ambiguities in the extraction of the longitudinal response from experimental data will be removed. Clearly, new data coming from Thomas Jefferson National Accelerator Facility

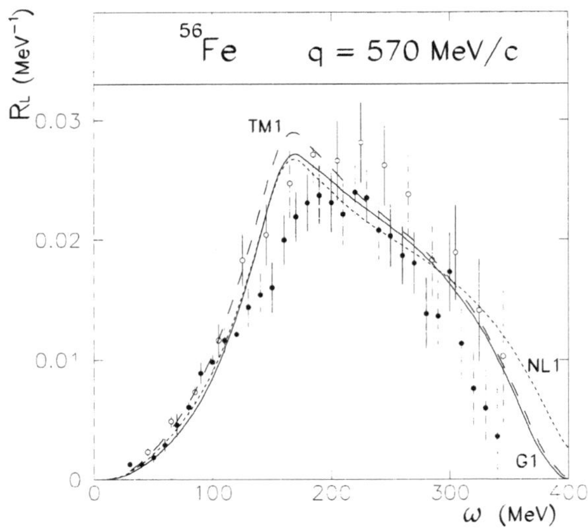

Figure 4. Longitudinal response functions for ^{56}Fe as a function of energy transfer ω for a three-momentum transfer of 570 MeV/c. The short dashed, long dashed and solid curves have been obtained by taking into account relativistic RPA correlations using respectively models NL1, TM1 and G1. The experimental data are taken from ref. [5] (open circles) and ref. [6] (filled circles).

will help the community to progress in this direction.

References

1. B.D. Serot and J.D. Walecka: Advances in Nuclear Physics **16**, 1 (1986)
2. P.G. Reinhard et al.: Z. Phys. **A 323**, (1986) 13
3. Y. Sugahara and H. Toki: Nucl. Phys. **A 579**, 557 (1994)
4. R.J. Furnstahl, B.D. Serot, H.-B. Tang: Nucl. Phys. **A 615**, 441 (1997)
5. J. Jourdan: Nucl. Phys. **A603**, 117 (1996)
6. J. Morgenstern and Z.-E. Meziani: Phys. Lett. **B** (in press), nucl-ex/0105016
7. Jlab experiment E01-016; Spokespeople: Seonho Choi, J.P. Chen and Z.-E. Meziani
8. Zhong-yu Ma et al.: Nucl. Phys. **A 686**, (2001) 173

Few-Body Systems Suppl. 15, 171–181 (2003)

Few-Body Systems

Photo and Electrodisintegration of Few-Body Systems Revisited

J.-M. Laget*

CEA/Saclay, DAPNIA-SPhN, F91191 Gif-sur-Yvette, France

Abstract. Twenty years after P. Sauer released the state of the art Faddeev solution of the bound state three nucleon systems, I revisit photo and electrodisengration of few body systems with a special emphasis on the prospects opened at Jefferson Laboratory.

1 Introduction

In the 80' P. Sauer, and his group at the University of Hanover, solved the Faddeev equations for the three nucleon systems [1, 2] with the state of the art realistic NN potentials [3]. Simultaneously a vigorous experimental program was developed at Saclay. Thanks to the availability of an ^{3}He and a tritium targets, the isoscalar and isovector components of the charge and the magnetic elastic form factors of the three nucleon systems were determined [4]. They remarkably agreed with the predictions of the Hanover group. The key to this success was not only the "tour de force" in operating the radioactive target, which was only made possible by an environment of high technology, but also the close collaboration between a theoretical and an experimental team.

I met Peter at this occasion and, starting from his wave functions, I extended my diagrammatic approach [5] to the disintegration channels of the three nucleon systems [6]. An experimental program was ongoing not only at Saclay but also at NIKHEF and Bates, in order to determine the wave function of the few body systems. Again, the experiments confirmed that Peter's wave functions were doing a good job up to nucleon momentum of 500 MeV/c.

However, it became evident that the energy and the duty factor of the available electron accelerators were not sufficient to map out the wave function at higher momentum. Kinematical limits prevented to increase significantly the momentum transfer. Rescattering and interaction effects were not negligible at such a low energy. Finally the low duty factor prevented to single out

* *E-mail address:* jlaget@cea.fr

the very small cross sections, associated with high momentum transfers, from background.

One of the reasons to built CEBAF at Jefferson Laboratory (JLab) was to overcome these drawbacks. The energy was increased by one order of magnitude (600 MeV to 6 GeV) and the duty factor by two orders of magnitude (1% to 100%). About seven years of operation have produced a few jewels in the disintegration of few body systems.

This meeting in honor of P. Sauer provides me with a good opportunity to revisit the field in view of recent achievements and to evaluate future developments.

2 The Two Body Disintegration Channels

The primary goal of the study of the (e,e'p) reaction on nuclei was, and still is, the determination of the high momentum components of the nuclear wave function, in view of the study of short range correlations and possible exotic configurations.

2.1 The Low Energy Regime

In the past the spectral functions measured at Saclay or Amsterdam suffered for large corrections (about a factor two or more) due to Final State Interactions (FSI) and Meson Exchange Currents (MEC). A survey of the state of the art at that time can be found in ref. [6]. The corresponding experiments were performed at low values ($\sim$ 0.3 GeV^2) of the virtuality Q^2 of the exchanged photon.

When CEBAF was decided, a common belief was that increasing Q^2 was the way to suppress FSI and MEC contributions. This was partly true, since both the FSI and MEC amplitudes involve a loop integral, which connects the nuclear bound and scattering states and which is expected to decrease when Q^2 increases as form factors do. But this was partly wrong, since the singular part of the FSI integral does not depends on Q^2, besides the trivial momentum dependency of the elementary operators. It comes from unitarity, and corresponds to the propagation of an on-shell nucleon. It involves on-shell elementary matrix elements and it is maximum when the kinematics allows for rescattering on a nucleon at rest. This happens in the quasi-free kinematics, when $X = Q^2/2m\nu = 1$ (being ν the energy of the virtual photon).

Figure 1 exhibits these features. It shows the angular distribution, against the neutron angle θ_R, of the ratio between the full cross section of the $\mathrm{D}(e, e'p)n$ reaction and the quasi free contribution, when the momentum P_R of the recoiling neutron is kept constant. FSI are maximum near $\theta_R = 70°$ where $X = 1$ and on-shell rescattering is maximized. At low values of the recoil momentum ($P_R = 200$ MeV/c), on-shell nucleon rescattering reduces the quasi free contribution, as expected from unitarity (a part of the strength of the quasi elastic channel is transferred to inelastic ones). At high values of the recoil momentum ($P_R = 400$ MeV/c) the quasi free contribution strongly decreases as the nucleon

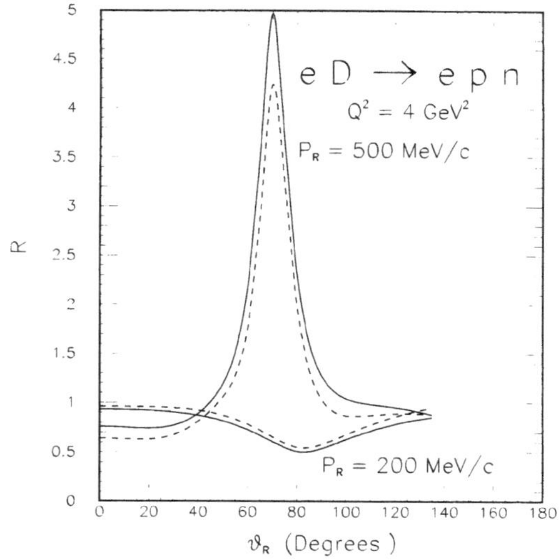

Figure 1. The ratio between the full cross section and the contribution of the quasi-free scattering.

momentum distribution: on-shell rescattering takes over and dominates.

An experiment [7] recently performed at JLab confirms this behavior. Figure 2 shows the recoil momentum distribution up to 600 MeV/c, in the quasi elastic kinematic (X=1). Above $P_R = 400$ MeV/c, FSI (dashed line) dominate while MEC (full line) contribute to a lesser extent.

Although the virtuality ($Q^2 = 0.665$ GeV^2) of the exchanged photon is about twice as much than what was achieved twenty years ago, the relative kinetic energy of the two outgoing nucleons ($T_L = Q^2/2m = 337$ MeV) is still low enough to rely on the partial wave expansion of the np scattering amplitude. The curves are the results of the original model [8], where both the on-shell and half off-shell np scattering amplitudes are solution of the Lippman-Schwinger equation with the same potential (Paris) as for the bound state.

2.2 The High Energy Regime

At higher energies (let's say when the relative kinetic energy of the outgiong fragment exceeds 500 MeV or so), too many partial waves enter into the game and their growing inelasticities prevent to compute the scattering amplitude from a potential. It is better to use a global parameterization of the NN scattering amplitude. On general ground, it can be expanded as follows

$$T_{NN} = \alpha + i\gamma(\boldsymbol{\sigma_1} + \boldsymbol{\sigma_2}) \cdot \boldsymbol{n} + \text{ spin} - \text{spin terms} \tag{1}$$

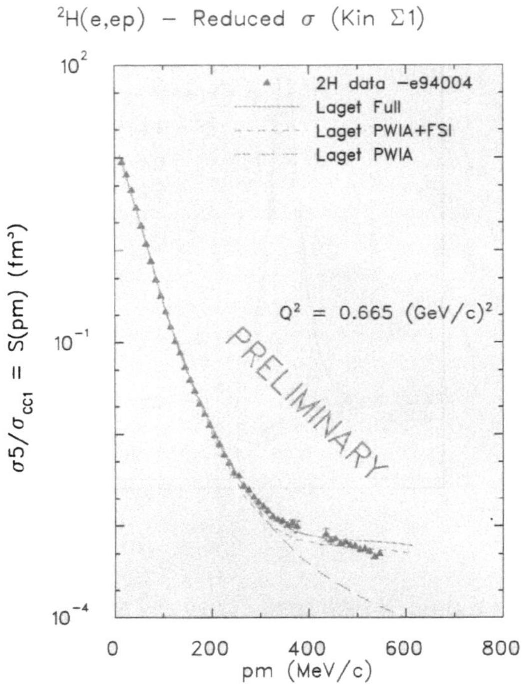

Figure 2. The momentum distribution in the $D(e, e'p)n$ reaction at X=1.

Above 500 MeV, the central part α dominates and is almost entirely absorptive, and takes the simple form

$$\alpha = \frac{k}{4\pi}\,(\epsilon + i)\,\sigma_{\rm NN}\,\exp[\frac{b}{2}t] \tag{2}$$

In the forward direction its imaginary part is related to the total cross section $\sigma_{\rm NN}$, while the slope parameter b is related to the angular distribution of NN scattering. Both can be determined from the experiments performed at Los Alamos, Saturne and Cosy. The ratio ϵ, between the real and imaginary part of the amplitude, is small and does not exceed 0.2. The spin-orbit and spin-spin terms are related to polarization observables, but I have not yet implemented them in the three body codes.

Such a parameterization is very convenient to compute the rescattering amplitude, but leads only to an accurate prediction of its singular part (on-shell scattering). Contrary to low energy, there is unfortunately no way to constrain the half-off-shell part of the NN scattering amplitude, and one can get only estimate of the principal part of the rescattering amplitude. It turns out that it does not dominate at high energy. So, the method is founded on solid grounds in the quasi-elastic kinematics (X∼1). Away, it tells us in which kinematics FSI are minimized.

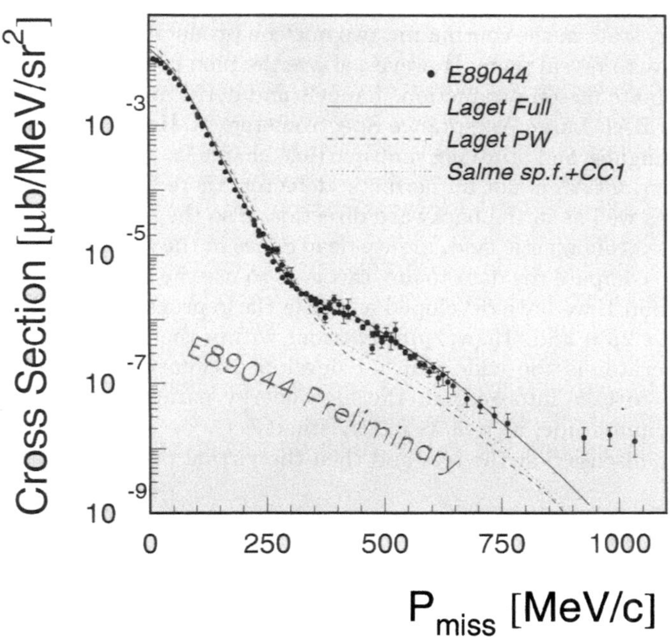

Figure 3. The momentum distribution in the ^{3}He$(e,e'p)d$ reaction at X=1 and $Q^2 =$ 1.55 GeV^2.

Figure 3 shows how well this method reproduces the cross section of the ^{3}He$(e,e'p)d$ reaction recently measured at JLab [9] at $Q^2 = 1.55$ GeV^2, in the quasi-free kinematics (X=1). At such a high virtuality, the relative kinetic energy between the outgoing proton and deuteron is $T_L = 850$ MeV, where the NN cross section reaches its maximun and becomes flat around $\sigma_{\rm NN} =$ 47mb. Again, FSI reduces the quasi-free contribution below 300 MeV/c and overwhelms it by more than a factor five around 500 MeV/c. Details on the model are given in ref. [10].

To summarize, a fair agreement with the recent JLab data has been reached in the quasi-elastic regime, up to recoil momentum of the order of 1 GeV/c, provided that the NN scattering amplitude relevant to the same energy range is used. In order to determine the high momentum components of the nuclear wave function, one has to go away the quasi-elastic kinematics: as demonstrated in Fig. 1, this occurs in parallel or anti-parallel kinematics, where on-shell nucleon rescattering is suppressed.

3 The Three Body Disintegration Channels

Two body short range correlations are the primary source of high momentum components in the nuclear wave function. Since they are strongly coupled to high energy state in the continuum, two nucleon production experiments are the natural way to reveal them. However, above the pion production threshold, one must eliminate meson production channels and perform exclusive experiments. CLAS (CEBAF Large Acceptance Spectrometer) at JLab is the ideal detector for disentangling and studying multiparticle channels.

However, CLAS is not an hermetic detector: there is a hole in the forward direction as well as in the backward direction; also the coils, which provide the toroidal electromagnetic field, induce dead zones in the acceptance. So the only fair way to compare the data to any theory is to use the same cuts. To that end, G. Audit and I, we have developed a Monte Carlo programme which simulates the ^{3}He$(e, e'2p)n$ and ^{3}He$(\gamma, 2p)n$ reactions within the CLAS acceptance. The event generator is the code which I developed twenty years ago and which I updated to take into account the high energy parameterization of the NN scattering amplitude, as well as relativistic effects.

Let me discuss first the real and then the virtual photon sectors.

3.1 Real Photons

Figure 4 shows the result of such a simulation within the CLAS acceptance. The model [11] has been calibrated against previous data recorded below 800 MeV in restricted parts of the phase space: two magnetic spectrometers [12, 13] in the two body disintegration sector (neutron at rest); two magnetic spectrometers [14] and DAPHNE detector [15] in the three body disintegration sector.

CLAS results [16] enlarge the available data set, not only by covering the full phase space but also by extending the energy range in a single shot. Cuts can be made in various places of the Dalitz plot, in order to emphazis different mechanisms and access different aspects of the three body dynamics. For instance, the peak at bottom of the Dalitz plot (small T_n) corresponds to the disintegration of a pair of protons at rest. Since this pair has no dipole moment, MEC associated with the Δ resonance are suppressed and one can probe higher order MEC. The peaks on the left and right edges of the Dalitz plot correspond to rescattering within a proton neutron pair, while the other proton takes almost all the available energy. Finally, in the center of the Dalitz plot the energy is shared by the three nucleons: this is the place where three nucleon effects dominate. Preliminary results show a clear distinction between two and three body effects and follow the trends of the model. I refer to ref. [16] for a more detailed account of the preliminary CLAS data.

3.2 Virtual Photons

The ^{3}He$(e, e'2p)n$ reaction was advocated [8] as the best way to access two proton correlations in few body systems. In particular, in kinematics where the

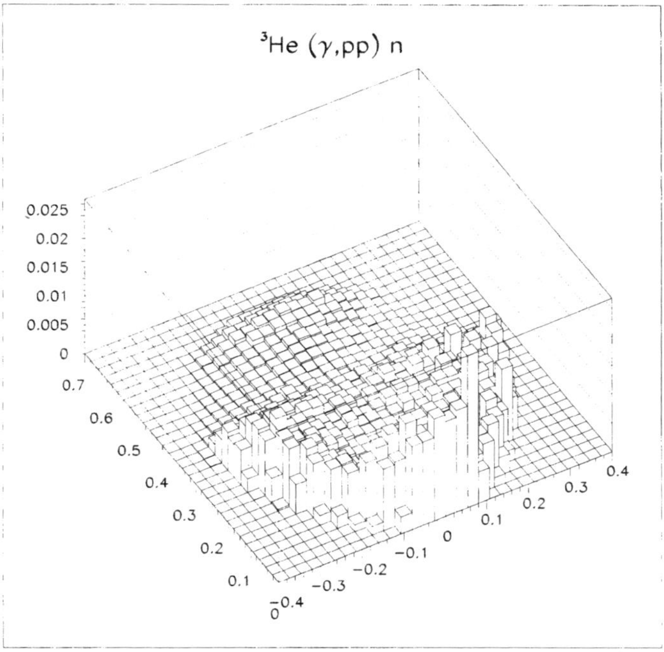

Figure 4. The Dalitz plot of the $^3\mathrm{He}(\gamma, 2p)n$ reaction at 550 MeV. The bottom-right axis is proportional to the difference $T_1 - T_2$ between the kinetic energy of the two protons, while the left axis is proportional to the kinetic energy of the neutron T_n.

neutron is almost at rest, the transverse part of the amplitude is suppressed since the pair of proton has no dipole moment. It also turns out that three body mechanisms are suppressed, since they prefer kinematics where the momentum transfer is shared by the three nucleon. On the contrary, the coupling of a longitudinal photon the the proton pair is not suppressed, and the only sizeable correction is due to FSI between the two outgoing protons.

CLAS provides us with the first comprehensive study of this channel [18, 19]. Figure 5 shows the momentum distribution of the pair of protons almost at rest in ^{3}He. The plane wave calculation exhibits the characteristic node of the S-wave part of its wave function, which is the only one which survives when the pair is at rest. Of course, FSI fill in this hole (nature does not like holes!), and MEC contribute moderatly at high momentum. Both the theory and the experiment are integrated within the same cuts (for instance, the fall off of the cross section below 0.2 GeV^2 reflects the CLAS acceptance) and there is no normalization factor (absolute comparison). The wave function of P. Sauer does a good job in a configuration where it has never been checked before.

Increasing the virtuality from $Q^2 \sim 1$ GeV^2 up to a few GeV^2 will provide

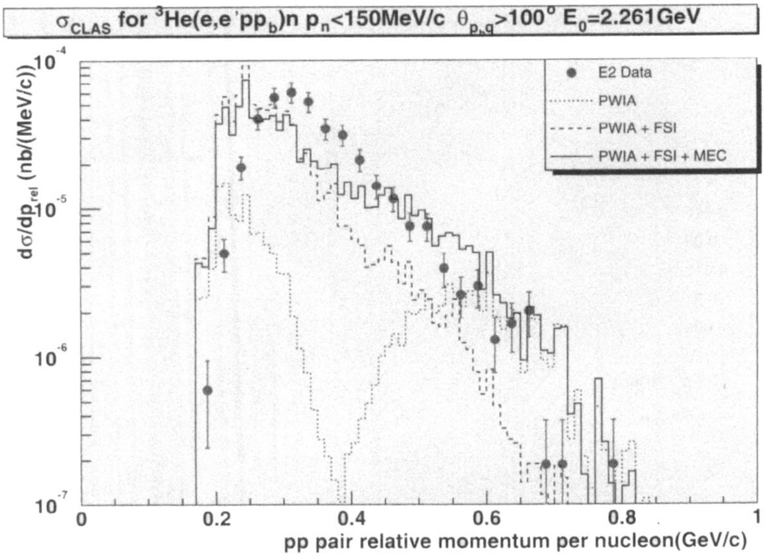

Figure 5. The cross section of the $^3\text{He}(e, e'2p)n$ reaction, when the neutron is almost at rest (CLAS preliminary). The fast proton is emitted forward, while the slow proton is emitted backward. The dotted line corresponds to one body processes. The dashed line includes FSI. The full line includes also MEC.

us with a more stringent test. Since this kinematic is far from the quasi-elastic one, on-shell nucleon rescattering is not dominant, and FSI and MEC will be suppressed.

On the contrary, on-shell nucleon rescattering dominates the cross section in the kinematics reported in Fig. 6. Here the two protons are required to be emitted in a symmetric way around the direction of the virtual photon, and the cross section is plotted against the energy of the virtual photon. The good agreement between the theory and the experiment tells us that our description of the on-shell rescattering matrix element is fine over a wide energy range ($0.2 < T_L < 2$ GeV).

Other cuts have been applied to CLAS data. For instance, one can select a very fast neutron and look for a proton pair almost at rest in ^{3}He. In a plane wave picture, this gives access to the relative wave function of the two protons. However, in the interesting situation where the relative momentum between the two protons is large, the three nucleons share the total energy and momentum and three body mechanisms contribute significantly. I refer to ref. [20] for a more detailed accounts of such a study.

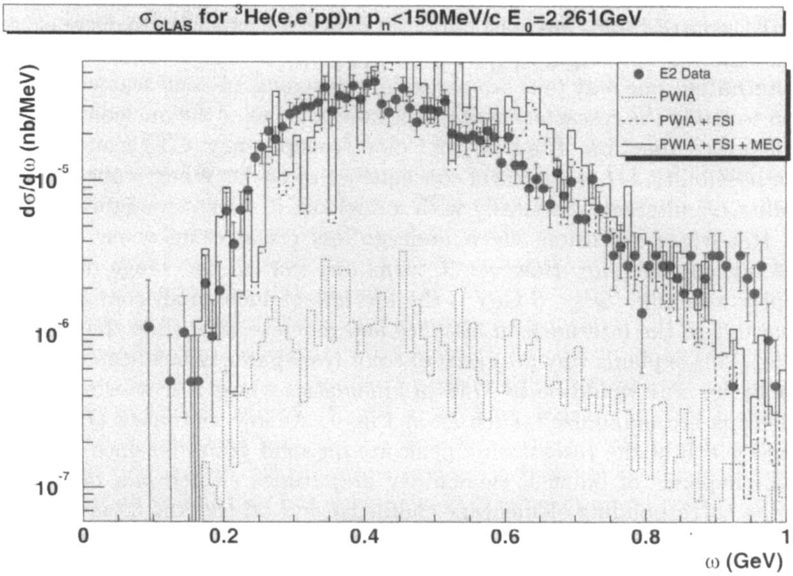

Figure 6. The cross section of the $^3\mathrm{He}(e, e'2p)n$ reaction, when the neutron is almost at rest(CLAS preliminary). The two protons are emitted in a symmetric way against the virtual photon direction. The dotted line corresponds to one body processes. The dashed line includes FSI. The full line includes also MEC.

4 Perspectives

The operation of CEBAF at Jefferson Laboratory, over the past few years, has confirmed the expectations which we had twenty years ago, but has enlarged in a considerable way the data set, both in accuracy and in energy as well as momentum range.

In the real photon sector, the three body photo disintegration of ^{3}He is dominated by three body mechanisms related to the on-shell propagation of a pion between the three nucleons. The extension to the virtual photon sector gives access to three body Meson Exchange Currents, which may be related to three body forces.

In the virtual photon sector, the two-body electrodisintegration of ^{3}He, and of a pair of two nucleons bound in ^{3}He, opens windows on the momentum distribution at short distance. However special care has to be taken in order to avoid on-shell nucleon scattering which dominates the electrodisintegration cross section in a well defined part of the phase space, namely near the quasi free kinematics. The best candidate to study the short range components of the nuclear wave function is the parallel kinematics, where the outgoing nucleons are emitted along the direction of the virtual photon. In that kinematics FSI and MEC contributions are reasonably small and are expected to decrease when the virtuality Q^2 increases in the range of several GeV^2.

Alternately, one may take advantage of the strong on-shell scattering contribution to study the rescattering of exotic components of the nucleon (and other hadrons) wave function. The study of Color Transparency (CT) is an example of such a possibility. CT is a natural consequence of QCD. When a photon of high virtuality Q^2 interacts elastically with a nucleon, it selects configurations with small transverse extension which undergo less rescattering when they travel in the nuclear medium. However it turns out that in the range of virtuality currently available ($Q^2 < 6\ \mathrm{GeV}^2$) the lifetime of such small configurations is comparable to the internucleon distance and much smaller than the radius of a nucleus. This explains why no signal of color transparency has been reported so far. A better way would be to work in kinematics where rescatterings between to nucleons are maximized, such as in Fig. 1. At low virtuality Q^2 the hight and the width of the rescattering peak are on solid grounds, since the matrix element depends of on-shell elementary amplitudes (which can be borrowed from the corresponding elementary channels) and on the low momentum part of the nuclear wave function. At higher virtuality, CT will cause the peak to decrease and its width to increase. I refer to my contribution in ref. [21] for a more detailed discussion.

While the present energy of CEBAF (6 GeV) allows to test this conjecture at low virtually ($Q^2 < 6\ \mathrm{GeV}^2$), only its energy upgrade to 12 GeV will allows to reach high enough virtuality ($Q^2 \sim 12\ \mathrm{GeV}^2$) to see the effects of color transparency.

Acknowledgement. This work was supported in part by the European Commission under Contract HPRN-CT-2000-00130.

References

1. Ch. Hajduk, P. Sauer: Nucl. Phys. **A369**, 321 (1981)
2. Ch. Hajduk, P. Sauer, W. Strueve: Nucl. Phys. **A405**, 581 (1983)
3. Lacombe et al.: Phys. Lett. **B101**, 139 (1981)
4. B. Frois, S. Platchkov: In: *Modern Topics in Electron Scattering* (Eds. B. Frois and I. Sick). Singapore: World Scientific 1991
5. J.-M. Laget: Phys. Rep. **69**, 1 (1981)
6. J.-M. Laget: In: *Modern Topics in Electron Scattering* (Eds. B. Frois and I. Sick) p.290. Singapore: World Scientific 1991
7. P. Ulmer et al.: Phys. Rev. Lett. **89**, 062301 (2002)
8. J.-M. Laget: Phys. Lett. **B199**, 493 (1987)

9. D. Higinbotham: In: *International Symposium on Electromagnetic Interactions in Nuclear and Hadron Physics*, Osaka, Japan (2001). Singapore: World Scientific (to appear)

10. J.-M. Laget: Nucl. Phys. **A579**, 333 (1994)

11. J.-M. Laget: J. Phys. **G14**, 1445 (1988)

12. G. Audit et al.: Phys. Lett. **B227**, 331 (1989)

13. G. Audit et al.: Phys. Lett. **B312**, 57 (1993)

14. G. Audit et al.: Phys. Rev. **C44**, R575 (1991)

15. G. Audit et al.: Nucl. Phys. **A614**, 461 (1997)

16. S. Niccolai: Few Body Systems Suppl. (to appear); In *XVIII Conference on Few-Body Problems in Physics*, Bled, Slovenia (2002)

17. J.-M. Laget: Phys. Rev. **C35**, 832 (1987)

18. B. Zhang: Thesis. MIT 2002 (unpublished)

19. S. Gilad et al.: In: *5th Workshop on Electromagnetically induced Two Hadron Emission*, Lund, Sweden (2001)

20. L. Weinstein: Eur. Phys. J. (to appear); nucl-ex/0209014

21. J.-M. Laget:In: *Exclusive and Semi-Exclusive Processes at High Momentum Transfer*, Newport-News, USA (1999), p.13. Singapore: World Scientific 2000

Few-Body Systems Suppl. 15, 183–194 (2003)

Few-Body Systems

Dynamical Model of Electroweak Pion Production Reactions

T.-S.H. Lee,[1,*] T. Sato[2]

[1] Physics Division, Argonne National Laboratory, Argonne, Illinois 60439, USA

[2] Department of Physics, Osaka University, Toyonaka, Osaka 560-0043, Japan

Abstract. The dynamical model we have developed in the past few years for investigating pion photo- and electroproduction reactions is reviewed. Some preliminary results from extending the model to investigate weak pion production reactions are presented.

1 Introduction

It is now well recognized that an important challenge in nuclear research is to understand the hadron structure and nuclear dynamics within Quantum Chromodynamics(QCD). In this contribution, we would like to report on an advance which we have made recently in this direction, by investigating the electroweak pion production reactions in the Δ excitation region. Our objective is to extract the N-Δ transition form factors. Such information is important for testing hadron models and Lattice QCD calculations. In particular, we can address the question about whether N and Δ are deformed. If they are deformed, is it due to the gluon interactions between quarks or due to the pion cloud which results from the spontaneously breaking of the chiral symmetry of QCD?

We started with a very extensive investigation of pion photo- and electroproduction reactions. There exist three different approaches for analyzing the data of these reactions. The approach based on dispersion relations was introduced by Chew, Goldberger, Low, and Nambu in 1950's [1]. This approach has recently been revived [2, 3] for investigating pion photoproduction up to the Δ energy region. The second approach is based on the effective lagrangian

* *E-mail address:lee@theory.phy.anl.gov*

method [4, 5]. In this approach, the nucleon resonance(N^*) excitations are described by the isobar model and the non-resonant (background) amplitudes are determined by the tree diagrams of effective lagrangians and appropriate phenomenological terms. The K-matrix method is used to unitarize the amplitude. The widely used data-fitting codes SAID [7] and MAID [6] are also based on this approach.

The third approach is the dynamical approach which has been pursued by several groups [8, 9, 10, 11, 12, 13, 14, 15]. This approach solves dynamical scattering equations with interactions defined by effective lagrangians and/or hadron models. Compared with the other two approaches, the main feature of the dynamical approach is to include the off-shell dynamics. Therefore, the contributions from reaction mechanisms can be separated from the excitation of the internal structure of hadrons.

In this contribution, we will only report on the dynamical approach developed by Sato and Lee [14]. In Sect. 2, we will briefly describe the dynamical model of ref. [14]. The results obtained for pion photo- and electroproduction reactions are reviewed in Sect. 3. By considering the data from LEGS [16], Mainz [17, 18, 19], Jlab [20, 21], and MIT-Bates [22], we have extracted vector N-Δ form factors. In Sect. 4, we present our recent results [23] from extending the dynamical approach to investigate weak pion production reactions. The objective is to provide a framework for extracting the axial vector N-Δ from factors from the new neutrino reaction data which could be obtained at Fermilab, KEK and Japanese Hadron Facility in the near future.

2 The Dynamical Approach

In the dynamical approach developed in ref. [14] (called SL model), the starting interaction Hamiltonian is $H_I = \Gamma_{MB \leftrightarrow B'}$ which describes the absorption and emission of a meson(M) by a baryon(B). In the SL model, such a Hamiltonian is obtained from phenomenological lagrangians. In fact, the approach is more general and this Hamiltonian can be defined in terms of an hadron model, as attempted, for example, in ref. [24], or by using the vertex functions predicted by Lattice QCD calculations.

It is a non-trivial many-body problem to calculate πN scattering and $\gamma N \to \pi N$ reaction amplitudes from the interaction Hamiltonian $H_I = \Gamma_{MB \leftrightarrow B'}$. To obtain a manageable reaction model, a unitary transformation method [14, 25] is used up to second order in H_I to derive an effective Hamiltonian. The essential idea of the employed unitary transformation method is to eliminate the unphysical vertex interactions $MB \to B'$ with $m_M + m_B < m_{B'}$ from the Hamiltonian and absorb their effects into $MB \to M'B'$ two-body interactions. For the pion production processes in the Δ region, the resulting effective Hamiltonian is defined in a subspace spanned by the πN, γN and Δ states and has the following form

$$H_{\text{eff}} = H_0 + v_{\pi N} + v_{\gamma\pi} + \Gamma_{\pi N \leftrightarrow \Delta} + \Gamma_{\gamma N \leftrightarrow \Delta}, \tag{1}$$

where $v_{\pi N}$ is a non-resonant πN potential, and $v_{\gamma\pi}$ describes the non-resonant

$\gamma N \leftrightarrow \pi N$ transition. The Δ excitation is described by the vertex interactions $\Gamma_{\gamma N \leftrightarrow \Delta}$ for the $\gamma N \leftrightarrow \Delta$ transition and $\Gamma_{\pi N \leftrightarrow \Delta}$ for the $\pi N \leftrightarrow \Delta$ transition. The non-resonant $v_{\gamma\pi}$ consists of the usual pseudo-vector Born terms, ρ and ω exchanges, and the crossed Δ term.

From the effective Hamiltonian Eq. (1), it is straightforward to derive a set of coupled equations for πN and γN reactions. The resulting pion photoproduction amplitude can be written as

$$T_{\gamma\pi}(E) = t_{\gamma\pi}(E) + \frac{\bar{\Gamma}_{\Delta \to \pi N} \bar{\Gamma}_{\gamma N \to \Delta}}{E - m_\Delta - \Sigma_\Delta(E)}. \tag{2}$$

The non-resonant amplitude $t_{\gamma\pi}$ is calculated from $v_{\gamma\pi}$ by

$$t_{\gamma\pi}(E) = v_{\gamma\pi} + t_{\pi N}(E) G_{\pi N}(E) v_{\gamma\pi}, \tag{3}$$

where $G_{\pi N}$ is the πN free propagator , and $t_{\pi N}$ is calculated from the non-resonant πN interaction $v_{\pi N}$.

The dressed vertices in Eq. (2) are defined by

$$\bar{\Gamma}_{\gamma N \to \Delta} = \Gamma_{\gamma N \to \Delta} + \bar{\Gamma}_{\pi N \to \Delta} G_{\pi N}(E) v_{\gamma\pi}, \tag{4}$$

$$\bar{\Gamma}_{\Delta \to \pi N} = [1 + t_{\pi N}(E) G_{\pi N}(E)] \Gamma_{\Delta \to \pi N}. \tag{5}$$

The Δ self-energy in Eq. (2) is then calculated from

$$\Sigma_\Delta(E) = \Gamma_{\pi N \to \Delta} G_{\pi N}(E) \bar{\Gamma}_{\Delta \to \pi N}. \tag{6}$$

As seen in the above equations, an important feature of the dynamical model is that the bare vertex $\Gamma_{\gamma N \to \Delta}$ is modified by the non-resonant interaction $v_{\gamma\pi}$ to give the dressed vertex $\bar{\Gamma}_{\gamma N \to \Delta}$, as defined by Eq. (4). The second terms in the right-hand sides of Eqs. (4) and (5) can be interpreted as the *dynamical* pion cloud effects on the N-Δ transitions. We can then identify the parameters of the bare vertex $\Gamma_{\gamma N \to \Delta}$ with the predictions from hadron models within which the π-baryon states in *continuum* are excluded. Such a separation of the reaction mechanisms from the excitations of hadron internal structure is an essential ingredient of a dynamical approach, but is not an objective of the approaches based on either dispersion relations or K-matrix method.

3 Results of Pion Photo- and Electroproduction Reactions

In ref. [14], we have found that the dynamical model briefly described above can describe very extensive data of pion photo- and electroproduction reactions in the Δ excitation region. Our typical results for pion photoproduction reactions are shown in Fig.1.

We see that the best fit to the photon asymmetry(Σ) data favors a non-zero ratio R_{EM} between the electric E2 and magnetic M1 transitions of the bare vertex $\Gamma_{\gamma N \to \Delta}$, indicating that the Δ is deformed. The calculated $\gamma N \to \Delta$ transition amplitudes of helicity $\lambda = 3/2$ and R_{EM} are listed in Table 1 and

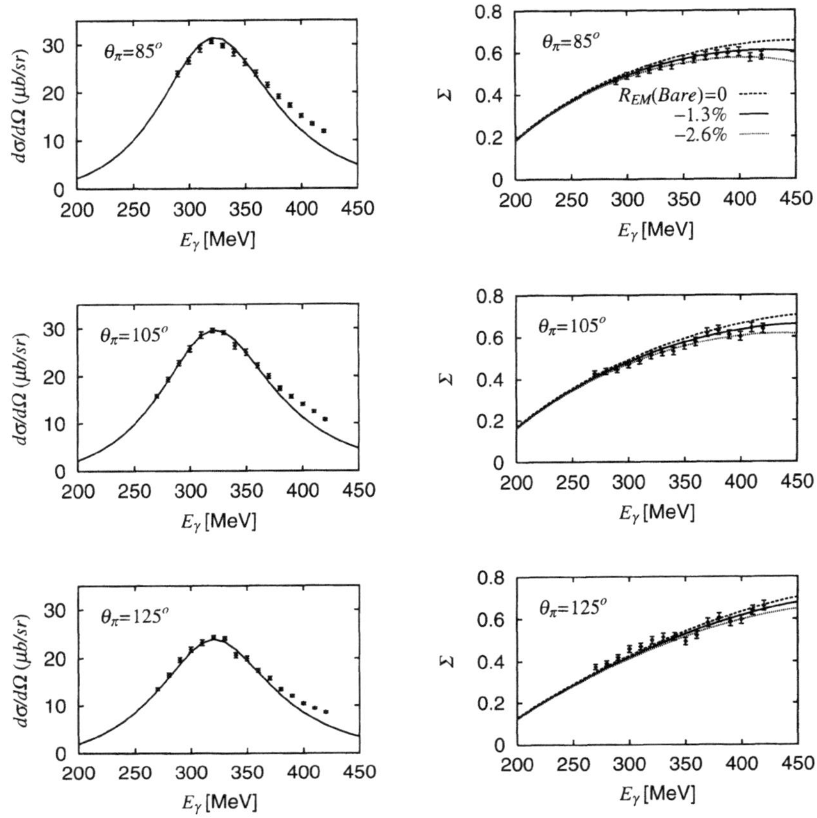

Figure 1. Differential cross section($d\sigma/d\Omega$) and photon asymmetry(Σ) of $\gamma p \to \pi^0 p$ reaction. $R_{\rm EM}(Bare)$ is the E2/M1 ratio of the bare vertex $\Gamma_{\gamma N \to \Delta}$. The data are from Refs. [16] and [17].

compared with the predictions from several constituent quark models [26, 27, 28].

The large differences between the dressed and bare values clearly indicate that the dynamical pion cloud effect is very important. It accounts for about 40 % of the M1 transition amplitude (which is proportional to $A_{3/2}$), and about a factor of two in changing the ratio $R_{\rm EM}$. We also note that the bare values are closer than the dressed values to the quark model predictions and our dressed values are close to the values listed by the Particle Data Group(PDG) [29]. This provides an explanation of a long-standing puzzle of the constituent quark model. Namely, the empirically determined PDG values contain the dynamical pion cloud effects which are included in our dressed vertex, but are excluded in all of constituent quark model calculations.

Table 1. Helicity amplitude $A_{3/2}$ and E2/M1 ratio $R_{\rm EM}$ for the $\gamma N \rightarrow \Delta$ transition at $Q^2 = 0$ photon point. $A_{3/2}$ is in unit of $10^{-3} GeV^{-1/2}$ and $R_{\rm EM}$ in %.

	$A_{3/2}$		$R_{\rm EM}$		
	Dressed	Bare	Dressed	Bare	Refs.
Dynamical Model	-228	-153	-2.7	-1.3	14
Quark Model	–	-186	–	~ 0	26
	–	-157	–	~ 0	27
	–	-182	–	-3.5	28

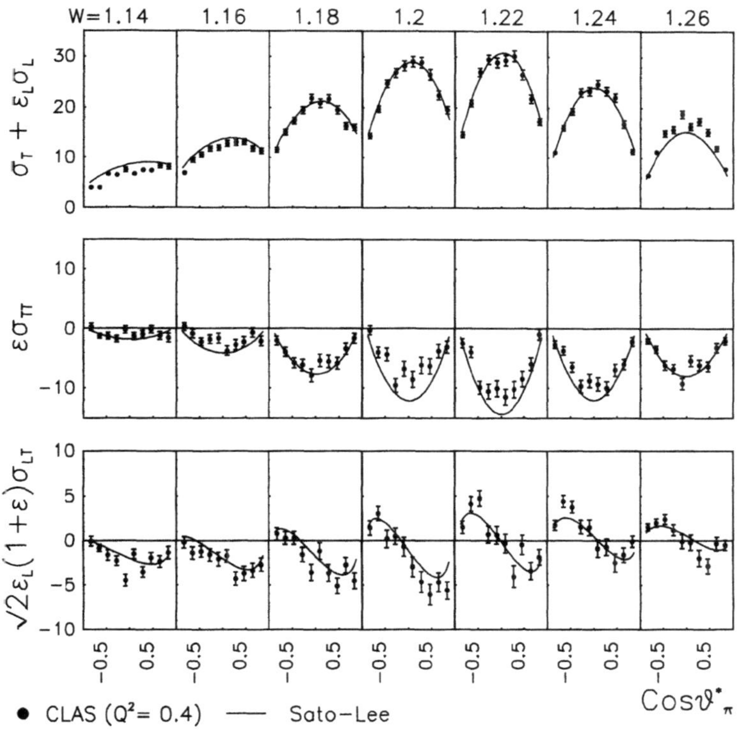

Figure 2. The differential cross sections of $p(e,e'\pi^0)$ at $Q^2 = 0.4$ (GeV/c)^2. The data are from ref. [21].

Next we turn to reviewing our results for electroproduction reactions. The electroproduction cross section has several components

$$\frac{d\sigma^v}{d\Omega} = A + B\cos\phi + C\cos 2\phi + h_e D \sin\phi, \tag{7}$$

where ϕ is the angle between the e-e' plane and the π-N plane, h_e is the helicity of the incoming electron, and

$$\begin{aligned} A &= \frac{d\sigma_T}{d\Omega} + \epsilon \frac{d\sigma_L}{d\Omega}, \\ B &= \sqrt{2\epsilon(\epsilon+1)} \frac{d\sigma_{LT}}{d\Omega}, \\ C &= \epsilon \frac{d\sigma_{TT}}{d\Omega}, \\ D &= \sqrt{2\epsilon(1-\epsilon)} \frac{d\sigma_{LT'}}{d\Omega}. \end{aligned}$$

By using sophisticated detectors, such as CLAS at Jlab [20, 21] and OOP at MIT-Bates [22], the data for the components A, B and C have been extracted from unpolarized $(e, e'\pi)$ measurements. With the polarization measurements [18, 19], we now also have data for the component D.

Our model can describe the Jlab data in a rather wide range of Q^2 up to 4 (GeV/c)^2. One example is shown Fig. 2. We see that each component of the differential cross section can be described very well.

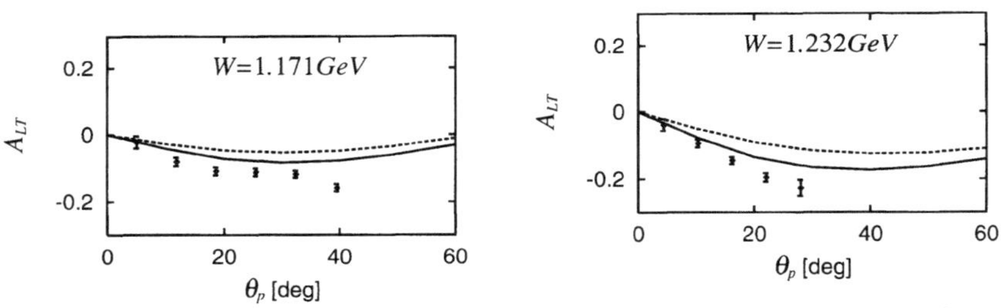

Figure 3. The predicted $A_{\rm LT}$ (as defined in ref. [14]) of the $p(e, e'\pi^0)$ reaction at $Q^2 = 0.126$ (GeV/c)^2 are compared with the data from MIT-Bates [22] The dotted curves are obtained from setting the electric E2 and Coulomb C2 transitions of the $\gamma N \to \Delta$ interaction to zero.

We also get good agreement with the data of L-T interference cross sections from MIT-Bates. This is shown in Fig. 3. Our predictions of various polarization observables are also in fare agreement with the recent data from and Mainz [18, 19]. This is illustrated in Fig. 4 for the beam-helicity asymmetry of $p(\vec{e}, e', \boldsymbol{p})\pi^0$.

The main outcome of our fits to the data is a determination of $\gamma N \to \Delta$ form factors. The extracted magnetic M1 form factors for both the dressed (solid curve) and bare (dotted curve) vertices are shown Fig. 5. We see that they drop faster than the proton form factor $G_D(Q^2)$ as Q^2 increases. The differences between the solid and dot curves indicate a very large pion cloud effect in the low Q^2 region, and much less at high Q^2.

The extracted electric E2 and Coulomb C2 form factors of the $\gamma N \to \Delta$ transition are compared in Fig. 6 with the data from ref. [21]. Here we plot their ratios with the magnetic M1 form factor. These ratios measure the deformation of the Δ.

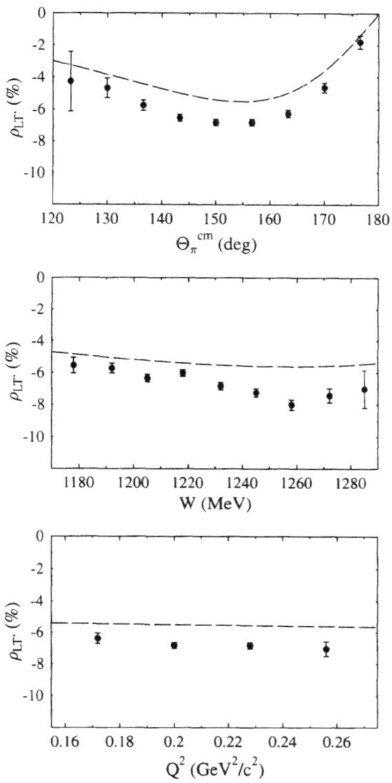

Figure 4. Beam-helicity asymmetry of $p(\vec{e}, e'p)\pi^0$. The data are from ref. [19].

Clearly, our results agree well with the data, and we can definitely conclude that N and Δ are deformed. In Fig. 7 we also show the imaginary parts of the electroproduction multipole amplitudes E_{1+} and S_{1+} which are proportional to the E2 and C2 form factors of the $\gamma N \rightarrow \Delta$ transition. The dotted curves are from the bare form factor. The differences between the solid and dotted curves are due to the pion cloud. We therefore conclude that the Δ deformation is largely due to pion cloud, not the intrinsic quark core. We also see that the pion cloud effect has a very strong energy dependence. More experiments to be performed will further test our predictions in the very low Q region.

4 Weak Pion Production Reaction

Since weak currents and electromagnetic currents are closely related within the Standard Model, it is straightforward to extend our dynamical approach to study weak pion production reactions. This is our current focus. Our objective

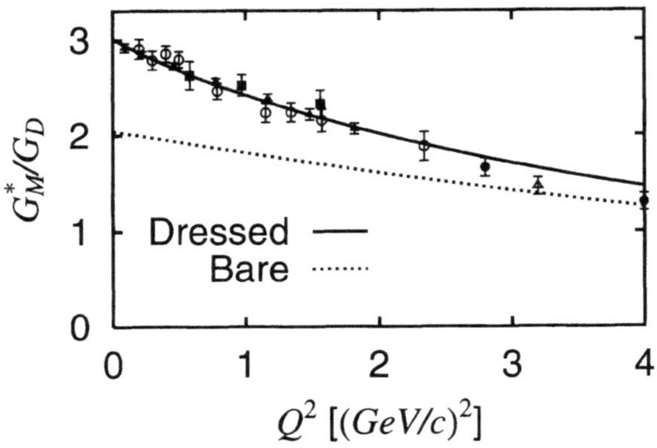

Figure 5. The ratio between the M1 form factor of the $\gamma N \to \Delta$ transition and the proton dipole form factor (G_D). The dotted curve is obtained when the dynamical pion cloud effects, as defined by the second term of Eq.(4), are switched off.

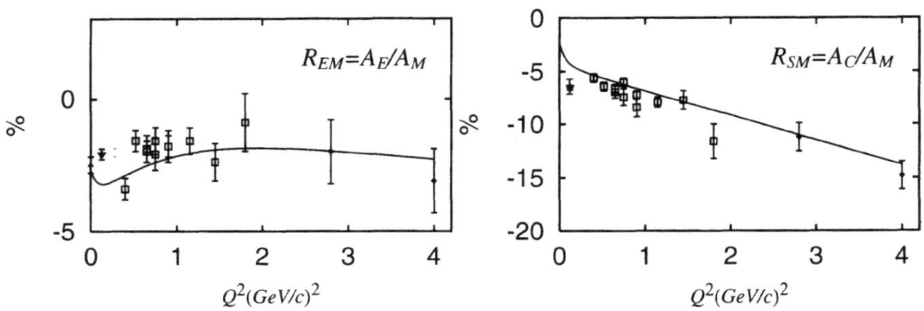

Figure 6. $R_{\rm EM}(R_{\rm SM})$ is the ratio between the electric E2(Coulomb C2) and magnetic M1 transitions of the $\gamma N \to \Delta$ interaction. The data are from ref. [21].

is to develop a framework for extracting the axial vector N-Δ form factors from the neutrino-induced pion production reaction data which could be obtained at Fermilab, KEK and Japanese Hadron Facility in the near future.

The first step is to extend our dynamical formulation, briefly reviewed in Sect. 2, to include the coupling with W and Z bosons. By Conserved Vector Current (CVC) relations, the vector(V) parts of the weak currents can be obtained from the electromagnetic currents by isospin rotations. By using procedures similar to the SL model, we can construct the axial (A) vector currents

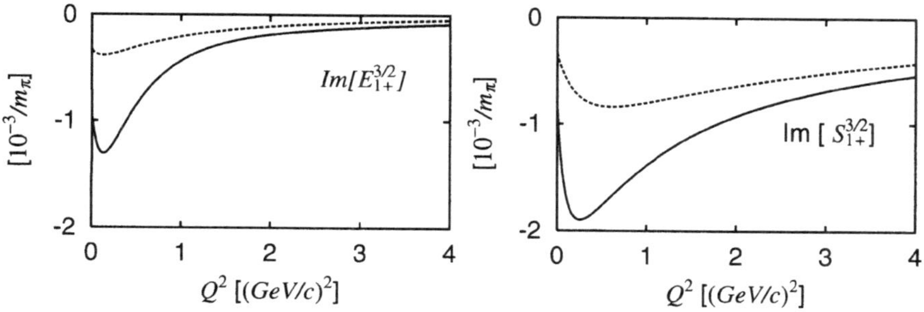

Figure 7. Q^2 dependence of the imaginary(Im) parts of $\gamma^* N \to \pi N$ multipole amplitudes at $W = 1232$ MeV. The dotted curves are from turning off the pion cloud effects.

for pion production reactions. The resulting axial current amplitude consists of nucleon-Born, rho-exchange, pion-pole, and Δ excitation terms. As a start, this amplitude is calculated by using the parameters from the previous works. In particular, the bare axial N-Δ form factors are assumed to be those calculated by Hemmer, Holstein, and Mukhopadhyay [30] within the constituent quark model. The dressed axial form factors are obtained by including the dynamical pion cloud effects, as defined by the dynamical equations similar to those given in Sect. 2. Thus our first calculations do not have any adjustable parameters.

Similar to the electroproduction case, we are able to examine the relative importance between the non-resonant amplitude, the bare N-Δ transition, and the dynamical pion cloud effect. Our results for the total cross sections of $N(\nu_\mu, \mu^- \pi)N$ reactions are displayed in Fig. 8. Clearly, our full calculations(solid curves) are close to the very limited existing data [31]. The differences between the solid curves and dotted curves indicate the pion cloud effects. This implies that our assumption that the bare axial vector N-Δ form factors can be identified with those predicted by the constituent quark model is consistent with the data. The 30 % discrepancy with the data observed in ref. [30] can be accounted for by including the dynamical pion cloud effects. This is consistent with our finding in the study of pion electroproduction, as reviewed in Sect. 3. In Fig. 8, we also see that the non-resonant contributions (dot-dashed curves), although much weaker, are essential in getting an overall good agreement with the data. More detailed results of our investigation will be published elsewhere [23].

5 Summary

As reviewed in this contribution, the dynamical model developed in ref. [14] can describe very extensive data of pion photo- and electroproduction reactions. We have extracted the vector N-Δ form factors and found that their E2/M1

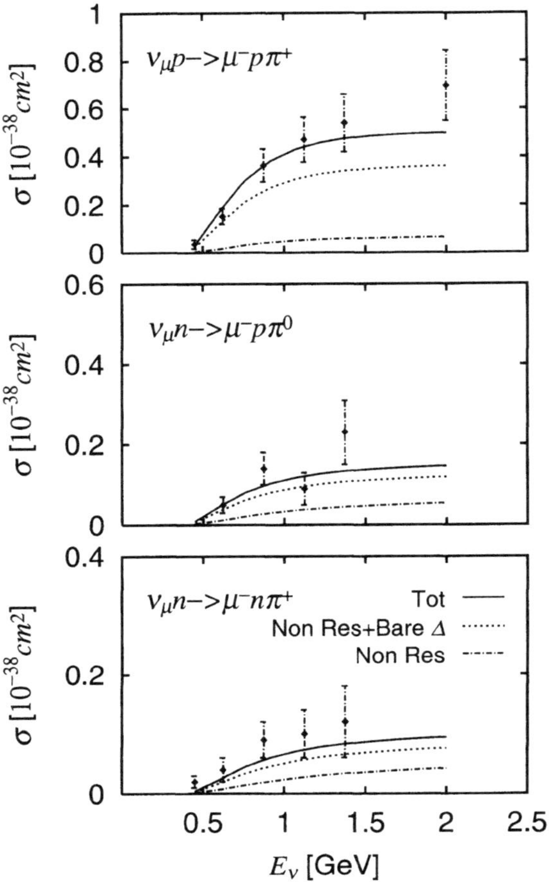

Figure 8. Total cross sections of $N(\nu_\mu, \mu^- \pi)N$ reactions. The data are from ref. [31]. "Non Res" means the contribution from non-resonant amplitude. "Bare Δ" means the contribution from bare N-Δ transition.

and C2/M1 ratios are negative and nonzero. We also have shown that the quadropole N-Δ transition is mainly due to the dynamical pion cloud.

The determined bare N-Δ strengths are consistent with the constituent quark model calculations. This has provided an explanation of a long-standing puzzle of the constituent quark model.

We have extended our dynamical model to investigate weak pion production reactions. The predicted total cross sections of neutrino-induced reactions are in fair agreement with the data. Our approach can be used to pin down the axial vector N-Δ form factors from the new data that could be obtained at Fermilab, KEK, and Japanese Hadron Facility in the near future.

Acknowledgement. This work was supported by U.S. DOE Nuclear Physics Division, Contract No. W-31-109-ENG and Japan Society for the Promotion of Science, Grant-in-Aid for Scientific Research (C) 12640273.

References

1. G.F. Chew, M.L. Goldberger, F.E. Low and Y. Nambu: Phys. Rev. **106**, 1345 (1957)
2. O. Hanstein, D. Drechsel and L. Tiator: Nucl. Phys. **A632**, 561 (1998)
3. I. G. Aznauryan: Phys. Rev. D **57**, 2727 (1998)
4. R. M. Davidson and N. C. Mukhopadyay: Phys. Rev. D **42**, 20 (1990); R. M. Davidson, N. C. Mukhopadyay and R. S. Wittman: Phys. Rev. D **43**, 71 (1990)
5. R. M. Davidson et al.: Phys. Rev. C **59**, 1059 (1999)
6. D. Drechsel, O. Hanstein, S.S. Kamalov and L. Tiator: Nucl. Phys. **A465**, 145 (1999)
7. R. A. Arndt, I. I. Strakovsky and R. L. Workman: Phys. Rev. C **53**, 430 (1996)
8. H. Tanabe, K. Ohta: Phys. Rev. **C31**, 1876 (1985)
9. S. N. Yang: J. Phys. G **11**, L205 (1985)
10. S. Nozawa, B. Blankleider, T.-S. H. Lee: Nucl. Phys. **A513**, 459 (1990)
11. B. Pearce, T.-S. H. Lee: Nucl. Phys. **A528**, 655 (1991)
12. C. van Antwerpen, I.R. Afnan: Phys. Rev. **C52**, 554 (1995)
13. Y. Surya, F. Gross: Phys. Rev. **C53**, 2422 (1996)
14. T. Sato, T.-S. H. Lee: Phys. Rev. **C54**, 2660 (1996); Phys. Rev. C **63**, 055201 (2001)
15. S. S. Kamalov, S. N. Yang: Phys. Rev. Lett. **83**, 4494 (1999)
16. G. Blanpied et al.: Phys. Rev. Lett. **79**, 4337 (1997)
17. R. Beck et al.: Phys. Rev. **C61**, 035204 (2000)
18. Th. Pospischil et al.:Phys. Rev. Lett. **86**, 2959 (2001)
19. P. Bartsch et al.: Phys. Rev. Lett. **88**, 142001 (2002)
20. V. V. Frolov et al.: Phys. Rev. Lett. **82**, 45 (1999)
21. K. Joo et al: Phys. Rev. Lett **88**, 12001 (2002)

22. C. Mertz et al.: Phys. Rev. Lett. **86**, 2963 (2001)

23. T. Sato, D. Uno, T.-S. H. Lee: in preparation.

24. T. Yoshimoto, T. Sato, M. Arima, T. -S. H. Lee: Phys. Rev. **C61** 065203 (2000)

25. M. Kobayashi, T. Sato, H. Ohtsubo: Prog. Theor. Phys. **98**, 927 (1997)

26. S. Capstick: Phys. Rev. D **46**, 1965 (1992); **D46**, 2864 (1992)

27. R. Bijker, F. Iachello, A. Leviatan: Ann. Phys. (N.Y.) **236**, 69 (1994); Phys. Rev. **D54**, 1935 (1996)

28. A. Buchmann, E. Hernandez, A. Faessler: Phys. Rev. **C55**, 448 (1997)

29. Particle Data Group; D.E. Groom, et al.: Eur. Phys. J. C **15**, 1 (2000)

30. T.R. Hemmer, B.R. Holstein, and N.C. Mukhopadhyay: Phys. Rev. **D51**, 158 (1995)

31. S.J. Barish et al.: Phys. Rev. **D19**, 2521 (1979)

Few-Body Systems Suppl. 15, 195–200 (2003)

Few-Body Systems

't Hooft Determinant: Fluctuations and Multiple Vacua

Brigitte Hiller*,†, Alexander A. Osipov**,††*

Centro de Física Teórica, Departamento de Física da Universidade de Coimbra, 3004-516 Coimbra, Portugal

Abstract. The 't Hooft six quark flavor mixing interaction ($N_f = 3$) is bosonized by the path integral method. The considered complete Lagrangian is constructed on the basis of the combined 't Hooft and $U(3) \times U(3)$ extended chiral four fermion Nambu – Jona-Lasinio interactions. The method of steepest descents is used to derive the effective mesonic Lagrangian. Additionally to the known lowest order stationary phase (SP) result of Reinhardt and Alkofer we obtain the contribution from the small quantum fluctuations of bosonic configurations around their stationary phase trajectories. Fluctuations give rise to multivalued solutions of the gap equations, marked at instances by drastic changes in the quark condensates, a novel scenarion for the vacuum state of hadrons at low energies.

1 Introduction

The global $U_L(3) \times U_R(3)$ chiral symmetry of the QCD Lagrangian (for massless quarks) is broken by the $U_A(1)$ Adler-Bell-Jackiw anomaly of the $SU(3)$ singlet axial current $\bar{q}\gamma_\mu\gamma_5 q$. Through the study of instantons [1, 2], it became clear that effective $2N_f$ quark interactions, known as 't Hooft interactions, violate $U_A(1)$, but are still invariant under chiral $SU(N_f) \times SU(N_f)$. In the case of two flavors they are four-fermion interactions, and the resulting low-energy theory resembles the old Vaks Larkin Nambu Jona-Lasinio model [3]. In the case of three flavors they are six-fermion interactions which are responsible for the correct description of η and η' physics, and additionally lead to the OZI-violating effects [4, 5],

$$\mathcal{L}_{\mathrm{det}} = \kappa(\det \bar{q}P_R q + \det \bar{q}P_L q) \tag{1}$$

*This work is supported by Fundação para a Ciência e a Tecnologia, POCTI/35304/FIS/2000 and NATO "Outreach" Cooperation program
† *E-mail address*: brigitte@teor.fis.uc.pt
** *On leave from:* Joint Institute for Nuclear Research, Laboratory of Nuclear Problems, 141980 Dubna, Moscow Region, Russia.
†† *E-mail address:* alexguest@teor.fis.uc.pt

where the matrices $P_{R,L} = (1 \pm \gamma_5)/2$ are projectors and the determinant is over flavor indices.

The physical degrees of freedom of QCD at low-energies are mesons. The bosonization of the effective quark interaction (1) by the path integral approach has been considered in [6], where the lowest order stationary phase approximation (SPA) has been used to estimate the leading contribution from the 't Hooft determinant. In this approximation the functional integral is dominated by the stationary trajectories $r_{\rm st}(x)$, determined by the extremum condition $\delta S(r) = 0$ of the action $S(r)$. The lowest order SPA corresponds to the case in which the integrals associated with $\delta^2 S(r)$, for the path $r_{\rm st}(x)$ are neglected and only $S(r_{\rm st})$ contributes to the generating functional. The subject of the present work is to include the contribution associated with $\delta^2 S(r)$. Our motivation is the following: let us consider the one-dimensional integrals of the form

$$F(\lambda) = \int_{-\infty}^{\infty} dt \exp[\mathrm{i}\lambda f(t)]\,. \tag{2}$$

What is sought by the method of SP is the dominant contribution to F as $\lambda \to \infty$. The dominant contribution to the integral comes from regions of t where f' vanishes. Supposing that f' vanishes at only one point t_0 and neglecting contributions to the integral from regions far from t_0, one can obtain the result

$$F(\lambda) = \sqrt{\frac{2\pi \mathrm{i}}{\lambda f''(t_0)}}\; e^{\mathrm{i}\lambda f(t_0)}. \tag{3}$$

Hence $F(\lambda)$ goes to zero like $1/\sqrt{\lambda}$. The lowest order SPA would correspond to the result $F(\lambda) = \mathrm{const} \cdot \exp[\mathrm{i}\lambda f(t_0)]$, which has the incorrect asymptotic behavior $F(\lambda) = \mathcal{O}(1)$. If results related to finite-dimensional integrals, such as $F(\lambda)$, mean anything with regard to corresponding functional integrals, one can conclude that the leading SPA should include the contribution from the integrals associated with the second functional derivative $\delta^2 S(r_{\rm st})$.

2 Path Integral Bosonization

To be definite, let us consider the theory of the quark fields in four dimensional Minkowski space, with dynamics defined by the Lagrangian density

$$\mathcal{L} = \mathcal{L}_{\rm NJL} + \mathcal{L}_{\rm det}. \tag{4}$$

The first term here is the extended version of the Nambu – Jona-Lasinio (NJL) Lagrangian $\mathcal{L}_{\rm NJL} = \mathcal{L}_0 + \mathcal{L}_{\rm int}$, consisting of the free field part

$$\mathcal{L}_0 = \bar{q}(\mathrm{i}\gamma^\mu \partial_\mu - \hat{m})q, \tag{5}$$

and the $U(3)_L \times U(3)_R$ chiral symmetric four-quark interaction

$$\mathcal{L}_{\rm int} = \frac{G}{2}[(\bar{q}\lambda_a q)^2 + (\bar{q}\mathrm{i}\gamma_5\lambda_a q)^2]. \tag{6}$$

We assume that quark fields have color and flavor indices running through the set $i = 1, 2, 3$; λ_a are the standard $U(3)$ Gell-Mann matrices with $a = 0, 1, \ldots, 8$. The current quark mass, $\hat{m}$, is a nondegenerate diagonal matrix with elements $\mathrm{diag}(\hat{m}_u, \hat{m}_d, \hat{m}_s)$. It explicitly breaks the global chiral $U(3)_L \times U(3)_R$ symmetry of the $\mathcal{L}_{\mathrm{NJL}}$ Lagrangian. The second term in (4) is given by (1). Letting

$$s_a = -\bar{q}\lambda_a q, \quad p_a = \bar{q}\mathrm{i}\gamma_5\lambda_a q, \quad s = s_a\lambda_a, \quad p = p_a\lambda_a\,, \tag{7}$$

yields

$$\mathcal{L}_{\mathrm{det}} = -\frac{\kappa}{64}\left[\det(s + \mathrm{i}p) + \det(s - \mathrm{i}p)\right]. \tag{8}$$

The dynamics of the system is described by the vacuum transition amplitude in the form of the path integral

$$Z = \int \mathcal{D}q\mathcal{D}\bar{q}\exp\left(\mathrm{i}\int d^4x\mathcal{L}\right), \tag{9}$$

which can be written in the equivalent form [6]

$$Z = \int \mathcal{D}\sigma_a\mathcal{D}\phi_a\mathcal{D}q\mathcal{D}\bar{q}\exp\left(\mathrm{i}\int d^4x\mathcal{L}_q\right)\int \mathcal{D}r_{1a}\mathcal{D}r_{2a}\exp\left(\mathrm{i}\int d^4x\mathcal{L}_r\right), \tag{10}$$

where

$$\begin{aligned} \mathcal{L}_q(\bar{q}, q, \sigma, \phi) &= \bar{q}(\mathrm{i}\gamma^\mu\partial_\mu - \hat{m} - \sigma + \mathrm{i}\gamma_5\phi)q, & (11)\\ \mathcal{L}_r(\sigma, \phi, r_1, r_2) &= 2G\left[(r_{1a})^2 + (r_{2a})^2\right] - 2(r_{1a}\sigma_a + r_{2a}\phi_a) \\ &- \frac{\kappa}{8}\left[\det(r_1 + \mathrm{i}r_2) + \det(r_1 - \mathrm{i}r_2)\right]. & (12) \end{aligned}$$

The Fermi fields enter the action bilinearly, we can always integrate over them. At this stage one should also shift the scalar fields $\sigma_a \to \sigma_a + \Delta_a$ by demanding that the vacuum expectation values of the shifted fields vanish $< 0|\sigma_a|0 >= 0$, yielding gap equations to fix parameters $\Delta_a = m_a - \hat{m}_a$, where m_a denotes the constituent quark masses [1]. To evaluate path integrals over $r_{1,2}$ one has to use the method of stationary phase, or, after the formal analytic continuation in the time coordinate $x_4 = \mathrm{i}x_0$, the method of steepest descents. The analytical continuation of the Euclidean version of the path integral under consideration is (see [8] for details),

$$J(\sigma, \phi) = \int_{-\mathrm{i}\infty + r_{\mathrm{st}}}^{+\mathrm{i}\infty + r_{\mathrm{st}}} \mathcal{D}r_{1a}\mathcal{D}r_{2a}\exp\left(\int d^4x\mathcal{L}_r(\sigma, \phi, r_1, r_2)\right). \tag{13}$$

Near the saddle point r^a_{st},

$$\mathcal{L}_r \approx \mathcal{L}_r(r_{\mathrm{st}}) + \frac{1}{2}\sum_{\alpha,\beta}\tilde{r}_\alpha\mathcal{L}''_{\alpha\beta}(r_{\mathrm{st}})\tilde{r}_\beta\,, \tag{14}$$

[1] The shift by the current quark mass is needed to hit the correct vacuum state, see e.g. [7].

where $r^a_{\rm st}$ is a solution of the equations $\mathcal{L}'_r(r_1, r_2) = 0$ determining a flat spot of the surface $\mathcal{L}_r(r_1, r_2)$,

$$\left\{ \begin{array}{rcl} 2Gr^a_1 - (\sigma + \Delta)_a - \frac{3\kappa}{8} A_{abc}(r^b_1 r^c_1 - r^b_2 r^c_2) & = & 0\,, \\ 2Gr^a_2 - \phi_a + \frac{3\kappa}{4} A_{abc} r^b_1 r^c_2 & = & 0\,, \end{array} \right. \tag{15}$$

with totally symmetric constants, A_{abc}, closely related with the $U(3)$ constants d_{abc}. We use in (14) symbols $\tilde{r}^a$ for the differences $(r^a - r^a_{\rm st})$. To deal with the multitude of integrals in (13) we define a column vector $\tilde{r}$ with eighteen components $\tilde{r}_\alpha = (\tilde{r}^a_1, \tilde{r}^a_2)$ and with the matrix $\mathcal{L}''_{\alpha\beta}(r_{\rm st})$ being equal to

$$\mathcal{L}''_{\alpha\beta}(r_{\rm st}) = 4GQ_{\alpha\beta}, \quad Q_{\alpha\beta} = \left(\begin{array}{cc} \delta_{ab} - \frac{3\kappa}{8G} A_{abc} r^c_{1\rm st} & \frac{3\kappa}{8G} A_{abc} r^c_{2\rm st} \\ \frac{3\kappa}{8G} A_{abc} r^c_{2\rm st} & \delta_{ab} + \frac{3\kappa}{8G} A_{abc} r^c_{1\rm st} \end{array} \right). \tag{16}$$

Equation (13) can now be concisely written as

$$J(\sigma, \phi) = \exp\left(\int d^4x \mathcal{L}_r(r_{\rm st}) \right) \int_{-\mathrm{i}\infty}^{+\mathrm{i}\infty} \mathcal{D}\tilde{r}_\alpha \exp\left(2G \int d^4x \tilde{r}^{\rm t} Q(r_{\rm st}) \tilde{r} \right) [1 + \mathcal{O}(\hbar)]\,. \tag{17}$$

The solutions of Eqs. (15) are the following even and odd parity combinations $r^a_{1\rm st}$ and $r^a_{2\rm st}$ expressed in the form of increasing powers in σ_a, ϕ_a

$$r^a_{1\rm st} = h_a + h^{(1)}_{ab}\sigma_b + h^{(1)}_{abc}\sigma_b\sigma_c + h^{(2)}_{abc}\phi_b\phi_c + \ldots \tag{18}$$

$$r^a_{2\rm st} = h^{(2)}_{ab}\phi_b + h^{(3)}_{abc}\phi_b\sigma_c + \ldots \tag{19}$$

Putting these expansions in Eqs.(15) one obtains a series of selfconsistent equations to determine the constants h_a, $h^{(1)}_{ab}$, $h^{(2)}_{ab}$, etc. [8]. As a result we get

$$\mathcal{L}_r(r_{\rm st}) = -2h_a\sigma_a - h^{(1)}_{ab}\sigma_a\sigma_b - h^{(2)}_{ab}\phi_a\phi_b + \mathcal{O}(\text{field}^3). \tag{20}$$

We now turn to the evaluation of the path integral in Eq. (17). In order to define the measure $\mathcal{D}\tilde{r}_\alpha$ more accurately we expand $\tilde{r}_\alpha$ in a Fourier series

$$\tilde{r}_\alpha(x) = \sum_{n=1}^{\infty} c_{n,\alpha}\varphi_n(x), \tag{21}$$

assuming that suitable boundary conditions are imposed. The set of the real functions $\{\varphi_n(x)\}$ form an orthonormal and complete sequence, therefore

$$\int \mathcal{D}\tilde{r}_\alpha \exp\left(2G \int d^4x \tilde{r}^{\rm t} Q(r_{\rm st}) \tilde{r} \right) = \frac{C}{\sqrt{\det(2G\lambda^{\alpha\beta}_{nm})}}, \tag{22}$$

with

$$2G\lambda^{\alpha\beta}_{nm} = \left(\begin{array}{cc} h^{(1)-1}_{ac} & 0 \\ 0 & h^{(2)-1}_{ac} \end{array} \right)_{\alpha\sigma} \left(\delta_{\sigma\beta}\delta_{nm} + \int d^4x \varphi_n(x) F_{\sigma\beta}(x) \varphi_m(x) \right) \tag{23}$$

and

$$F_{\sigma\beta} = \frac{3\kappa}{4} A_{eba} \begin{pmatrix} -h_{ce}^{(1)}(r_{1\mathrm{st}}^a - h_a) & h_{ce}^{(1)} r_{2\mathrm{st}}^a \\ h_{ce}^{(2)} r_{2\mathrm{st}}^a & h_{ce}^{(2)}(r_{1\mathrm{st}}^a - h_a) \end{pmatrix}_{\sigma\beta} \tag{24}$$

Only the matrix $F_{\sigma\beta}$ depends here on fields σ, ϕ. By absorbing in C the irrelevant field independent part of $2G\lambda_{nm}^{\alpha\beta}$, and expanding the logarithm in the representation $\det(1+F) = \exp[\mathrm{tr}\ln(1+F)]$, one can obtain finally for the complete action

$$S_r = \int d^4x \left\{ \mathcal{L}_r(r_{\mathrm{st}}) + \frac{a}{2G^2} \sum_{n=1}^{\infty} \frac{(-1)^n}{n} \mathrm{tr}\left[F_{\alpha\beta}^n(r_{\mathrm{st}})\right] \right\} \tag{25}$$

proposing that the undetermined dimensionless constant a will be fixed by confronting the model with experiment afterwards [9].

3 The Ground State

Let us study the ground state of the model under consideration, then properties of the excitations will follow naturally. In Eq. (12) the field coefficients h_i obey

$$2Gh_i - \Delta_i = \frac{\kappa}{8} t_{ijk} h_j h_k \tag{26}$$

with the totally symmetric coefficients t_{ijk} equal to zero except for $t_{uds} = 1$, and with order parameters $\Delta_i \neq 0 \quad (i = u, d, s)$.

A tadpole graphs calculation gives for the gap equations the following result

$$2h_i + \frac{3a\kappa}{8G^2}\left(h_{ab}^{(2)} - h_{ab}^{(1)}\right) A_{abc} h_{ci}^{(1)} = \frac{N_c}{2\pi^2} m_i J_0(m_i^2)\,, \tag{27}$$

where the left hand side is the contribution from (25) and the right hand side is the contribution of the quark loop from (10), [8].

The second term on the left hand side of Eq. (27) is the correction resulting from the Gaussian integrals of the steepest descent method, comprising the effects of small fluctuations around the stationary path. If one puts for a moment $a = 0$ in Eq. (27), and combines the result with Eqs. (26), one finds gap equations which are very similar to the ones obtained in [4]. For this case one obtains values of (m_u, m_s) which are uniquely related to values of (G, κ).

The general case, which we have when $a > 0$ in Eq. (27), yields in turn a region for m_u, m_s, where three values of couplings (G, κ) are possible.

Conversely, one can study the solutions $m_u = m_u(G,\kappa)$, $m_s = m_s(G,\kappa)$, at fixed values of input parameters: $\Lambda, \hat{m}_u = \hat{m}_d, \hat{m}_s$. Again we find several extremal solutions. The minima will be identified in a future analysis of the effective potential. We refer to [8] for graphical displays.

Multivalued solutions of the gap equations have been obtained in a different context in [10].

4 Concluding remarks

The purpose of this work has been twofold. Firstly we have developed the technique which is necessary to go beyound the lowest order SPA in the problem of the path integral bosonization of the 't Hooft six quark interaction. This technique is rather general and can be readily used in other applications. Second, we have explored in considerable detail the implications of taking the quantum fluctuations into account in the description of the hadronic vacuum. We encountered several classes of solutions. These multiple vacua may have very interesting physical consequences and applications.

References

1. A.M. Polyakov: Phys. Lett. **B 59** 82 (1975); Nucl. Phys. **B 121** 429 (1977). A.A. Belavin, A.M. Polyakov, A. Schwartz, Y. Tyupkin: Phys. Lett. **B 59** 85 (1975). G. 't Hooft: Phys. Rev. Lett. **37** 8 (1976); Phys. Rev. **D 14** 3432 (1976). C. Callan, R. Dashen, D.J. Gross: Phys. Lett. **B 63** 334 (1976). R. Jackiw, C. Rebbi: Phys. Rev. Lett. **37** 172 (1976). S. Coleman, *The uses of instantons*, In: *Aspects of Symmetry: Selected Erice Lectures*. Cambridge: Cambridge University Press 1985
2. D. Diakonov: *Chiral symmetry breaking by instantons*. In: *Selected Topics in Non Perturbative QCD*, Lectures at the International School of Physics Enrico Fermi, Varenna, June 27 - July 7 (1995). Amsterdam: IOS Press 1997; `hep-ph/9602375`
3. V.G. Vaks, A.I. Larkin: ZhETF **40** 282 (1961). Y. Nambu, G. Jona-Lasinio: Phys. Rev. **122** 345 (1961); **124** 246 (1961)
4. V. Bernard, R.L. Jaffe, U.-G. Meißner: Nucl. Phys. **B 308** 753 (1988)
5. T. Kunihiro, T. Hatsuda: Phys. Lett. **B 206** 385 (1988). T. Hatsuda: Phys. Lett. **B 213** 361 (1988). Y. Kohyama, K. Kubodera, M. Takizawa: Phys. Lett. **B 208** 165 (1988). M. Takizawa, Y. Kohyama, K. Kubodera: Prog. Theor. Phys. **82** 481 (1989)
6. H. Reinhardt, R. Alkofer: Phys. Lett. **B 207** 482 (1988)
7. A.A. Osipov, B. Hiller: Phys. Rev. **D 62** 114013 (2000); Phys. Rev. **D 63** 094009 (2001)
8. A.A. Osipov, B. Hiller: Phys. Lett. **B 539** 76 (2002)
9. R. Jackiw: Int. J. Mod. Phys. **B 14** 2011 (2000); `hep-th/9903044`
10. P.J. Bicudo, J.E. Ribeiro, A.V. Nefediev: Phys. Rev. **D 65** 085026 (2002)

Few-Body Systems Suppl. 15, 201–206 (2003)

Few-Body Systems

Does it make any sense to talk about a Δ-isobar?

Frieder Kleefeld*

Centro de Física das Interacções Fundamentais, Instituto Superior Técnico, Edifício Ciência, Piso 3, Av. Rovisco Pais, P-1049-001 LISBOA, Portugal

Abstract. It is shortly investigated, on what basis an experimentally observed resonance like the $\Delta(1232)$-isobar can be embedded into the framework of Quantum Theory (QT), i.e. Quantum Field Theory (QFT) and Quantum Mechanics (QM). After a short discussion of the particle concept in the context of the "bootstrap" idea of G. Chew and S. Mandelstam we will focus on the theoretical formalism being necessary to describe resonances in the Lagrangian or Hamiltonian formulation of QT.

1 Why this title?

During a workshop on the (anticipated) occasion of the 65th birthday of Peter U. Sauer we want to recall the intriguing and revealing question raised by Peter U. Sauer [1] and collegues in the year 1990 on the resonance having excited him most of all: *"Does it make any sense to talk about $N\Delta$ phase shifts?"*. In order to be allowed to ask this question we first have to explore the underlying question: *"Does it make any sense to talk about a Δ-isobar?"* or *"Do unstable particles in physics have experimentally and theoretically a well defined meaning?"*.

The positive answer to the last question sketched in this short presentation will be based on previous work [2, 3, 4] (and references therein). Certain technical aspects will be displayed in more detail.

2 Particle Concept and Bootstrap

Experimentally it seems to be a not well defined problem to classify what is a physical "particle" and to specify most decisively its mass properties in physical data tables. The most accurate theoretical definition of a particle has been given by G. Chew (see ref. [5]) who associated "particles" with poles of the scattering matrix. According to this definition the *Review of Particle Properties* [6] contains not only a small number of "stable", i.e. "elementary" particles

* *E-mail address:* kleefeld@cfif.ist.utl.pt

(which play the quantum theoretical role of "asymptotic states") yet lists also a vast majority of "particles", which are not elementary. Such particles are either considered to be "unstable" or — which is intimately related — "composite" [2, 7] (topical examples are e.g. scalar mesons [8, 9, 10]). As the parametrization of data in terms of (improved) non-relativistic Breit-Wigner fits being close to the idea of G. Chew fails in the description of processes of relativistic and strongly interfering unstable "particles" the developement of an adequate consistent relativistic formalism for such "particles" has been imperative. Consider a simple non-relativistic scattering problem at an energy at which the T-matrix can be described to a good approximation by a Breit-Wigner amplitude. In this case the Breit-Wigner amplitude describes the full T-matrix already at "tree-level". The Breit-Wigner amplitude itself being determined by the solution of a Lippmann-Schwinger equation is perturbatively expanded into a Born series. Correspondingly, in the S-matrix picture of G. Chew the "effective action" which may be descibed by "particles" in the sense of G. Chew (which need not all to be stable) describes a scattering process already at "tree-level" while the "effective action" is usually perturbatively obtained from the generating functional of an interacting theory of asymptotic fields[1]. The scenario that G. Chew's "particles" may be (at least approximately) chosen such that they describe scattering problems already (at least approximately) at "tree-level" may be summarized under the term "bootstrap" introduced by G. Chew and S. Mandelstam [11] (see also ref. [5]). A nice feature of such quasi-bootstrap theories of (local) Chew "particles" is that even for large coupling constants physics is already exhaustively described at "tree-level"[2]. Yet there emerges the theoretically demanding consequence that not only self-energies, but also coupling constants may be complex valued. In order to handle such theories without getting in conflict with unitarity, causality, Lorentz covariance, locality, positivity, renormalizability etc. it is not only convenient, but imperative to achieve a consistent non-Hermitian prescription of QT in the context of a Lagrangian or Hamiltonian framework. The solution to this problem has been illuminated to a certain extent in refs. [2, 3, 4]. It consists of the replacement of standard Hermitian QT by an (anti)causal formulation of QT, in which the underlying causal and anticausal Lagrangians (or Hamiltonians) are non-Hermitian. As a consequence the non-Hermitian causal and anticausal fields being solutions of the causal and anticausal non-Hermitian Lagrange equations of motion represent elements of a generalized (anti)causal, i.e. biorthogonal Fock space. This Fock space contains elementary ("quasi-Hermitian") *and* unstable ("non-Hermitian") fields. As stated above we should bear in mind that the calculation of scattering or boundstate problems performed only in the "stable" particle basis is usually highly non-perturbative, while the inclusion of unstable particles might bring us by bootstrapping close to tree-level.

[1]Here we assume the existence of a Lagrangian and fields which is in general not guaranteed.
[2]A typical example for a bootstrap theory is the "quark-level linear sigma model" (see e.g. ref. [10, 12] and references therein). Even for Hermitian fields nature is well described at tree-level and loop-corrections are estimated to be very small [13].

3 The (anti)causal Klein-Gordon (KG) field ("Nakanishi-field")

In order to tackle a spin-3/2-isospin-3/2-resonance like the $\Delta(1232)$ we first have to understand the properties of underlying (iso)spin 0, 1/2, 1 resonances. The (anti)causal spin 0 field has been studied for the first time[3] by N. Nakanishi [14, 15] in the context of his "Complex-Ghost Relativistic Field Theory" defining a *complex* (resonance) *mass* $M = m - i\,(\Gamma/2)$ and denoting the "free" Lagrangian for the respective causal (anticausal) KG field $\phi_r(x)$ ($\phi_r^+(x)$) with isospin $(N-1)/2$ as $(r = 1, \ldots, N)$:

$$\mathcal{L}(x) = \sum_r \frac{1}{2}\Big\{(\partial\phi_r(x))^2 - M^2(\phi_r(x))^2 + (\partial\phi_r^+(x))^2 - M^{*\,2}(\phi_r^+(x))^2\Big\}$$

By variation of the action one obtains the respective causal and anticausal KG-equations $(\,\partial^2 + M^2)\,\phi_r(x) = 0$ and $(\,\partial^2 + M^{*\,2})\,\phi_r^+(x) = 0$, respectively. As $\mathrm{Im}\,[M] = -\Gamma/2 \neq 0$ they are solved by a *Laplace transform*[4,5]:

$$\begin{aligned}
\phi_r(x) &= \int \frac{d^4p}{(2\pi)^3}\ \text{``}\,\delta(p^2 - M^2)\,\text{''}\ a_r\,(p)\,e^{-ipx} \\
&\stackrel{!}{=} \int \frac{d^3p}{(2\pi)^3\,2\,\omega(\boldsymbol{p})}\ \Big[a_r(\boldsymbol{p})\ e^{-ipx} + c_r^+(\boldsymbol{p})\ e^{+ipx}\Big]\Big|_{p^0=\omega(\boldsymbol{p})} \\
\phi_r^+(x) &\stackrel{!}{=} \int \frac{d^3p}{(2\pi)^3\,2\,\omega^*(\boldsymbol{p})}\ \Big[c_r(\boldsymbol{p})\ e^{-ip^*x} + a_r^+(\boldsymbol{p})\ e^{+ip^*x}\Big]\Big|_{p^0=\omega(\boldsymbol{p})}
\end{aligned}$$

The Canonical conjugate momenta to $\phi_r(x)$ and $\phi_r^+(x)$ are:

$$\Pi_r(x) := \frac{\delta\,\mathcal{L}(x)}{\delta\,(\partial_0\,\phi_r(x))} \stackrel{!}{=} \partial_0\,\phi_r(x)\ , \quad \Pi_r^+(x) := \frac{\delta\,\mathcal{L}(x)}{\delta\,(\partial_0\,\phi_r^+(x))} \stackrel{!}{=} \partial_0\,\phi_r^+(x)$$

The consistent (anti)causal quantization requires *even* for *resonance* fields *equal-time* commutation relations. The respective non-vanishing relations are:

$$[\,\phi_r(\boldsymbol{x},t), \Pi_s(\boldsymbol{y},t)\,] = i\,\delta^3(\boldsymbol{x}-\boldsymbol{y}\,)\,\delta_{rs}\ ,\ [\,\phi_r^+(\boldsymbol{x},t), \Pi_s^+(\boldsymbol{y},t)\,] = i\,\delta^3(\boldsymbol{x}-\boldsymbol{y}\,)\,\delta_{rs}$$

Hence the non-vanishing commutation relations in momentum space are:

$$\begin{aligned}
[\,a_r(\boldsymbol{p}), c_s^+(\boldsymbol{p}')\,] &= (2\pi)^3\,2\omega\,(\boldsymbol{p})\ \delta^3(\boldsymbol{p}-\boldsymbol{p}'\,)\,\delta_{rs} \\
[\,c_r(\boldsymbol{p}), a_s^+(\boldsymbol{p}')\,] &= (2\pi)^3\,2\omega^*(\boldsymbol{p})\,\delta^3(\boldsymbol{p}-\boldsymbol{p}')\,\delta_{rs}
\end{aligned}$$

The Hamilton operator is derived in the standard way from the Lagrangian[6,7]:

$$H = \sum_r \int d^3p\,\Big(\,\frac{1}{2}\,\omega(\boldsymbol{p})\,\Big\{c_r^+(\boldsymbol{p}), a_r(\boldsymbol{p})\Big\} + \frac{1}{2}\,\omega^*(\boldsymbol{p})\,\Big\{a_r^+(\boldsymbol{p}), c_r(\boldsymbol{p})\Big\}\,\Big)$$

---

[3]The considerations in the 2nd article of ref. [2] have been done without knowing refs. [14, 15].

[4]Let $\omega(\boldsymbol{p}) := \sqrt{\boldsymbol{p}^2 + M^2}$ ($\omega(\boldsymbol{0}) = M$) and $a_r(\boldsymbol{p}) := a_r(p)|_{p^0=\omega(\boldsymbol{p})}$, $c_r^+(\boldsymbol{p}) := a_r(-p)|_{p^0=\omega(\boldsymbol{p})}$.

[5]For details with respect to the Laplace transform involving a symbolic δ-distribution denoted as "$\delta(p^2 - M^2)$" for the complex resonance mass M see refs. [16, 15, 2]!

[6]$H = \int d^3x\,\{\sum_r(\Pi_r(x)\,(\partial_0\,\phi_r(x)) + (\partial_0\,\phi_r^+(x))\,\Pi_r^+(x)\,) - \mathcal{L}(x)\}$

[7]In 1-dim. QM we have $H = \frac{1}{2}\,\omega\,[c^+, a]_\pm + \frac{1}{2}\,\omega^*\,[a^+, c]_\pm$ ($\pm$ for Bosons/Fermions with $[c, a^+]_\mp = 1$). This (anti)causal Harmonic Oscillator is diagonalized by the (normalized) right eigenstates $|n, m\rangle = \frac{1}{\sqrt{n!\,m!}}\,(c^+)^n(a^+)^m\,|0\rangle$ with $E_{nm} = \omega\,(n \pm \frac{1}{2}) + \omega^*\,(m \pm \frac{1}{2})$.

Standard real-time ordering leads to the causal "Nakanishi" propagator:

$$\langle 0|\, T\,(\,\phi_r\,(x)\,\phi_s\,(y)\,)\,|0\rangle \stackrel{!}{=} i\int \frac{d^4p}{(2\,\pi)^4}\,\frac{e^{-i\,p(x-y)}}{p^2-M^2}\;\delta_{rs}$$

The anticausal propagator is obtained by Hermitian conjugation or by a vacuum expectation value of an *anti-real-time ordered* product of two anticausal fields.

4 Lorentz boost, spinors and polarization vectors

The definition of spinors and polarization vectors for spin 1/2 and 1 resonances requires the introduction of the respective concept of a Lorentz boost of unstable fields (see e.g. refs. [2]). For any symmetric metric $g_{\mu\nu}$ Lorentz transformations (LTs) $\Lambda^{\mu}{}_{\nu}$ are defined by $\Lambda^{\mu}{}_{\rho}\, g_{\mu\nu}\, \Lambda^{\nu}{}_{\sigma} = g_{\rho\sigma}$. Let n^{μ} be a timelike unit 4-vector ($n^2 = 1$) and ξ^{μ} an *arbitrary complex* 4-vector with $\xi^2 \neq 0$. The 4 independent LTs[8] relating ξ^{μ} with its "restframe" (i.e. $\xi^{\,\mu} = \Lambda^{\mu}{}_{\nu}(\xi)\; n^{\,\nu} \sqrt{\xi^2}$ and $n_{\,\nu}\sqrt{\xi^2} = \xi_{\,\mu}\, \Lambda^{\mu}{}_{\nu}(\xi)$) are ($P^{\,\mu}_{\;\;\nu} := 2\, n^{\,\mu}\, n_{\,\nu} - g^{\,\mu}{}_{\nu}$ = reflection matrix):

$$\begin{aligned}
\Lambda^{\mu}{}_{\nu}(\xi) &= \pm\left\{ g^{\,\mu}{}_{\rho} - \frac{\sqrt{\xi^2}}{\sqrt{\xi^2}\mp\xi\cdot n}\left[n^{\,\mu}\mp\frac{\xi^{\,\mu}}{\sqrt{\xi^2}}\right]\left[n_{\,\rho}\mp\frac{\xi_{\,\rho}}{\sqrt{\xi^2}}\right]\right\} P^{\,\rho}{}_{\nu} \\
\Lambda^{\mu}{}_{\nu}(\xi) &= \pm\left\{ g^{\,\mu}{}_{\nu} - \frac{\sqrt{\xi^2}}{\sqrt{\xi^2}\mp\xi\cdot n}\left[n^{\,\mu}\mp\frac{\xi^{\,\mu}}{\sqrt{\xi^2}}\right]\left[n_{\,\nu}\mp\frac{\xi_{\,\nu}}{\sqrt{\xi^2}}\right]\right\}
\end{aligned}$$

Let ξ^{μ} be e.g.[9] the 4-momentum p^{μ} of a"force-free"[10] *unstable* particle with complex mass M in a $(+,-,-,-)$ metric. Its restframe is defined by $n^{\mu} = (1,\mathbf{0})^{\mu}$ [11]. For such a spin 1/2 Fermion we define spinors $u(p) \equiv v(-p)$ and their transpose (i.e. $\overline{u^c}(p) := u^T(p)\; C$ with $C = i\,\gamma^{\,2}\,\gamma^{\,0}$) by the generalized Dirac equation $(\not p - \sqrt{p^2}\,)u(p) = 0$ [12]. Lorentz covariance of this Dirac equation requires $S^{-1}(\Lambda(p))\,\gamma^{\,\mu}\, S(\Lambda(p)) = \Lambda^{\mu}{}_{\nu}(p)\,\gamma^{\,\nu}$ for $u(p,s) = S(\Lambda(p))\; u(\sqrt{p^2}\, n, s)$ resulting for the metric $(+,-,-,-)$ and the proper orthochronous LTs in

$$\begin{aligned}
u(p,s)|_{\mathrm{Re}[p^0]>0} &= \frac{\not p + \sqrt{p^2}}{\sqrt{2\sqrt{p^2}\,(p^0+\sqrt{p^2})}}\, u(\sqrt{p^2}\, n, s) = \frac{\not p + \sqrt{p^2}}{\sqrt{p^0+\sqrt{p^2}}}\begin{pmatrix}\chi_s\\ 0\end{pmatrix} \\
\overline{v^c}(p,s)|_{\mathrm{Re}[p^0]>0} &= \overline{v^c}(\sqrt{p^2}\, n, s)\frac{\not p + \sqrt{p^2}}{\sqrt{2\sqrt{p^2}\,(p^0+\sqrt{p^2})}} = (\chi_s^+, 0)\,\frac{\not p + \sqrt{p^2}}{\sqrt{p^0+\sqrt{p^2}}}.
\end{aligned}$$

---

[8] They are well known as ortho-chronous/non-ortho-chronous proper/improper LTs.

[9] In "Thermal Field Theory" the inverse temperature $\beta = 1/T$ may be treated as an *imaginary time*. Hence $\xi^{\mu} = (t - i\,\beta, \boldsymbol{x}\,)^{\mu}$ and $\sqrt{\xi^2}\, n^{\mu} = (\sqrt{(t-i\,\beta)^2 - \boldsymbol{x}^{\,2}}, \mathbf{0})^{\mu}$ are related by a LT!

[10] A remarkable "bootstrap" feature of (anti)causal QT is that a boost of such resonances does *not* produce particles — in contrast to the "traditional" boost of interacting particles!

[11] The space-time trajectories of such a particle are determined by the condition $p\cdot x = const$, or — in 1 dimension — by $x^1 = (\omega(\mathbf{p})/p^1)\, t + const$. If the movement proceeds along the *real time axis*, the quantity $\omega(\mathbf{p})/p^1$ has the interpretation of a *complex* velocity.

[12] Note that $u^c(p) = u(-p^*) = v(p^*)$. The spin projections are introduced such that for $\mathrm{Re}[p^0] \neq 0$ there holds the spinor *normalization* condition $\mathrm{sgn}(\mathrm{Re}[p^0])\, \sum_s u(p,s)\, \overline{v^c}(p,s) = \not p + \sqrt{p^2}$. The restframe 2-spinors used above obey $\chi_s^+ \chi_{s'} = \delta_{ss'}$ and $\sum_s \chi_s \chi_s^+ = 1_2$.

The spin-projection vector $s^{\mu}(p) = \Lambda^{\mu}_{\ \nu}(p)\ s^{\nu}(\sqrt{p^2},\mathbf{0}) = \Lambda^{\mu}_{\ \nu}(p)\,(0,\boldsymbol{s})^{\nu}$ with $|\boldsymbol{s}|^2 = 1$ obeys $s^{\mu}(p)s_{\mu}(p) = -1$ and $p^{\mu}\,s_{\mu}(p) = 0$. For Re$[p^0] > 0$ there holds $\gamma_5\ \not{s}(p)\,u(p,s) = 2s\ u(p,s)$ and $\gamma_5\ \not{s}(p^*)\,v^c(p,s) = 2s\ v^c(p,s)$ (with $s = \pm 1/2$). Similarly, defining 3 orthonormal Cartesian unitvectors $\boldsymbol{e}^{(i)}$ $(i = x,y,z)$ by $\boldsymbol{e}^{(i)} \cdot \boldsymbol{e}^{(j)} = \delta^{ij}$ we may introduce polarization vectors by $\varepsilon^{\mu(i)}(p) := \Lambda^{\mu}_{\ \nu}(p)\ \ \varepsilon^{\nu\,(i)}(\sqrt{p^2},\mathbf{0}) = \Lambda^{\mu}_{\ \nu}(p)\,(0,\boldsymbol{e}^{(i)})^{\nu}$ fulfilling $\varepsilon^{\mu\,(i)}(p)\,\varepsilon_{\mu}^{(j)}(p) = -\,\delta^{ij}$, $p^{\mu}\varepsilon_{\mu}^{(i)}(p) = 0$ and $\sum_i \varepsilon^{\mu\,(i)}(p)\,\varepsilon^{\nu\,(i)}(p) = -g^{\mu\nu} + (p^{\mu}p^{\nu})/p^2$. All these identities can be projected on the complex mass shell $p^2 = M^2$, e.g. for $p^0 = \omega(\boldsymbol{p})$.

5 The (anti)causal Dirac and Proca field

As shown in refs. [2, 3, 4] the (anti)causal Lagrangian for an isospin $(N-1)/2$ spin $1/2$ Fermionic resonance is given by $(r = 1,\ldots,N)(\bar{M} := \gamma_0\ M^+\gamma_0)$:

$$\mathcal{L}(x) = \sum_r \frac{1}{2}\left(\overline{\psi^c_r}(x)\,(\frac{1}{2}\,i\,\overleftrightarrow{\not{\partial}} - M)\,\psi_r(x)\ +\ \overline{\psi}_r(x)\,(\frac{1}{2}\,i\,\overleftrightarrow{\not{\partial}} - \bar{M})\,\psi^c_r(x)\,\right)$$

As a whole we have four Lagrange equations of motion, i.e. $(i\,\not{\partial} - M)\,\psi_r(x) = 0$ and $(i\,\not{\partial} - \bar{M})\,\psi^c_r(x) = 0$ and the two *transposed* equations for $\overline{\psi^c_r}(x)$ and $\overline{\psi}_r(x)$. As for the KG field their solution is obtained by a *Laplace transform* yielding[13]:

$$\begin{aligned}
\psi_r(x) &= \sum_s \int \frac{d^3p}{(2\pi)^3\,2\omega(\boldsymbol{p})}\ [e^{-iqx}b_r(\boldsymbol{p},s)\,u(\boldsymbol{p},s) + e^{iqx}d^+_r(\boldsymbol{p},s)\,v(\boldsymbol{p},s)] \\
\psi^c_r(x) &= \sum_s \int \frac{d^3p}{(2\pi)^3\,2\omega^*(\boldsymbol{p})}\ [e^{iq^*x}b^+_r(\boldsymbol{p},s)\,u^c(\boldsymbol{p},s) + e^{-iq^*\cdot x}d_r(\boldsymbol{p},s)\,v^c(\boldsymbol{p},s)] \\
\overline{\psi_r}(x) &= \sum_s \int \frac{d^3p}{(2\pi)^3\,2\omega^*(\boldsymbol{p})}\ [e^{iq^*x}b^+_r(\boldsymbol{p},s)\,\bar{u}(\boldsymbol{p},s) + e^{-iq^*x}d_r(\boldsymbol{p},s)\,\bar{v}(\boldsymbol{p},s)] \\
\overline{\psi^c_r}(x) &= \sum_s \int \frac{d^3p}{(2\pi)^3\,2\omega(\boldsymbol{p})}\ [e^{-iqx}b_r(\boldsymbol{p},s)\,\overline{u^c}(\boldsymbol{p},s) + e^{iqx}d^+_r(\boldsymbol{p},s)\,\overline{v^c}(\boldsymbol{p},s)]
\end{aligned}$$

Canonical ("*real-equal time*") quantization leads — due to Fermi-statistics — to the following non-vanishing momentum-space anticommutation relations:

$$\begin{aligned}
\{\,b_r(\boldsymbol{p},s), d^+_{r'}(\boldsymbol{p}',s')\,\} &= (2\pi)^3\,2\,\omega\,(\boldsymbol{p})\ \ \delta^3(\boldsymbol{p}-\boldsymbol{p}'\,)\ \delta_{rr'}\ \delta_{ss'} \\
\{\,d_r(\boldsymbol{p},s), b^+_{r'}(\boldsymbol{p}',s')\,\} &= (2\pi)^3\,2\,\omega^*(\boldsymbol{p})\ \delta^3(\boldsymbol{p}-\boldsymbol{p}'\,)\ \delta_{rr'}\ \delta_{ss'}
\end{aligned}$$

Convince yourself that — as for the causal KG field — the propagator of a causal spin 1/2 Fermion is obtained by *standard Fermionic real-time ordering*[14]:

$$\langle 0|\,T\,(\,(\psi_r(x))_\alpha\,(\overline{\psi^c_s}(y))_\beta\,)\,|0\rangle \stackrel{!}{=} i\int \frac{d^4p}{(2\,\pi)^4}\ \frac{e^{-i\,p(x-y)}}{p^2 - M^2}\ (\not{p} + M)_{\alpha\beta}\ \delta_{rs}$$

---

[13] Here we have defined $q^{\mu} := (\omega(\boldsymbol{p}),\boldsymbol{p})^{\mu} = p^{\mu}|_{p^0=\omega(\boldsymbol{p})}$ and $b_r(\boldsymbol{p},s) := b_r(p,s)|_{p^0=\omega(\boldsymbol{p})}$, $d^+_r(\boldsymbol{p},s) := b_r(-p,s)|_{p^0=\omega(\boldsymbol{p})}$ and $u(\boldsymbol{p},s) := u(p,s)|_{p^0=\omega(\boldsymbol{p})}$, $v(\boldsymbol{p},s) := v(p,s)|_{p^0=\omega(\boldsymbol{p})}$.

[14] The anticausal propagator is obtained by Hermitian conjugation or by a vacuum expectation value of a *anti-real-time ordered* product of two anticausal fields.

As discussed in ref. [2] we may use the formalism of Jun-Chen Su [17] to introduce — with the polarization vectors derived above — a renormalizable and unitary Lagrangian for a massive (anti)causal vector field. In combining our results for a spin 0, 1/2 and 1 resonance the door of (anti)causal QT is opened widely for an (anti)causal Δ-isobar.

Acknowledgement. This work has been supported by the *Fundação para a Ciência e a Tecnologia* (FCT) of the *Ministério da Ciência e da Tecnologia (e do Ensinio Superior)* of Portugal, under Grant no. PRAXIS XXI/BPD/20186/99 and SFRH/BDP/9480/2002.

References

1. H. Garcilazo et al.: Phys. Rev. **C 42**, 2315 (1990)
2. F. Kleefeld: hep-ph/0211460; F. Kleefeld, E. van Beveren, G. Rupp: Nucl. Phys. **A 694**, 470 (2001)
3. F. Kleefeld: Doctoral Thesis (University of Erlangen-Nürnberg, Germany, 1999); F. Kleefeld: Acta Physica Polonica **B 30**, 981 (1999)
4. F. Kleefeld: pp. 69-77 in Vol. I of Proc. XIV Int. Seminar on High En. Phys. Probl. (ISHEPP 98), 17.-22.8.1998, JINR, Dubna, Russia, nucl-th/9811032
5. F. Capra: "Bootstrap physics: a conversation with Geoffrey Chew", in Berkeley Chew Jubilee 1984:247 (LBL-18372), KEK-library 198412075; B. Diu: "Qu'est-ce qu'une particule êlêmentaire?", collection de monographies de physique publiée sous la direction de P. Aigrain, A. Blanc-Lapierre, J. Friedel, M. Lévy, ©1965 par Masson et Cie – Éditeurs, Paris
6. K. Hagiwara et al. (Particle Data Group): Phys. Phys. **D66**, 010001 (2002)
7. N.D. Hari Dass and V. Soni: hep-th/0204177
8. E. van Beveren et al.: hep-ph/0211411
9. F. Kleefeld et al.: Phys. Rev. **D 66**, 034007 (2002)
10. E. van Beveren et al.: Mod. Phys. Lett. **A 17**, 1673 (2002)
11. G.F. Chew and S. Mandelstam: Nuovo Cim. **19**, 752 (1961)
12. M.D. Scadron et al.: hep-ph/0211275
13. L.-H. Chan, R. W. Haymaker: Phys. Rev. **D 7**, 402 (1973)
14. N. Nakanishi: Phys. Rev. **D 5**, 1968 (1972)
15. N. Nakanishi: Prog. Theor. Phys. Suppl. **51**, 1 (1972)
16. N. Nakanishi: Prog. Theor. Phys. **19**, 607 (1958)
17. Jun-Chen Su: hep-th/9805195, 9805192, 9805193, 9805194.

Few-Body Systems Suppl. 15, 207–218 (2003)

Few-Body Systems

Relativity, Chiral Symmetry, and the Nucleon Electromagnetic Form Factors

Gerald A. Miller*

Department of Physics, University of Washington Seattle, WA 98195-1560

Abstract. A relativistic interpretation for why the proton's G_E/G_M falls and QF_2/F_1 is approximately constant is presented. Reproducing the observed G_E^n mandates the inclusion of the effects of the pion cloud. The full relativistic model with a pion cloud provides a good reproduction of all of the nucleon electromagnetic form factors.

1 Introduction

An alternate title could be "Surprises in the Proton". This talk owes its existence to the precise, stunning and exciting recent experimental work on measuring G_E/G_M (or QF_2/F_1) for the proton and G_E, G_M for the neutron. My goal here is to interpret the data. Symmetries including Poincaré invariance and chiral symmetry will be the principal tool I'll use. This talk is based on three papers [1],[2],[3].

If, a few years ago, one had asked participants at a meeting like this about the Q^2 dependence of the proton's G_E/G_M or QF_2/F_1. Almost everyone one have answered that for large enough values of Q^2, G_E/G_M would be flat and QF_2/F_1 would fall with increasing Q^2. The reason for the latter fall being conservation of hadron helicity. Indeed, the shapes of the curves have been obtained in the new measurements, except for the mis-labeling of the ordinate axes. The expected flatness of G_E/G_M holds for QF_2/F_1, and the quantity G_E/G_M falls rapidly and linearly with Q^2. This behavior needs to be understood!

2 Outline

I will begin with a brief introduction to Light Front Physics. Then I will discuss a particular relativistic model of the nucleon, and proceed to apply it to the proton form factors, with the aim of providing a qualitative understanding

* *E-mail address:* miller@phys.washington.edu

of the salient experimental features. The same model fails to reproduce the neutron $G_{\rm E}$ unless the effects of the pion cloud are included. The combination of relativistic effects with those of the pion cloud lead to a model that is able to reproduce all of the nucleon electromagnetic form factors.

3 Light Front

Light-front dynamics is a relativistic many-body dynamics in which fields are quantized at a "time"$=\tau = x^0 + x^3 \equiv x^+$. The τ-development operator is then given by $P^0 - P^3 \equiv P^-$. These equations show the notation that a four-vector A^μ is expressed as $A^\pm \equiv A^0 \pm A^3$. One quantizes at $x^+ = 0$ which is a light-front, hence the name "light front dynamics". The canonical spatial variable must be orthogonal to the time variable, and this is given by $x^- = x^0 - x^3$. The canonical momentum is then $P^+ = P^0 + P^3$. The other coordinates are $\mathbf{x}_\perp$ and $\mathbf{P}_\perp$.

The most important consequence of this is that the relation between energy and momentum of a free particle is given by: $p_\mu p^\mu = m^2 = p^+p^- - p_\perp^2 \rightarrow p^- = (p_\perp^2 + m^2)/p^+$, a relativistic kinetic energy which does not contain a square root operator. This allows the separation of center of mass and relative coordinates, so that the computed wave functions are frame independent.

The use of the light front is particularly relevant for calculating form factors, which are probability amplitudes for an nucleon to absorb a four momentum q and remain a nucleon. The initial and final nucleons have different total momenta. This means that the final nucleon is a boosted nucleon, with different wave function than the initial nucleon. In general, performing the boost is difficult for large values of $Q^2 = -q^2$. However the light front technique allows one to set up the calculation so that the boosts are independent of interactions. Indeed, the wave functions are functions of relative variables and are independent of frame.

4 Definitions

Let us define the basic quantities concerning us here. These are the independent form factors defined by

$$\langle N, \lambda' p' \,|J^\mu|\, N, \lambda p\rangle = \bar{u}_{\lambda'}(p') \left[F_1(Q^2)\gamma^\mu + \frac{\kappa F_2(Q^2)}{2M_N} i\sigma^{\mu\nu}(p'-p)_\nu \right] u_\lambda(p). \quad (1)$$

The Sachs form factors are defined by the equations:

$$G_{\rm E} = F_1 - \frac{Q^2}{4M_N^2}\kappa F_2, \; G_{\rm M} = F_1 + \kappa F_2. \quad (2)$$

There is an alternate light front interpretation, based on field theory, in which one uses the "good" component of the current, J^+, to suppress the effects of quark-pair terms. Then, using nucleon light-cone spinors:

$$F_1(Q^2) = \frac{1}{2P^+}\langle N, \uparrow |J^+| N, \uparrow\rangle, \quad Q\kappa F_2(Q^2) = \frac{-2M_N}{2P^+}\langle N, \uparrow |J^+| N, \downarrow\rangle. \quad (3)$$

The form factor F_1 is obtained from the non-spin flip matrix element, while F_2 is obtained from the spin-flip term.

5 Three-Body Variables and Boost

We use light front coordinates for the momentum of each of the i quarks, such that $\mathbf{p}_i = (p_i^+, \mathbf{p}_{i\perp}), p^- = (p_\perp^2 + m^2)/p^+$. The total (perp)-momentum is $\mathbf{P} = \mathbf{p_1} + \mathbf{p_2} + \mathbf{p_3}$, the plus components of the momenta are denoted as

$$\xi = \frac{p_1^+}{p_1^+ + p_2^+}, \qquad \eta = \frac{p_1^+ + p_2^+}{P^+}, \tag{4}$$

and the perpendicular relative coordinates are given by

$$\mathbf{k}_\perp = (1-\xi)\mathbf{p}_{1\perp} - \xi\mathbf{p}_{2\perp}, \quad \mathbf{K}_\perp = (1-\eta)(\mathbf{p}_{1\perp} + \mathbf{p}_{2\perp}) - \eta\mathbf{p}_{3\perp}. \tag{5}$$

In the center of mass frame we find:

$$\mathbf{p}_{1\perp} = \mathbf{k}_\perp + \xi\mathbf{K}_\perp, \qquad \mathbf{p}_{2\perp} = -\mathbf{k}_\perp + (1-\xi)\mathbf{K}_\perp, \quad \mathbf{p}_{3\perp} = -\mathbf{K}_\perp. \tag{6}$$

The coordinates $\xi, \eta, \mathbf{k}, \mathbf{K}$ are all relative coordinates so that one obtains a frame independent wave function $\Psi(\mathbf{k}_\perp, \mathbf{K}_\perp, \xi, \eta)$.

Now consider the computation of a form factor, taking quark 3 to be the one struck by the photon. One works in a special set of frames with $q^+ = 0$ and $Q^2 = \mathbf{q}_\perp^2$, so that the value of $1-\eta$ is not changed by the photon. The coordinate $\mathbf{p}_{3\perp}$ is changed to $\mathbf{p}_{3\perp} + \mathbf{q}_\perp$, so only one relative momentum, $\mathbf{K}_\perp$ is changed:

$$\mathbf{K'}_\perp = (1-\eta)(\mathbf{p}_{1\perp} + \mathbf{p}_{2\perp}) - \eta(\mathbf{p}_{3\perp} + \mathbf{q}_\perp) = \mathbf{K}_\perp - \eta\mathbf{q}_\perp, \quad \mathbf{k}'_\perp = \mathbf{k}_\perp, \tag{7}$$

The arguments of the spatial wave function are taken as the mass-squared operator for a non-interacting system:

$$M_0^2 \equiv \sum_{i=1,3} p_i^- P^+ - P_\perp^2 = \frac{K_\perp^2}{\eta(1-\eta)} + \frac{k_\perp^2 + m^2}{\eta\xi(1-\xi)} + \frac{m^2}{1-\eta}. \tag{8}$$

This is a relativistic version of the square of the center-of-mass kinetic energy, expressed in terms of light-front variables. Note that the absorption of a photon changes the value to:

$${M_0'}^2 = \frac{(K_\perp - \eta q_\perp)^2}{\eta(1-\eta)} + \frac{k_\perp^2 + m^2}{\eta\xi(1-\xi)} + \frac{m^2}{1-\eta}. \tag{9}$$

6 Wave Function

Our wave function is based on symmetries. The wave function is anti-symmetric, a function of relative momenta, independent of reference frame, an eigenstate of the spin operator and rotationally invariant (in a specific well-defined sense). The use of symmetries is manifested in the construction of such wave functions, as originally described by Terent'ev [4], Coester [5] and their collaborators. A schematic form of the wave function is

$$\Psi(p_i) = \Phi(M_0^2)u(p_1)u(p_2)u(p_3)\psi(p_1,p_2,p_3), \quad p_i = \mathbf{p}_i s_i, \tau_i \tag{10}$$

where ψ is a spin-isospin color amplitude factor, the p_i are expressed in terms of relative coordinates (for example, $\mathbf{p}_{3\perp} = -\mathbf{K}_\perp$), the $u(p_i)$ are ordinary Dirac spinors and Φ is a spatial wave function. The ordinary Dirac spinors depend on the third or z component of the momenta and this is given, for example, by $p_{3z} = \frac{1}{2}[M_0(1-\eta) - \frac{m^2+K_\perp^2}{(1-\eta)M_0}]$.

We take the the spatial wave function from Schlumpf [6]:

$$\Phi(M_0) = \frac{N}{(M_0^2+\beta^2)^\gamma}, \beta = 0.607 \text{ GeV}, \ \gamma = 3.5, \ m = 0.267 \text{ GeV}. \tag{11}$$

The value of γ is chosen that $Q^4 G_\mathrm{M}(Q^2)$ is approximately constant for $Q^2 > 4 \text{ GeV}^2$ in accord with experimental data. The parameter β helps govern the values of the perp-momenta allowed by the wave function Φ and is closely related to the rms charge radius, and m is mainly determined by the magnetic moment of the proton.

At this point the wave function and the calculation are completely defined. One could evaluate the form factors as $\langle\Psi|J^+|\Psi\rangle$ and obtain the results. Actually the first results were obtained in 1995[1], and are given as Figs. 10,11 in that reference. It is clear that the model predicted that G_E falls off much more rapidly with increasing Q^2 than does G_M. The result that $Q^4 G_\mathrm{M}$ is flat at high Q^2 is a consequence of using Schlumpf's wave function, but our values of G_E for large Q^2 were a prediction.

So we had this early prediction, which was confirmed by experimental measurements. But we didn't present an explanation of the numerical results. This is the next task.

7 Simplify Calculation- Light Cone Spinors

The evaluation of the operator $J^+ \sim \gamma^+$ is simplified by using light cone spinors. These solutions of the free Dirac equation, related to ordinary Dirac spinors by a unitary transformation, conveniently satisfy:

$$\bar{u}_\mathrm{L}(p^+,\mathbf{p}',\lambda')\gamma^+ u_\mathrm{L}(p^+,\mathbf{p},\lambda) = 2\delta_{\lambda\lambda'}p^+. \tag{12}$$

To take advantage of this, re-express the wave function in terms of light-front spinors using the completeness relation: $1 = \sum_\lambda u_\mathrm{L}(p,\lambda)\bar{u}_\mathrm{L}(p,\lambda)$. We then find

$$\Psi(p_i) = u_\mathrm{L}(p_1,\lambda_1)u_\mathrm{L}(p_2,\lambda_2)u_\mathrm{L}(p_3,\lambda_3)\psi_\mathrm{L}(p_i,\lambda_i), \tag{13}$$

$$\psi_\mathrm{L}(p_i,\lambda_i) \equiv [\bar{u}_\mathrm{L}(\mathbf{p}_1,\lambda_1)u(\mathbf{p}_1,s_1)][\bar{u}_\mathrm{L}(\mathbf{p}_2,\lambda_2)u(\mathbf{p}_2,s_2)]$$

$$\times \quad [\bar{u}_\mathrm{L}(\mathbf{p}_3,\lambda_3)u(\mathbf{p}_3,s_3)]\ \psi(p_1,p_2,p_3). \tag{14}$$

This is the very same Ψ as before, it is just that now it is easy to compute the matrix elements of the γ^+ operator.

The unitary transformation is also known as the Melosh rotation. The basic point is that one may evaluate the coefficients in terms of Pauli spinors: $|\lambda_i\rangle, |s_i\rangle$, with $\langle\lambda_i|R_{\rm M}^\dagger(\mathbf{p}_i)|s_i\rangle \equiv \bar{u}_{\rm L}(\mathbf{p}_i,\lambda_i)u(\mathbf{p}_i,s_i)$. It is easy to show that

$$\langle\lambda_3|R_{\rm M}^\dagger(\mathbf{p}_3)|s_3\rangle = \langle\lambda_3|\left[\frac{m+(1-\eta)M_0+i\boldsymbol{\sigma}\cdot(\mathbf{n}\times\mathbf{p}_3)}{\sqrt{(m+(1-\eta)M_0)^2+p_{3\perp}^2}}\right]|s_3\rangle. \tag{15}$$

The important effect resides in the term $(\mathbf{n}\times\mathbf{p}_3)$ which originates from the lower components of the Dirac spinors. This large relativistic spin effect can be summarized: the effects of relativity are to replace Pauli spinors by Melosh rotation operators acting on Pauli spinors. Thus

$$|\uparrow \mathbf{p}_i\rangle \equiv R_{\rm M}^\dagger(\mathbf{p}_i)\begin{pmatrix}1\\0\end{pmatrix}, \quad |\uparrow\mathbf{p}_3\rangle \neq |\uparrow\rangle. \tag{16}$$

In the non-relativistic limit, the Melosh rotation matrices become unit operators and one recovers the familiar $SU(6)$ quark model.

8 Proton F_1, F_2-Analytic Insight

The analytic insight is based on Eq. (15). Consider high momentum transfer such that $Q = \sqrt{\mathbf{q}_\perp^2} \gg \beta = 560$ MeV. *Each* of the quantities: M_0, M_0', $\mathbf{p}_{3\perp}$, $\mathbf{p}_{3\perp}$ can be of order $q_\perp$, so the spin-flip term is as large as the non-spin flip term. In particular, $(s_3 = +1/2)$ may correspond to $(\lambda_3 = -1/2)$, so the spin of the struck quark $\neq$ proton spin. This means that there is no hadron helicity selection rule [7, 8].

The effects of the lower components of Dirac spinors, which cause the spin flip term $\boldsymbol{\sigma}\times\mathbf{p}_3$, are the same as having a non-zero L_z, if the wave functions are expressed in the light-front basis.

We may now qualitatively understand the numerical results, since

$$F_1(Q^2) = \int\frac{d^2q_\perp d\xi}{\xi(1-\xi)}\frac{d^2K_\perp d\eta}{\eta(1-\eta)}\cdots\langle\uparrow\mathbf{p}_3'|\uparrow\mathbf{p}_3\rangle \tag{17}$$

$$Q\kappa F_2(Q^2) = 2M_N\int\frac{d^2q_\perp d\xi}{\xi(1-\xi)}\frac{d^2K_\perp d\eta}{\eta(1-\eta)}\cdots\langle\uparrow\mathbf{p}_3'|\downarrow\mathbf{p}_3\rangle, \tag{18}$$

where the $\cdots$ represents common factors. The term $F_1 \sim \langle\uparrow\mathbf{p}_3'|\uparrow\mathbf{p}_3\rangle$ is a spin-non-flip term and $QF_2 \sim \langle\uparrow\mathbf{p}_3'|\downarrow\mathbf{p}_3\rangle$ depends on the spin-flip term. In doing the integral each of the momenta, and M_0, M_0' can take the large value Q for some regions of the integration. Thus in the integral

$$\langle\uparrow\mathbf{p}_3'|\uparrow\mathbf{p}_3\rangle \sim \frac{Q}{Q}, \quad \langle\uparrow\mathbf{p}_3'|\downarrow\mathbf{p}_3\rangle \sim \frac{Q}{Q}, \tag{19}$$

so that F_1 and QF_2 have the same Q^2 dependence. This is shown in Fig. 1. Indeed, for Q^2 greater than about 2 GeV^2, the ratio QF_2/F_1 varies very little with increasing Q^2.

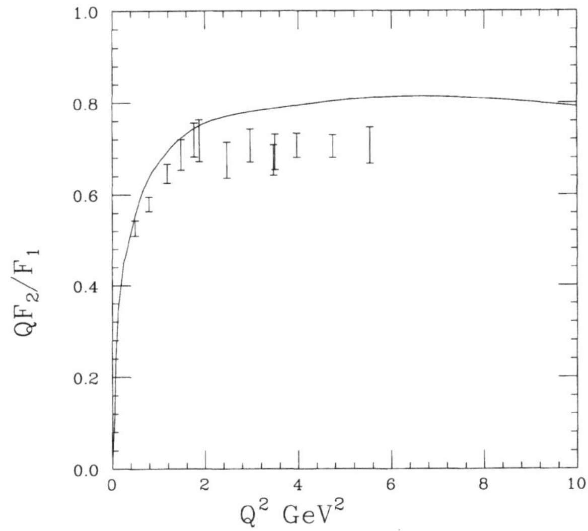

Figure 1. Calculation of refs. [1, 2], data are from Jones et al. [9] for $2 \leq Q^2 \leq 3.5$ GeV^2 and from Gayou et al. [10]. for $3.5 \leq Q^2 \leq 5.5$ GeV^2.

9 Neutron Charge Form Factor

The neutron has no charge, $G_{\rm E}^n(Q^2 = 0) = 0$, and the square of its charge radius is determined from the low Q^2 limit as $G_{\rm E}^n(Q^2) \to -Q^2R^2/6$. The quantity R^2 is well-measured [11] as $R^2 = -0.113 \pm 0.005$ fm^2. The Galster parameterization [12] has been used to represent the data for $Q^2 < 0.7$ GeV^2.

Our proton respects charge symmetry, the interchange of u and d quarks, so it contains a prediction for neutron form factors. This is shown in Fig. 2. The resulting curve labeled relativistic quarks is both large and small. It is very small at low values of Q^2. Its slope at $Q^2 = 0$ is too small by a factor of five, if one compares with the straight line. But at larger values of Q^2 the prediction is relatively large.

Our model gives $R^2_{\rm model} = -0.025$ fm^2, about five times smaller than the data. The small value can be understood in terms of $F_{1,2}$. Taking the definition (2) for small values of Q^2 gives

$$-Q^2R^2/6 = -Q^2R_1^2/6 - \kappa_n Q^2/4M^2 = -Q^2R_1^2/6 - Q^2R_{\rm F}^2/6, \tag{20}$$

where the Foldy contribution, $R_{\rm F}^2 = 6\kappa_n/4M^2 = -0.111$ fm^2, is in good agreement with the experimental data. That a point particle with a magnetic moment can explain the charge radius has led some to state that $G_{\rm E}$ is not a measure of the structure of the neutron. However, one must include the Q^2 dependence of F_1 which gives R_1^2. In our model $R_1^2 = +0.086$ fm^2 which nearly

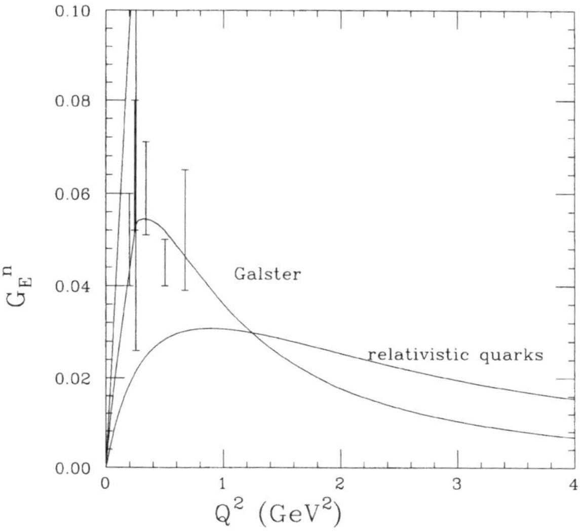

Figure 2. Calculation of G_E^n. The data are from ref. [13], with more expected soon [14].

cancels the effects of R_F^2. Isgur [15] showed that this cancellation is a natural consequence of including the relativistic effects of the lower components of the Dirac spinors. Thus our relativistic effects are standard. We need another source of R^2. This is the pion cloud.

10 Pion Cloud and the Light Front Cloudy Bag Model

The effects of chiral symmetry require that sometimes a physical nucleon can be a bare nucleon immersed in a pion cloud. An incident photon can interact electromagnetically with a bar nucleon, a pion in flight or with a nucleon while a pion is present. These effects were included in the cloudy bag model [16], and are especially pronounced for the neutron. Sometimes the neutron can be a proton plus a negatively charged pion. The tail of the pion distribution extends far out into space (see Figs. 10 and 11) of ref. [16], so that the square of the charge radius is negative.

It is necessary to modernize the cloudy bag model, so as to make it relativistic. This involves using photon-nucleon form factors from our model, using a relativistic π-nucleon form factor, and treating the pionic contributions relativistically by doing a light front calculation. We define the resulting model as the light-front cloudy bag model LFCBM.

The calculation is implemented by evaluating the relevant Feynman di-

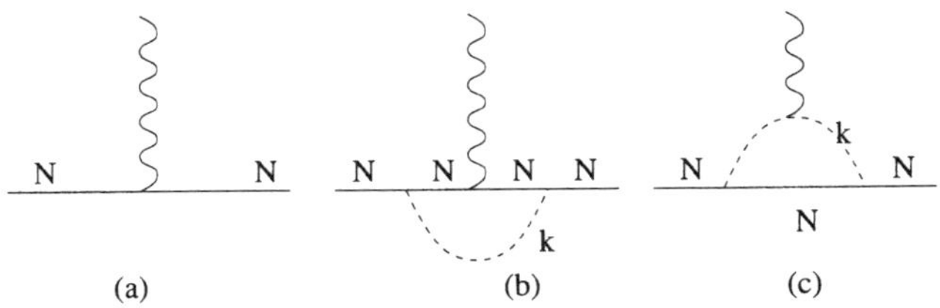

Figure 3. Diagrams. The dashed line represents the virtual pion of momentum k^μ.

agrams of Fig. 3 by integrating over k^- analytically (k^μ is the momentum of the emitted virtual pion) and the other three components numerically; see ref. [3]. Thus the Feynman graphs, Fig. 3, are represented by a single τ-ordered diagram. The use of J^+ and the Yan identity [17] $S_{\rm F}(p) = \sum_s u(p,s)\overline{u}(p,s)/(p^2-m^2+{\rm i}\epsilon)+\gamma^+/2p^+$ allows one see that the nucleon current operators appearing in Fig.3b act between on-mass-shell spinors.

There are four model parameters: $m, \beta, \gamma, \Lambda_{\pi N}$, with $\Lambda_{\pi N}$ representing the cut-off in the pion-nucleon form factor. Including the effects of the pion cloud gives contributions to the magnetic moments of the proton and neutron, so it is necessary to re-fit the parameters. A sample of the values of the parameters is given in Table 1.

The result is termed the light front cloudy bag model [3] (even though there is no bag), and results are shown in Fig. 4. We see that the pion cloud effects are important for small values of Q^2 and, when combined with those of the relativistic quarks coming from the bare nucleon, leads to a good description of the low Q^2 data. The total value of $G_{\rm E}$ is substantial for large values of Q^2.

One might ask how the effects of the pion cloud influence the results shown in Fig. 1. They do not change the picture very much. The pion cloud effects influence the low momentum transfer properties, but vanish at large momentum transfer. Fig. 1 shows QF_2/F_1 which vanishes at $Q = 0$, so the influence of pion cloud effects is hidden.

We have shown one ratio and one form factor. But there are four nucleon

Table 1. Different parameter sets, units in terms of fm

Set(legend)	m	β	Λ	γ	$-R_n^2$	$-\mu_n$	μ_p
1 solid	1.8	3.65	3.1	4.1	0.111	1.73	2.88
2 dot-dash	1.7	3.4	3.1	3.9	0.110	1.79	2.95
3 dash	1.7	2.65	3.1	3.7	0.109	1.79	2.95

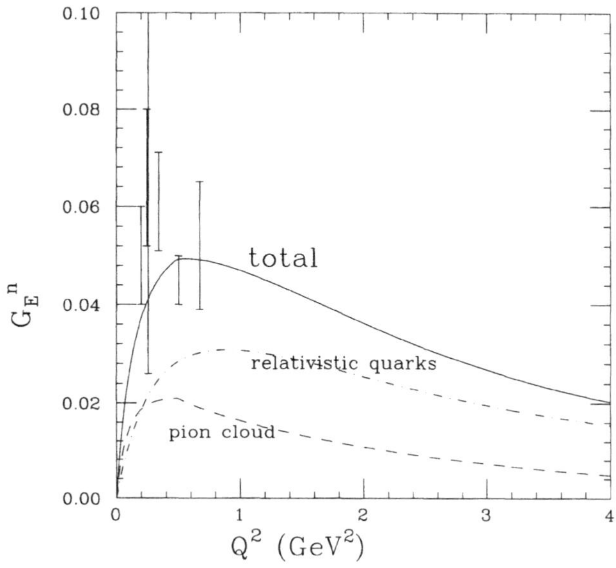

Figure 4. Light front cloudy bag model LFCBM Calculation of G_E^n [3].

electromagnetic form factors. The remaining two are shown in Figs. 5 and 6.

Speaking in a general fashion, I can say that good descriptions of $G_M^{p,n}$ are obtained. If one looks closely there are disagreements with the data. One sees, at very large values of Q^2, that our value of G_M^p is too small, and this allows room for the effects of perturbative QCD [18]. that have been neglected. The wiggles at low values of Q^2 are also worthy of comment. These arise because the effects of the pion field fall off faster with increasing Q^2 than the dipole form used for comparison. Our result for G_M^n shows too deep a dip at low Q^2 in comparison with recent data [19].

The axial form factor, $G_A(Q^2)$ is not calculated here. Schlumpf's model obtained an excellent reproduction of existing data[6], and our parameters are similar to his. The lowest-order effect of the pion cloud vanishes, so the principal difference between G_A of our model and that of ref. [6] is that our quark masses have larger values. This increases the computed value of g_A for a bare nucleon by about $10-15\%$. But this is opposed by the need to multiply the bare nucleon result by the renormalization factor Z of about $0.85-0.9$. Thus our results for $G_A(Q^2)$ should be similar to those of [6].

11 Summary

These calculations show that the combination of Poincaré invariance and pion cloud effects, is sufficient to describe the existing experimental data up to about

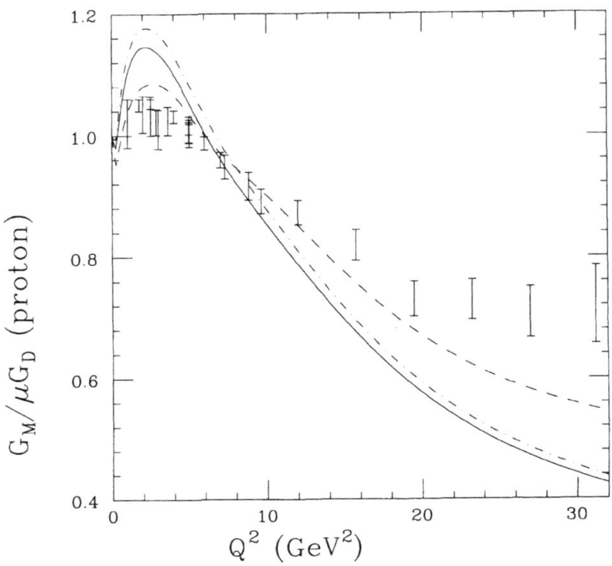

Figure 5. Light front cloudy bag model LFCBM Calculation of G_M^p [3].

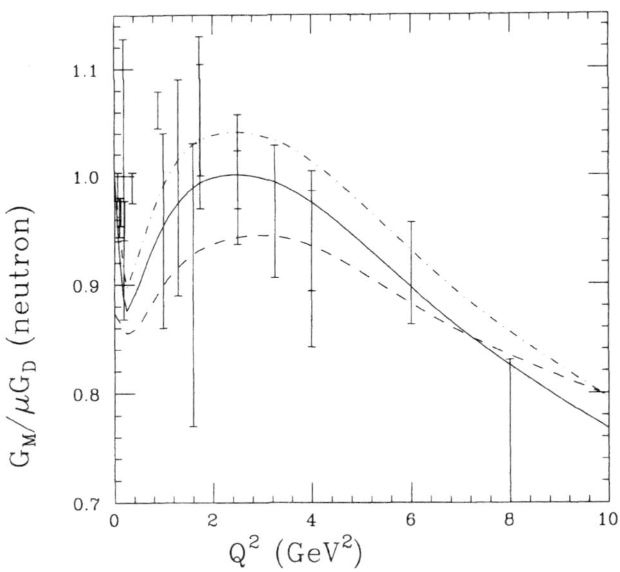

Figure 6. Light front cloudy bag model LFCBM Calculation of G_M^n [3].

$Q^2 = 20\ \text{GeV}^2$. This is somewhat surprising as the model keeps only two necessary effects. Configuration mixing of quark [20], the variation of the quark mass with Q^2 [21], exchange currents [22] and an intermediate Δ [16] have not been included. These effects either have modest influence, are incorporated implicitly through the choice of parameters, or will help to remove any remaining differences with experiment.

Acknowledgement. I thank the USDOE for partial support of this work.

References

1. M.R. Frank, B.K. Jennings and G.A. Miller: Phys. Rev. C **54**, 920 (1996)
2. G.A. Miller and M.R. Frank: Phys. Rev. C **65**, 065205 (2002)
3. G.A. Miller: Phys. Rev. C **66**, 032201 (2002) [nucl-th/0207007]
4. V.B. Berestetskii and M.V. Terent'ev: Sov. J. Nucl. Phys. **25**, 347 (1977)
5. P.L. Chung and F. Coester: Phys. Rev. D **44**, 229, (1991)
6. F. Schlumpf, Ph. D. Thesis (Uni Zurich) [hep-ph/9211255]
7. P. Jain, B. Pire and J.P. Ralston: Phys. Rept. **271**, 67 (1996);
   T. Gousset, B. Pire and J.P. Ralston: Phys. Rev. D **53**, 1202 (1996)
8. V.M. Braun et al.: Phys. Rev. D **65**, 074011 (2002)
9. M.K. Jones et al. [Jefferson Lab Hall A Collaboration]: Phys. Rev. Lett. **84**, 1398 (2000) [nucl-ex/9910005]
10. O. Gayou et al.: Phys. Rev. Lett. **88**, 092301 (2002)
11. S. Kopecky et al.: Phys. Rev. Lett. **74**, 2427 (1995)
12. S. Galster et al.: Nucl. Phys. B **32**, 221 (1971)
13. T. Eden et al.: Phys. Rev. C **50**, 1749 (1994);
    M. Meyerhoff et al.: Phys. Lett. B **327**, 201 (1994);
    M. Ostrick et al.: Phys. Rev. Lett. **83**, 276 (1999);
    J. Becker et al.: Eur. Phys. J. A **6**, 329 (1999);
    I. Passchier et al.: Phys. Rev. Lett. **82**, 4988 (1999);
    D. Rohe et al.: Phys. Rev. Lett. **83**, 4257 (1999);
    H. Zhu et al.: Phys. Rev. Lett. **87**, 081801 (2001)
14. Jefferson Laboratory Experiment 93-038, R. Madey Spokesperson;
    R. Madey (Jlab E93-038 collaboration): "Neutron Electric Form Factor Via Recoil Polarimetry", contribution to Baryons 2002
15. N. Isgur: Phys. Rev. Lett. **83**, 272 (1999)

16. S. Théberge, A.W. Thomas and G.A. Miller: Phys. Rev. **D22** (1980) 2838; A.W. Thomas, S. Théberge, and G.A. Miller: Phys. Rev. **D24** (1981) 216; S. Théberge, G.A. Miller and A.W. Thomas: Can. J. Phys. **60**, 59 (1982) G.A. Miller, A.W. Thomas and S. Théberge, Phys. Lett. B **91**, 192 (1980)

17. S.J. Chang and T.M. Yan: Phys. Rev. D **7**, 1147 (1973)

18. G.P. Lepage and S.J. Brodsky: Phys. Rev. D **22**, 2157 (1980)

19. W. Xu et al.: Phys. Rev. Lett. **85**, 2900 (2000); G. Kubon et al.: Phys. Lett. B **524**, 26 (2002)

20. F. Cardarelli and S. Simula: Phys. Rev. C **62**, 065201 (2000)

21. J.C. Bloch et al.: Phys. Rev. C **60**, 062201 (1999); C.D. Roberts, and S.M. Schmidt: Prog. Part. Nucl. Phys. **45**, S1 (2000)

22. W.R. de Araujo et al.: Phys. Lett. B **478**, 86 (2000); E.F. Suisso et al.: Nucl. Phys. A **694**, 351 (2001); K. Bodoor et al.: Mod. Phys. Lett. A **15**, 2191 (2000)

Few-Body Systems Suppl. 15, 219–230 (2003)

Few-Body Systems

Current Density Operators in Relativistic Quantum Mechanics

F. Coester

Physics Division, Argonne National Lab., Argonne IL 60439, USA

Abstract. Within the framework of Poincaré covariant relativistic quantum mechanics the dynamics generates fully covariant conserved currents from current density operators that are covariant under the kinematic subgroup. Simple toy models provide convenient comparisons of the qualitative features generated by different forms of kinematics.

1 Introduction

Relativistic quantum mechanics [1, 2] is a mathematically well defined framework for the description of few nucleons, few hadrons and/or few constituent quarks for a limited range of the total mass spectrum. The Hilbert space of states consists of irreducible representation spaces of the Poincaré group (single-particle states), finite tensor products of single particle states, and finite direct sums of such tensor products. The Hilbert space of states is the same for interacting and noninteracting systems. The dynamics is specified by a unitary representation, $U(\Lambda, d)$ of the Poincaré group which differs from the representation, $U_0(\Lambda, d)$, of noninteracting constituents. The Poincaré Clebsch-Gordan coefficients relating reducible to irreducible representations are an essential element of the formal structure [3]. The form of kinematics depends on the choice of a subgroup for which the representations U and U_0 are identical.

Conserved current-density operators $I^\mu(x)$ must satisfy the covariant relations

$$U^\dagger(\Lambda, d) I^\mu(x) U(\Lambda, d) = \Lambda^\mu{}_\nu I^\nu(\Lambda^{-1}x - d) \,, \qquad [P_\mu, I^\mu(x)] = 0 \,. \tag{1}$$

This covariance together with the covariance of the eigenstates $|P\rangle$ of the momentum establishes the invariance of the form factors extracted from matrix elements $\langle P_f | I^\mu(0) | P_a \rangle$. In practice model currents are generated by the dynamics from input currents that are covariant only under a kinematic subgroup. In particular the input currents may be the currents of noninteracting constituents. Form-factor calculations depend on the assumed features of the input current, the form of kinematics and the structure of the bound-state wave functions.

Wave functions representing bound states are vectors in the representation space of the little group of the Poincaré group [4]. They are eigenfunctions of mass and spin operators. As such they are independent of the form of kinematics which depends on the parameters chosen to specify the spectrum of the four-momentum operator. The wave functions are not observable, but some features become observable by assumed features of electroweak currents which depend on the form of kinematics. In particular "impulse currents" depend on the definition of constituent-particle momenta which depends on the form of kinematics. Different forms of kinematics emphasize different structures. The instant form emphasizes spatial structure, the point form velocity structures and the front form combines features of both.

Recently, Desplanques et al. [5] emphasized the importance of comparing the effects of different forms of kinematics on the same system. The purpose of these notes is to illuminate some qualitative features of form factors depending on the choice of the kinematics. While the formal structure applies to all few-body systems the emphasis will be on properties of constituent-quark models of hadrons.

2 Representations of the Poincaré Generators

The ten generators, P^μ, $J^{\mu\nu}$, can be expressed as functions of the mass operator $\mathcal{M}$, the spin j^2 and eight other operators subject to choices of convenience, for instance

$$\{P^\mu, J^{\mu\nu}\} = \mathcal{G}[\mathcal{M}, j^2, j_x, j_z, \vec{P}, \mathrm{i}\nabla_P] \,, \tag{2}$$

with

$$\mathcal{M}^2 := -P^2 \,, \qquad j^2 := W^2 \mathcal{M}^{-2} \,, \qquad W_\tau := \tfrac{1}{2} P^\nu J^{\rho\sigma} \epsilon_{\nu\rho\sigma\tau} \,, \tag{3}$$

and the spin vector is related to the Poincaré generators by the canonical boost,

$$\{0, \vec{j}\} := B^{-1}(P\mathcal{M}^{-1}) W \mathcal{M}^{-1} \,. \tag{4}$$

With this particular choice the generators of space translations and rotat= ions are independent of the mass operator. Following Bakamjian and Thomas [6, 7] the dynamics is realized replacing the free-particle mass operator $\mathcal{M}_0$ by $\mathcal{M}$. The "instant-form" generators of the subgroup of space translations and rotations, E(3), are independent of $\mathcal{M}$.

Alternatively the generators of space-time translations could be specified by the mass operator and the velocity v, or by the mass operator and the momentum components $P^+, P_\perp$. The kinematic subgroups are then the Lorentz group SO(1,3), or the symmetry of the null-plane ISO(1,2). The boost relating the spin to the Pauli-Lubanski vector depend on the form of kinematics.

The generators depending on $\mathcal{M}$ are called "Hamiltonians" in Dirac's seminal paper [8], where the emphasis is on the choice of the dynamic generators which generate the evolution of an initial state.

2.1 Euclidean Kinematics ("Instant Form")

For E(3) kinematics explicit expressions of the Poincaré generators are represented by

$$\vec{P}\,, \qquad \vec{J} = -\mathrm{i}\vec{P} \times \nabla_P + \vec{j}\,, \tag{5}$$

for the kinematic generators and by

$$H := P^0 = \sqrt{\mathcal{M}^2 + \vec{P}^{\,2}}\,, \qquad \vec{K} = \frac{\mathrm{i}}{2}\{H\nabla_P + \nabla_P H\} - \frac{\vec{j} \times \vec{P}}{\mathcal{M} + H}\,, \tag{6}$$

for the dynamic generators.

2.2 Lorentz Kinematics ("Point Form")

For SO(1,3) kinematics the generators of the Lorentz transformations

$$\vec{J} = -i\vec{v} \times \nabla_v + \vec{j}\,, \qquad \vec{K} = \frac{\mathrm{i}}{2}\{v^0\nabla_v + \nabla_v v^0\} - \frac{\vec{j} \times \vec{v}}{(1 + v^0)}\,, \tag{7}$$

are independent of M, and all space-time translations are dynamic.

$$\{P^\mu\} = \mathcal{M}\,\{v^0\,, \vec{v}\}\,, \qquad v^0 := \sqrt{1 + \vec{v}^{\,2}}\,. \tag{8}$$

2.3 Null-Plane Kinematics ("Front Form")

For ISO(1,2) kinematics [9, 10, 11] the kinematic generators

$$\vec{P}_\perp := \{P_x, P_y\}\,, \qquad \vec{E} := \vec{K}_\perp + \hat{z} \times \vec{J} = \mathrm{i}P^+\nabla_{P_\perp}\,,$$

$$P^+ := P^0 + P_z\,, \qquad K_z = \mathrm{i}P^+\partial_{P^+}\,, \qquad J_z = \mathrm{i}P_\perp \times \nabla_{P_\perp} + j_z\,, \tag{9}$$

generate translations in the null-plane and boosts of the null-plane momenta, while the transverse rotations are dynamic.

$$H := P^- := P^0 - P_z = \frac{\mathcal{M}^2 + P_\perp^2}{P^+}\,, \qquad \hat{z} \times \vec{J} = \tfrac{1}{2}(\vec{E} - \vec{F})\,,$$

$$\vec{F} := \vec{K}_\perp - \hat{z} \times \vec{J}\,, \quad \vec{F} = \mathrm{i}P^- \nabla_{P_\perp} + 2\mathrm{i}P_\perp \partial_{P^+} + \frac{P_\perp \times \vec{j} - \mathcal{M}\,\hat{z} \times \vec{j}}{P^+}\,. \tag{10}$$

The principal virtue of this form of kinematics is that it shares important features with Galilean relativity.

3 Representations of Bound States

Vectors in the little Hilbert space [4] of n constituents can be represented by functions $\phi(\vec{k}_1, \sigma_1, \ldots \vec{k}_n, \sigma_n)$ of spin variables σ_i and internal momenta $\vec{k}_i$ which are subject to the constraint $\sum_i \vec{k}_i = 0$ and the normalization

$$\|\phi\|^2 = \sum_{\sigma_1 \ldots \sigma_n} \int d^3k_1 \cdots \int d^3k_n\, \delta^{(3)}\left(\sum_i \vec{k}_i\right) |\phi(\vec{k}_1, \sigma_1, \ldots, \vec{k}_n, \sigma_n)|^2\,. \tag{11}$$

The invariant inner-product measure of functions of the four-momentum P is

$$\begin{aligned} d^4P\,\delta(P^2 + M^2) &= \tfrac{1}{2}\left(M^2 + \vec{P}^2\right)^{-\frac{1}{2}} d^3\vec{P} = \frac{dP^+}{P^+}\, d^2P_\perp \\ &= M^2 d^4v \delta(v^2 + 1) = \frac{M^2 d^3v}{2\sqrt{1 + \vec{v}^2}} \end{aligned} \tag{12}$$

It is convenient to absorb the square root of the weight factors into the functions representing states. With this normalization wave functions representing bound states of mass M and spin j obtain in the representation

$$\begin{aligned} \Psi_{M,j,\sigma,\vec{P}_a}(\vec{k}_1, \sigma_1, \ldots, \vec{k}_n, \sigma_n; \vec{P}) &= \phi_{M,j,\sigma}(\vec{k}_1, \sigma_1, \ldots \vec{k}_n, \sigma_n)\delta^{(3)}(\vec{P} - \vec{P}_a) \\ &\equiv \phi_{M,j,\sigma}(\vec{k}_1, \sigma_1, \ldots \vec{k}_n, \sigma_n)\frac{\delta^{(3)}(\vec{v} - \vec{v}_a)}{M^3}\,, \end{aligned} \tag{13}$$

or

$$\Psi_{M,j,\sigma,\mathbf{P}_a}(\vec{k}_1, \sigma_1, \ldots, \vec{k}_n, \sigma_n; \mathbf{P}) = \phi_{M,j,\sigma}(\vec{k}_1, \sigma_1, \ldots \vec{k}_n, \sigma_n)\delta^{(3)}(\mathbf{P} - \mathbf{P}_a) \tag{14}$$

with the four-momentum specified by $\mathbf{P} := \{P^+, P_\perp\}$. Under Lorentz transformations the internal momenta $\vec{k}_i$ and the spins undergo Wigner rotations, $\mathcal{R}_W(\Lambda, v) := B(\Lambda^{-1}v)\Lambda B(v)$. Functions that are invariant under rotations are both Lorentz invariant and independent of a choice of boost.

Constituent momenta which are covariant under the kinematic subgroup, are related to the internal momenta $\vec{k}_i$ by different boosts which depend on the kinematics.

3.1 Euclidean Kinematics

With E(3) kinematics with momenta $\vec{p}_i$ are defined by

$$p_i = B(\vec{P}/M_0)\{\omega_i, \vec{k}_i\} , \qquad M_0 := \sum_i \omega_i , \qquad \sum_i \vec{p}_i = \vec{P} , \tag{15}$$

with $\omega_i := \sqrt{m^2 + |\vec{k}_i|^2}$. The canonical boost $B(\vec{P}/M_0)$ is a rotationless Lorentz trabsformation with the defining property

$$B(\vec{P}/M_0)\{M_0, 0\} = \{\sqrt{M_0^2 + \vec{P}^2}, \vec{P}\} . \tag{16}$$

3.2 Lorentz Kinematics

With SO(1,3) kinematics velocities $\vec{v}_i := \vec{p}_i/m$ are defined by

$$p_i = B(\vec{v})\{\omega_i, \vec{k}_i\} , \qquad \frac{\sum_i p_i}{\sqrt{-(\sum_i p_i)^2}} \equiv \frac{\sum_i v_i}{\sqrt{-(\sum_i v_i)^2}} = v . \tag{17}$$

3.3 Null-Plane Kinematics

With ISO(1,2) kinematics momenta $\mathbf{p}_i$ are defined by

$$p_i = B\left(\frac{P^+}{M_0}, \frac{P_\perp}{P^+}\right)\{\omega_i, \vec{k}_i\} , \qquad \sum_i \mathbf{p}_i = \mathbf{P} . \tag{18}$$

3.4 "Manifest" Poincaré Covariance

There are closely related representations that are "manifestly covariant" under the full Poincaré group, ISO(1,3). Four-momenta p_i related to the internal momenta by $B(v)$ can be defined such that $P = \sum_i p_i$.

For instance

$$p_i = B(\vec{v})\{\omega_i, \vec{k}_i\} , \quad \text{for } i = 2 \ldots n, \qquad p_1 := Mv - \sum_{i=2}^{n} p_i , \tag{19}$$

so that $\sum_i p_i = Mv$, or

$$\tilde{p}_i = B(\vec{v})\{0, \vec{k}_i\} , \qquad p_i = \tilde{p}_i + \frac{1}{n} P , \qquad \sum_i p_i = P . \tag{20}$$

Such representaions may be useful for comparisons with field theories. For two constituents they correspond to representations obtained by familiar quasi-potential reductions of Bethe-Salpeter amplitudes. Relations to quantum filed theory are beyond the scope of these notes.

4 Current Density Operators and Form Factors

Single particle currents provide the input from which the full current is generated by the dynamics. Given a Poincaré covariant current density operator $I^\mu(x)$ with the covariant matrix elements $\langle P_f|I^\mu(0)|P_a\rangle$ the Lorentz invariant matrix

$$\langle P_f|\mathcal{I}|P_a\rangle := \frac{(v_f + v_a)\cdot\langle P_f|I(0)|P_a\rangle}{\sqrt{-(v_f+v_a)^2}} \tag{21}$$

is related to the dimensionless charge form factor $G_e(\eta)$ by

$$G_e(\eta) := \langle P_f|\mathcal{I}|P_a\rangle M\sqrt{1+\eta}\,, \qquad \eta := \frac{(P_f - P_a)^2}{4M^2} \equiv \tfrac{1}{4}(v_f - v_a)^2\,. \tag{22}$$

In the Breit frame, $\vec{P}_f = M\{0,0,\sqrt{\eta}\} = -\vec{P}_a$, this reduces to the familiar expression

$$G_e(\eta) := \langle \vec{P}_f|I^0(0)|\vec{P}_a\rangle\,. \tag{23}$$

The covariance of the single-particle input currents is consistent with the covariance of the bound compound only for the kinematic subgroup. Since a given bound state is represented by different functions of the momenta $p_1,\ldots,p_n$ for different forms of kinematics, inequivalent current operators obtain with different kinematic subgroups. Distinctive features of current densities with different kinematics are sketched in in this Section.

With instant-form kinematics impulse current densities are defined by the representations

$$\langle \vec{p}'_n,\ldots,\vec{p}'_1|I^\mu(\vec{x}))|\vec{p}_1,\ldots,\vec{p}_n\rangle = \langle \vec{p}'_1|I^\mu_1(\vec{x})|\vec{p}_1\rangle\langle \vec{p}'_n,\ldots,\vec{p}'_2|\mathbf{1}|\vec{p}_2,\ldots,\vec{p}_n\rangle\,, \tag{24}$$

where

$$\langle \vec{p}'_1|I^\mu_1(\vec{x})|\vec{p}_1\rangle = \langle \vec{p}'_1|I^\mu_1(0)|\vec{p}_1\rangle e^{\mathrm{i}(\vec{p}'_1-\vec{p}_1)\cdot\vec{x}}\,, \tag{25}$$

and for spin 0

$$\langle \vec{p}'_1|I^\mu_1(0)|\vec{p}_1\rangle = \frac{p'^\mu_1 + p^\mu_1}{2\sqrt{p'^0_1 p^0_1}} f(\eta_1) \equiv \frac{v'^\mu_1 + v^\mu_1}{2\sqrt{v'^0_1 v^0_1}} f(\eta_1)\,, \tag{26}$$

where $\eta_1 := (p'_1 - p_1)^2/4m^2 \equiv \frac{1}{4}(v'_1 - v_1)^2$ and $f(0) = 1$.
For spin $\frac{1}{2}$ the corresponding Dirac current is

$$\langle \vec{p}'_1|I^\mu(0)|\vec{p}_1\rangle = \bar{u}(\vec{p}'_1)\gamma^\mu u(\vec{p}_1) f(\eta_1) \equiv \bar{u}(\vec{v}'_1)\gamma^\mu u(\vec{v}_1) f(\eta_1)\,. \tag{27}$$

However, with p_i defined by Eq. (15) the Lorentz covariance of p_1 is inconsistent with the Lorentz covariance of the compound. This problem could be avoided using dynamic Lorentz transformations to generate the velocity dependence of of the one-body current, $\langle \vec{p}'_1|I^\mu_1(0)|\vec{p}_1\rangle$.

Recently nucleon form factors of a constituent-quark model have been calculated using point-form kinematics [12]. With point-form kinematics the momentum p_1 does not generate translations consistent with the dynamics. In that case the spectator current density may be defined by

$$\langle \vec{p}'_n, \ldots, \vec{p}'_1 | I^\mu(\vec{x}) | \vec{p}_1, \ldots, \vec{p}_n \rangle :=$$

$$\langle \vec{p}'_1 | I^\mu_1(0 | \vec{p}_1 \rangle \langle \vec{p}'_n, \ldots, \vec{p}'_2 | \mathbf{1} | \vec{p}_2, \ldots, \vec{p}_n \rangle e^{\mathrm{i}M(\vec{v}'-\vec{v})\cdot\vec{x}} , \tag{28}$$

with the momenta p_i defined by Eq. (17).

Similar expressions obtain for null-plane kinematics with $\vec{p}_1$ replaced by $\mathbf{p}_1$ and canonical boosts replaced by null-plane boosts. The current matrix

$$\langle \mathbf{p}'_n, \ldots, \mathbf{p}'_1 | I^+(x^-, x_\perp) | \mathbf{p}_1, \ldots, \mathbf{p}_n \rangle$$

$$= \langle \mathbf{p}'_1 | I^+_1(x^-, x^+) | \mathbf{p}_1 \rangle \langle \mathbf{p}'_n, \ldots, \mathbf{p}'_2 | \mathbf{1} | \mathbf{p}_2, \ldots, \mathbf{p}_n \rangle \tag{29}$$

is covariant under kinematic subgroup. All values $\mathbf{p}_i$ are related by kinematic Lorentz transformations. For spin 0

$$\langle \mathbf{p}'_1 | I^+_1(0) | \mathbf{p}_1 \rangle = \frac{p_1'^+ + p_1^+}{\sqrt{p_1'^+ p_1^+}} f(\eta_1) \tag{30}$$

With the kinematically invariant constraint $p_1'^+ = p_1^+$ it follows that

$$\langle \mathbf{p}'_1 | I^+_1(0) | \mathbf{p}_1 \rangle = f(\eta_1) , \qquad \text{with} \qquad \eta_1 = \frac{(p'_{1\perp} - p_{1\perp})^2}{4m^2} . \tag{31}$$

However, rotations about $\vec{Q} = p'_{1\perp} - p_{1\perp}$ are not kinematic. They become kinematic when the orientation of the null-plane is treated as a global gauge variable [11, 13]. Rotational symmetry about the the direction of the momentum transfer is emphasized by the kinematic constraint $p'_{1\perp} = p_{1\perp}$ [14]. In that case

$$\eta_1 = \frac{1}{4}\left(1 + \frac{p^2_{1\perp}}{m^2}\right)\left(\sqrt{\frac{p_1'^+}{p_1^+}} - \sqrt{\frac{p_1^+}{p_1'^+}}\right)^2 , \tag{32}$$

and the spectator current does not provide an impulse current.

Comparative features of currents constructed in this manner are illustrated in the next Section by a simple toy model.

5 Toy Models of Two Constituents with Spin 0

For two particles with spin 0 a compound state with mass M and mom= entum $\vec{P}_A$ may be represented by

$$\Psi_{M,\vec{P}_a}(\vec{P}, \vec{k}) = \phi(\vec{k})\delta^3(\vec{P} - \vec{P}_a) , \qquad \int d^3k |\phi(\vec{k})|^2 = 1 , \tag{33}$$

or the equivalent representations

$$\Psi_{M,\vec{P}_a}(\vec{p}_1,\vec{p}_2) \quad := \quad \sqrt{\frac{\partial(\vec{P},\vec{k})}{\partial(\vec{p}_1,\vec{p}_2)}}\phi\Big(\vec{k}(\vec{p}_1,\vec{p}_2)\Big)\delta^{(3)}\Big(\vec{P}(\vec{p}_1,\vec{p}_2)-\vec{P}_a\Big)\,,$$

$$\Psi_{M,\vec{P}_a}(\vec{P},\vec{p}_2) \quad := \quad \sqrt{\frac{\partial(\vec{P},\vec{k})}{\partial(\vec{P},\vec{p}_2)}}\phi\Big(\vec{k}(\vec{P},\vec{p}_2)\Big)\delta^{(3)}(\vec{P}-\vec{P}_a)\,, \tag{34}$$

which depend on the boosts which define the the momenta $\vec{p}_1, \vec{p}_2$ as functions of $\vec{k}$ and $\vec{P}$.

A simple model may be specified by the bound-state function $\phi(\vec{k})$ with the Fourier transform $\phi(\vec{r})$,

$$\phi(\vec{k}) = \left(\frac{1}{b\sqrt{\pi}}\right)^{\frac{3}{2}} e^{-k^2/2b^2}\,, \qquad \phi(\vec{r}) = \left(\frac{b}{\sqrt{\pi}}\right)^{\frac{3}{2}} e^{-b^2r^2/2}\,. \tag{35}$$

The parameter b specifies the unobservable matter radius

$$\langle r^2\rangle := \int d^3r r^2|\phi(\vec{r})|^2 = \frac{3}{8b^2}\,. \tag{36}$$

The spectator definition of the charge density

$$\langle \vec{p}_2{}', \vec{P}_f|I^0(0)|\vec{P}_a,\vec{p}_2) = \delta^{(3)}(\vec{p}_2{}'-\vec{p}_2)\langle\vec{p}_1{}'|I^0_1(0)|\vec{p}_1\rangle \tag{37}$$

yields different form factors depending on the boost that relates $\vec{p}_2$ to $\vec{P}, \vec{k}$ and the momentum dependence of the single-particle charge I^0_1,

$$G_e(\eta) = \int d^3p_2\phi(\vec{k}')\langle\vec{p}_1{}'|I^0_1(0)|\vec{p}_1\rangle\sqrt{\frac{\partial(\vec{P}_f,\vec{k})}{\partial(\vec{P}_f,\vec{p}_2)}\frac{\partial(\vec{P}_a,\vec{k})}{\partial(\vec{P}_a,\vec{p}_2)}}\phi(\vec{k})\,, \tag{38}$$

where $\vec{k}' \equiv \vec{k}(P_f,p_2)$ and $\vec{k} \equiv \vec{k}(P_a,p_2)$.

5.1 *Euclidean Kinematics*

With "instant-form" kinematics the momenta $\vec{p}_1, \vec{p}_2$ are defined by the canonical boosts of $\{\omega, \pm\vec{k}\}$

$$p_1 := B(\vec{P}/2\omega)\{\omega,\vec{k}\}\,, \qquad p_2 := B(\vec{P}/2\omega)\{\omega,-\vec{k}\}\,. \tag{39}$$

That definition assures the relation $\vec{p}_1+\vec{p}_2 = \vec{P}$, which is required by Euclidean kinematics. The Jacobian is then

$$\frac{\partial(\vec{P},\vec{k})}{\partial(\vec{P},\vec{p}_2)} = \frac{\partial(\vec{P},\vec{k})}{\partial(\vec{p}_1,\vec{p}_2)} = \frac{(E_1+E_2)\,\omega}{2E_1E_2}\,, \qquad E_i := \sqrt{\vec{p}_i^{\,2}+m^2}\,. \tag{40}$$

The assumption of free-particle Lorentz covariance of the single-particle current implies

$$\langle \vec{p}_1{}' | I_1^0(0) | \vec{p}_1 \rangle := \frac{E_1' + E_1}{2\sqrt{E_1' E_1}} , \qquad \vec{p}_1 \equiv \vec{P}_a - \vec{p}_2 , \quad \vec{p}_1{}' \equiv \vec{P}_f - \vec{p}_2 . \quad (41)$$

This Lorentz covariance is not consistent with the Lorentz covariance of the compound and is not required by the Euclidean kinematics. Alternative assumptions consistent with Euclidean kinematics are no velocity structure [15] $\langle \vec{p}_1{}' | I_1^0(0) | \vec{p}_1 \rangle = f(|\vec{p}_1' - \vec{p}_1|^2)$ or a velocity structure consistent with Lorentz covariance of the compound,

$$\langle \vec{p}_1{}' | I_1^0(0) | \vec{p}_1 \rangle = \frac{v_1^{0\prime} + v_1^0}{2\sqrt{v_1^{0\prime} v_1^0}} , \qquad v_1 := B(v)\{\omega, \vec{k}\}/m . \quad (42)$$

5.2 *Lorentz Kinematics*

"Point-form" kinematics is implemented by the definitions

$$p_1 := B(v)\{\omega, \vec{k}\} , \qquad p_2 := B(v)\{\omega, -\vec{k}\} , \quad (43)$$

and $P := Mv$, which imply

$$\frac{p_1 + p_2}{\sqrt{-(p_1 + p_2)^2}} = v , \qquad p_1 + p_2 \neq P , \qquad \frac{\partial(\vec{P}, \vec{k})}{\partial(\vec{P}, \vec{p}_2)} = \frac{\omega}{p_2^0} . \quad (44)$$

5.3 *Null-Plane Kinematics*

With the four-momentum specified by the mass M and the null-plane components $\mathbf{P} := \{P^+, P_\perp\}$, constituent null-plane momenta $\mathbf{p}_1$ and $\mathbf{p}_2$ are related to $\vec{k}$ by null-plane boosts,

$$\{p_1^-, \mathbf{p}_1\} = B_f\Big(\frac{P^+}{M_0}, \frac{P_\perp}{P^+}\Big)\{\omega, \vec{k}\} , \qquad \{p_2^-, \mathbf{p}_2\} = B_f\Big(\frac{P^+}{M_0}, \frac{P_\perp}{P^+}\Big)\{\omega, -\vec{k}\} , \quad (45)$$

which give the explicit relations

$$\begin{aligned} p_1^+ &= \xi P^+ , \qquad p_2^+ = (1-\xi)P^+ , \\ p_{1\perp} &= k_\perp + \xi P_\perp , \qquad p_{2\perp} = -k_\perp + (1-\xi)P_\perp . \end{aligned} \quad (46)$$

where $\xi := (\omega + k_z)/2\omega$. The Jacobian

$$\frac{\partial(\vec{k}, \mathbf{P})}{\partial(\mathbf{p}_1, \mathbf{p}_2)} = \frac{\omega P^+}{2p_1^+ p_2^+} \quad (47)$$

relates the representations $\Psi_{M,\mathbf{P}_a}(\vec{k};\mathbf{P})$ and $\Psi_{M,\mathbf{P}_a}(\mathbf{p}_1,\mathbf{p}_2)$,

$$\Psi_{M,\mathbf{P}_a}(\mathbf{p}_1,\mathbf{p}_2) = \sqrt{\frac{\partial(\vec{k},\mathbf{P})}{\partial(\mathbf{p}_1,\mathbf{p}_2)}}\phi_M(k^2)\delta^{(3)}(\mathbf{p}_1+\mathbf{p}_2-\mathbf{P}_a) \ . \tag{48}$$

The spectator current is given by

$$\langle \mathbf{p}_2{}',\mathbf{P}_f|I^+(0)|\mathbf{P}_a,\mathbf{p}_2\rangle := \delta^{(3)}(\mathbf{p}_2{}'-\mathbf{p}_2)\langle \mathbf{p}_1{}'|I^+(0)|\mathbf{p}_1\rangle \ . \tag{49}$$

The spectator constraint $\mathbf{p}_2' = \mathbf{p}_2$ implies

$$(1-\xi')P_f^+ = (1-\xi)P_a^+ \ , \qquad k_\perp' - k_\perp = (1-\xi')P_{f\perp} - (1-\xi)P_{a\perp} \ . \tag{50}$$

There are two kinematically independent choices for $\mathbf{P}_a$ and $\mathbf{P}_f$:

5.3.1 Transverse Momentum Transfer, $P_a^+ = P_f^+$, $\eta = \frac{1}{4}Q^2/M^2$

In this case the spectator constraint (50) and the one body current reduce to

$$\xi' = \xi \ , \qquad k_\perp' - k_\perp = (1-\xi)Q_\perp \ , \quad \langle \mathbf{p}_1'|I^+(0)|\mathbf{p}_1\rangle = 1 \ , \tag{51}$$

and the form factor is

$$G_e(\eta) = \int d^2k_\perp \int_0^1 d\xi\phi(k'^2)\phi(k^2)\frac{\sqrt{\omega'\omega}}{2\xi(1-\xi)} \ . \tag{52}$$

5.3.2 Longitudinal Momentum Transfer, $P_{f\perp} = P_{a\perp} = 0$

In this case η is a function of P_f^+/P_a^+. From

$$P_f^+ = M\left(\sqrt{1+\eta}+\sqrt{\eta}\right) \ , \qquad P_a^+ = M\left(\sqrt{1+\eta}-\sqrt{\eta}\right) \ , \tag{53}$$

it follows that

$$\eta = \frac{1}{4}\left(\sqrt{\frac{P_f^+}{P_a^+}} - \sqrt{\frac{P_a^+}{P_f^+}}\right)^2 \ . \tag{54}$$

The spectator constraint (50) and the one body current reduce to

$$k_\perp' = k_\perp \ , \qquad \xi' - \xi = 2\left(1 - \frac{\xi'+\xi}{2}\right)\sqrt{\frac{\eta}{1+\eta}} \ ,$$

$$\mathcal{I}_1^+ := \langle \mathbf{p}_1'|I^+(0)|\mathbf{p}_1\rangle = \frac{(\xi'+\xi)\sqrt{1+\eta}+(\xi'-\xi)\sqrt{\eta}}{2\sqrt{\xi'\xi}} \ , \tag{55}$$

and the form factors is

$$G_e(\eta) = \int d^2k_\perp \int_0^1 d\bar{\xi}\,\phi(k'^2)\phi(k^2)\,\mathcal{I}_1^+\sqrt{\frac{\omega'\omega}{4\xi'(1-\xi')\xi(1-\xi)}} \ , \tag{56}$$

with $\bar{\xi} := \frac{1}{2}(\xi'+\xi)$.

5.4 Remark

The construction of current density operators illustrated by these examples does not require a "consistent" use of a kinematic subgroup. There is the relative spatial structure conjugate to the spacelike momentum $p := B(v)\{0, \vec{k}\}$. The boost implements the Lorentz contraction of the relative distance. For $\vec{P}_f = -\vec{P}_a$ the Jacobian of the transformation $\vec{P}, \vec{k} \to \vec{P}, \vec{p}$ is

$$\frac{\partial(\vec{P}_a, \vec{k})}{\partial(\vec{P}_a, \vec{p})} = \frac{\partial(\vec{P}_f, \vec{k})}{\partial(\vec{P}_f, \vec{k})} = \frac{1}{\sqrt{1+\eta}} \,. \tag{57}$$

This representation together with the velocity $v_1 := B(v)\{\omega, \vec{k}\}/m$ might useful in constructing simple model currents.

6 Summary Discussion

Historically constituent quark models are older than QCD. They are still a useful phenomenological framework for hadron spectroscopy and electroweak structure of hadrons. Consistency with symmetry requirements of the Poincaré group are an essential requirement which is satisfied by relativistic quantum mechanics. States of single hadrons are represented by rotationally invariant functions of spin, flavor, color and one relative momentum or two Jacobi momenta. These wave functions representing confined quark are not observable and they not necessarily involve a constituent quark mass [16]. Introduction of a constituent quark mass defines an internal velocity structure of the hadrons. Features of the quark structure of a hadron become observable by assumed properties of electroweak currents. The Poincaré covariance requirements of current density operators can be met generating the full current from an input current with limited symmetry. This process leaves considerable freedom of choice in actual model constructions. Different kinematic subgroups of the Poincaré group imply differences in the generation of a quark-particle structure. Independently of the ultimate agreement of a particular model with a limited set of data there are numerous questions one should ask even though the answers may involve subjective preferences: Is the consistent use of any one kinematic symmetry necessary or preferable? Why? Are the model structures more involved than necessary? Are the qualitative features of the currents plausible? Questions like these can be illuminated by toy models which allow quick and easy computation. The models described in Sec. 5 provide an illustration. They could be extended modifying the shape of the wave functions and providing for spin $\frac{1}{2}$ constituents. Instead of one internal momentum $\vec{k}$ there could be two Jacobi momenta $\vec{\kappa}$ and $\vec{k}$. The main point is that for a given compound model different current constructions are easily computable.

Acknowledgement. I have benefited from discussions with W. Plessas, W.H. Klink and B. Desplanques. Supported in part by the U.S. Department of Energy, Nuclear Physics Division, contract W-31-109-ENG-38.

References

1. E.P. Wigner: Ann. Math. **40**, 141 (1939)
2. B.D. Keister, W.N. Polyzou: Adv. Nucl. Phys. **20**, 225 (1991)
3. W.N. Polyzou: J. Math. Phys. (to be published)
4. R. Haag: *Local Quantum Physics.* New York: Springer Verlag 1992 see Sec. I.3.2, p. 29
5. B. Desplanques, L. Theußl: Eur. Phys. **A13**, 461 (2002); A. Amghar, B. Desplanques, L. Theußl: Nucl. Phys. A (to be published)
6. B. Bakamjian, L.H. Thomas: Phys. Rev. **92**, 1300 (1953)
7. B. Bakamjian: Phys. Rev. **121**, 1849 (1961)
8. P.A.M. Dirac: Rev. Mod. Phys. **21**, 392 (1949)
9. F. Coester: Prog. Part. Nucl. Phys. **29**, 1 (1992)
10. V.A. Karmanov: ZhETF **71**, 399 (1976) [transl.: JETP **44**, 210 (1976)]
11. J. Carbonell, B. Desplanques, V.A. Karmanov, J.-F Mathiot: Phys. Rep. **300**, 215 (1998)
12. W. Plessas: these proceedings; R.F. Wagenbrunn et al.: Phys. Lett. B **511**, 33 (2001); S. Boffi et al.: Eur. Phys. **A14**, 17 (2002)
13. F. Coester, W.H. Klink, W.N. Polyzou: Few-Body Syst. Suppl. **10**, 115 (1999)
14. F.M. Lev, E. Pace, G. Salmé: Phys. Rev. **C62** , 0640 04 (2000)
15. F. Coester, K. Dannbohm, D.O. Riska: Nucl. Phys. **A634**, 335 (1998)
16. R.P. Feynman, M. Kislinger, F. Ravndal: Phys. Rev. **D3**, 2706 (1971)

Few-Body Systems Suppl. 15, 231–236 (2003)

Few-Body Systems

Light-front Time Picture of the Bethe-Salpeter Equation

T. Frederico[1]*, J.H.O. Sales[2], B.V. Carlson[1], P.U. Sauer[3]

[1] Departamento de Física, Instituto Tecnológico de Aeronáutica, Centro Técnico Aeroespacial, 12.228-900, São José dos Campos, São Paulo, Brazil

[2] Instituto de Física Teórica, Universidade Estadual Paulista, 0145-900, São Paulo, São Paulo, Brazil

[3] Institute for Theoretical Physics, University Hannover, D-30167 Hannover, Germany

Abstract. We show the light-front representation of the field theoretical Bethe-Salpeter equation (BSE) in the ladder approximation using the quasi potential reduction. We discuss the equivalence of the covariant ladder Bethe-Salpeter equation with an infinite set of coupled equations for the Green's functions of the different light-front Fock-states.

1 Introduction

Recently, we have proposed a quasi-potential reduction of the field-theoretical Bethe-Salpeter equation to study two-particle bound states on the light-front for composite systems of bosons and fermions [1]. The reduction scheme eliminates the relative light-front time between the particles, and the global propagation of the intermediate system appears leading to Dirac's [2] concept of representing the dynamics of the quantum system for light-front times $x^+ = t + z$. We have derived two-body equations for which the effective interaction is irreducible with respect to the light-front two-body propagation. The global light-front time projection makes explicit the intermediate propagation within higher Fock-space sectors, and the mixing with the valence component of the wave-function. The effective two-body equation for the valence component of the light-front wave-function from the quasi-potential reduction has an effective interaction which contains the coupling with higher-Fock state components. In lowest order, the two-body bound-state equation is the one first derived by Weinberg [3].

* *E-mail address:* tobias@fis.ita.br

In view of the possible application of the proposed scheme to study the structure of the bound-state, for example, with electromagnetic probes [1, 4], it is timely to reveal the physical aspects of the dynamics generated by the quasi-potential reduction with respect to the mixing of the lowest Fock component with higher Fock components of the light-front wave function. We exemplify our procedure using a model for interacting bosons ϕ_1, ϕ_2 and σ for which the interaction Lagrangian is given by,

$$\mathcal{L}_I = g_S\,\phi_1^\dagger\phi_1\sigma + g_S\,\phi_2^\dagger\phi_2\sigma. \tag{1}$$

2 Projection on the Light-Front Hyperplane

The two-boson transition matrix $T(K)$ in relativistic field theory, for a given total four-momentum of the system, is the solution of

$$T(K) = V(K) + V(K)G_0(K)T(K), \tag{2}$$

where $V(K)$ is two-body irreducible. Neglecting self-energy parts the disconnected Green's function, written in *light-front* coordinates, e.g., $k_i = (k_i^- := k_i^0 - k_i^3\ ,\ k_i^+ := k_i^0 + k_i^3\ ,\ \boldsymbol{k}_{i\perp})$, is given by:

$$\langle k_1'^-|\,G_0(K)\,|k_1^-\rangle = -\frac{1}{2\pi}\frac{\delta\left(k_1'^- - k_1^-\right)}{\widehat{k}_1^+(K^+ - \widehat{k}_1^+)\left(k_1^- - \frac{\widehat{\boldsymbol{k}}_{1\perp}^2 + m_1^2 - io}{\widehat{k}_1^+}\right)\left(K^- - k_1^- - \frac{\widehat{\boldsymbol{k}}_{2\perp}^2 + m_2^2 - io}{K^+ - \widehat{k}_1^+}\right)}, \tag{3}$$

where the "hat" indicates operator form in the "kinematic" momentum $(k_1^+, \boldsymbol{k}_{1\perp})$, and $\widehat{k}_{1\mathrm{on}}^- = \frac{\widehat{\boldsymbol{k}}_{1\perp}^2 + m_1^2}{\widehat{k}_1^+}$ and $\widehat{k}_{2\mathrm{on}}^- = \frac{(\boldsymbol{K}_\perp - \widehat{\boldsymbol{k}}_{1\perp})^2 + m_2^2}{K^+ - \widehat{k}_1^+}$. The basis states for functions of the kinematical light-front momentum are defined by $\langle x_i^- \boldsymbol{x}_{i\perp}\,|k_i^+ \boldsymbol{k}_{i\perp}\rangle = \exp[-\mathrm{i}(\frac{1}{2}k_i^+ x_i^- - \boldsymbol{k}_{i\perp}.\boldsymbol{x}_{i\perp})]$, which are eigenfunctions of the momentum operators $(\widehat{k}_i^+, \widehat{\boldsymbol{k}}_{i\perp})$ and the free energy operator $\widehat{k}_{i\mathrm{on}}^-$. The states $|k^+\boldsymbol{k}_\perp\rangle$ form a complete basis in the space of functions of the kinematical variables, e.g., $\int \frac{dk^+ d^2k_\perp}{2(2\pi)^3}\,|k^+\boldsymbol{k}_\perp\rangle\langle k^+\boldsymbol{k}_\perp\,| = 1$ with normalization $\langle k'^+\boldsymbol{k}'_\perp|k^+\boldsymbol{k}_\perp\rangle = 2(2\pi)^3\delta(k'^+ - k^+)\delta(\boldsymbol{k}'_\perp - \boldsymbol{k}_\perp)$.

The global light-front time propagator of the free two-boson system between the hyperplanes $x_1^+ = x_2^+ = x^+$ and $x_1'^+ = x_2'^+ = x'^+$, is the Fourier transform of

$$g_0(K) = |G_0(K)| := \int dk_1'^- dk_1^-\,\langle k_1'^-|\,G_0(K)\,|k_1^-\rangle = \frac{\mathrm{i}g_0^{(2)}(K)}{\widehat{k}_1^+(K^+ - \widehat{k}_1^+)}, \tag{4}$$

where $K^+ > 0$ is used without any loss of generality. In Eq. (4) the vertical bar | indicates that the dependence on k_1^- is integrated out. The bar on the left of the Green's function represents integration on k_1^- in the bra-state, the bar on the right in the ket-state.

The free two-body light-front Green's function, $g_0^{(2)}(K)$ in Eq. (4) is a particular case of the light-front Green's function for N particles:

$$g_0^{(N)}(K) = \left[\prod_{j=1}^{N} \theta(\widehat{k}_j^+)\theta(K^+ - \widehat{k}_j^+)\right] \left(K^- - \widehat{K}_0^{(N)-} + \mathrm{io}\right)^{-1} , \tag{5}$$

where $\widehat{K}_0^{(N)-} = \sum_{j=1}^{N} \widehat{k}_{j\ on}^-$ is the free light-front Hamiltonian. The difference between the free two-body light-front Green's function and $g_0(K)$, Eq. (4), is the phase-space factor for particles 1 and 2.

The covariant two-body propagator is the solution of

$$G(K) = G_0(K) + G_0(K)V(K)G(K) = G_0(K) + G_0(K)T(K)G_0(K). \tag{6}$$

It gives the individual time propagation of the two-body system. The interacting propagator between hyperplanes of $x^+=$ const, is given by $g(K) \equiv |G(K)|$.

Our goal is the decomposition of the kernel of the integral equation of the global two-body propagator, $g(K)$, with an effective interaction irreducible with respect to the light-front propagation of the intermediate two-body system. For that purpose, we use the quasi-potential reduction of ref. [5], where the transition matrix $T(K)$ and the Bethe-Salpeter amplitude $|\Psi_\mathrm{B}\rangle$ of the covariant BSE can be obtained with the help of an auxiliary Green's function $\widetilde{G}_0(K)$, which we choose as

$$\widetilde{G}_0(K) := G_0(K)|g_0^{-1}(K)|G_0(K) , \tag{7}$$

because it allows to make explicit the two-boson propagation on the global light-front time in the intermediate states [1]. The transition matrix satisfies

$$\begin{aligned} T(K) &= W(K) + W(K)\widetilde{G}_0(K)T(K), & (8)\\ W(K) &= V(K) + V(K)[G_0(K) - \widetilde{G}_0(K)]W(K) . & (9)\end{aligned}$$

We introduce the auxiliary three-dimensional transition matrix

$$t(K) = g_0(K)^{-1}|G_0(K)T(K)G_0(K)|\, g_0(K)^{-1} . \tag{10}$$

The interaction in the integral equation for $t(K)$, has the desired property to be irreducible in respect light-front two-body propagation. The auxiliary transition matrix is the solution of a three-dimensional integral equation, derived from Eq. (8):

$$\begin{aligned} t(K) &= w(K) + w(K)g_0(K)t(K), & (11)\\ w(K) &= g_0(K)^{-1}|G_0(K)W(K)G_0(K)|\, g_0(K)^{-1} . & (12)\end{aligned}$$

From Eqs. (6) and (11), one has:

$$g(K) = g_0(K) + g_0(K)w(K)g(K) . \tag{13}$$

The dynamics of the interacting two-particle system can be fully described by its propagation between hyperplanes $x^+ = x^0 + x^3 =$ const in light-front dynamics [2], and covariant the transition matrix can be expressed in terms of $t(K)$ [1].

3 Light-Front Two-Particle Green's Function

The integral equation satisfied by the light-front Green's function, derived from Eq. (13), is

$$g^{(2)}(K) = g_0^{(2)}(K) + g_0^{(2)}(K)\nu(K)g^{(2)}(K) \ , \tag{14}$$

where $g^{(2)}(K) \equiv -\mathrm{i}\widehat{\Omega}g(K)\widehat{\Omega}$, $\nu(K) = \mathrm{i}\widehat{\Omega}^{-1}w(K)\widehat{\Omega}^{-1}$ and the phase space operator is $\widehat{\Omega} := \sqrt{\widehat{k}_1^+(K^+ - \widehat{k}_1^+)}$.

The interaction $w(K)$, Eq. (12), expanded according to Eq. (9) in first order of the driving term $V(K)$, is given by

$$w^{(2)}(K) = g_0(K)^{-1}|G_0(K)V(K)G_0(K)|\, g_0(K)^{-1} \ . \tag{15}$$

The matrix element $\langle k_1'^+ \boldsymbol{k}_{1\perp}'|w^{(2)}(K)|k_1^+ \boldsymbol{k}_{1\perp}\rangle$ is obtained from Eq. (15):

$$\begin{aligned} \langle k_1'^+ \boldsymbol{k}_{1\perp}'|w^{(2)}(K)|k_1^+ \boldsymbol{k}_{1\perp}\rangle &= (\mathrm{i}g_S)^2 \frac{\theta(k_1^+ - k_1'^+)}{(k_1^+ - k_1'^+)} \frac{\mathrm{i}}{K^- - K_0^{(3)-} + \mathrm{i}o} \\ &+ [k_1' \leftrightarrow k_1] \ , \end{aligned} \tag{16}$$

where $K_0^{(3)-} = \frac{\boldsymbol{k}_{1\perp}'^2 + m_1^2}{k_1'^+} - \frac{(\boldsymbol{K}_\perp - \boldsymbol{k}_{1\perp})^2 + m_2^2}{K^+ - k_1^+} - \frac{(\boldsymbol{k}_{1\perp}' - \boldsymbol{k}_{1\perp})^2 + \mu^2}{k_1^+ - k_1'^+}$.

The second order term in the expansion of $w(K)$, is given by

$$\begin{aligned} w^{(4)}(K) &= g_0(K)^{-1}|G_0(K)V(K)G_0(K)V(K)G_0(K)|\, g_0(K)^{-1} \\ &- g_0(K)^{-1}|G_0(K)V(K)\widetilde{G}_0(K)V(K)G_0(K)|\, g_0(K)^{-1} \ . \end{aligned} \tag{17}$$

The second term in the r.h.s of Eq. (17) comes from the iteration of $w^{(2)}(K)$, which is $w^{(2)}g_0(K)w^{(2)}$. The subtraction of the iterated term in Eq. (17) cancels the corresponding terms, such that the matrix element $\langle k_1'^+ \boldsymbol{k}_{1\perp}'|w^{(4)}(K)|k_1^+ \boldsymbol{k}_{1\perp}\rangle$ is two-body irreducible. The final form is:

$$\begin{aligned} \langle k_1'^+ \boldsymbol{k}_{1\perp}'|w^{(4)}(K)|k_1^+ \boldsymbol{k}_{1\perp}\rangle &= \frac{(\mathrm{i}g_S)^4}{2(2\pi)^3} \int dp_1^+ d^2p_{1\perp} \frac{\theta(p_1^+)}{p_1^+} \frac{\theta(K^+ - p_1^+)}{(K^+ - p_1^+)} \\ \times & \frac{\theta(k_1'^+ - p_1^+)}{(k_1'^+ - p_1^+)} \frac{\mathrm{i}}{K^- - \frac{\boldsymbol{p}_{1\perp}^2 + m_1^2}{p_1^+} - \frac{(\boldsymbol{K}_\perp - \boldsymbol{k}_{1\perp}')^2 + m_2^2}{K^+ - k_1'^+} - \frac{(\boldsymbol{k}_{1\perp}' - \boldsymbol{p}_{1\perp})^2 + \mu^2}{k_1'^+ - p_1^+} + \mathrm{i}o} \\ \times & \frac{\mathrm{i}}{K^- - \frac{\boldsymbol{k}_{1\perp}^2 + m_1^2}{k_1^+} - \frac{(\boldsymbol{K}_\perp - \boldsymbol{k}_{1\perp}')^2 + m_2^2}{K^+ - k_1'^+} - \frac{(\boldsymbol{k}_{1\perp}' - \boldsymbol{p}_{1\perp})^2 + \mu^2}{k_1'^+ - p_1^+} - \frac{(\boldsymbol{p}_{1\perp} - \boldsymbol{k}_{1\perp})^2 + \mu^2}{p_1^+ - k_1^+} + \mathrm{i}o} \\ \times & \frac{\theta(p_1^+ - k_1^+)}{(p_1^+ - k_1^+)} \frac{\mathrm{i}}{K^- - \frac{\boldsymbol{k}_{1\perp}^2 + m_1^2}{k_1^+} - \frac{(\boldsymbol{K}_\perp - \boldsymbol{p}_{1\perp})^2 + m_2^2}{K^+ - p_1^+} - \frac{(\boldsymbol{p}_{1\perp} - \boldsymbol{k}_{1\perp})^2 + \mu^2}{p_1^+ - k_1^+} + \mathrm{i}o} \\ + & [k_1' \leftrightarrow k_1] \ . \end{aligned} \tag{18}$$

Introducing the interaction between light-front states which creates or destroys a quantum of the intermediate boson, σ defined by the matrix elements

$$\begin{aligned} \langle q k_\sigma | v | k \rangle &= -2(2\pi)^3 \delta(q + k_\sigma - k) \frac{g_S}{\sqrt{q^+ k_\sigma^+ k^+}} \theta(k_\sigma^+) \\ \langle q | v | k_\sigma k \rangle &= -2(2\pi)^3 \delta(k + k_\sigma - q) \frac{g_S}{\sqrt{q^+ k_\sigma^+ k^+}} \theta(k_\sigma^+) , \end{aligned} \tag{19}$$

for the model defined by the Lagrangian of Eq. (1), we can rewrite the first and second order terms $w^{(2)}$ and $w^{(4)}$ in the form of the effective interaction terms $\nu^{(2)}(K)$ and $\nu^{(4)}(K)$, respectively, as:

$$\langle k_1'^+ \boldsymbol{k}_{1\perp}' | \nu^{(2)}(K) | k_1^+ \boldsymbol{k}_{1\perp} \rangle = \langle k_1'^+ \boldsymbol{k}_{1\perp}' | v g_0^{(3)}(K) v | k_1^+ \boldsymbol{k}_{1\perp} \rangle , \tag{20}$$

$$\langle k_1'^+ \boldsymbol{k}_{1\perp}' | \nu^{(4)}(K) | k_1^+ \boldsymbol{k}_{1\perp} \rangle = \langle k_1'^+ \boldsymbol{k}_{1\perp}' | v g_0^{(3)}(K) v g_0^{(4)}(K) v g_0^{(3)}(K) v | k_1^+ \boldsymbol{k}_{1\perp} \rangle . \tag{21}$$

Therefore, the effective interaction $\nu(K)$ up to second order in the expansion is the sum of Eqs. (20) and (21):

$$\nu(K) \approx v \left(g_0^{(3)}(K) + g_0^{(3)}(K) v g_0^{(4)}(K) v g_0^{(3)}(K) \right) v . \tag{22}$$

We identify in Eq. (22) the propagation of the intermediate three-particle system with corrections up to second order in the coupling constant. It is easy to imagine that the interacting three-particle Green's function should be obtained if the expansion of the auxiliary interaction $w(K)$ is performed to all orders in the ladder using Eqs. (9) and (12). Thus, the effective interaction should be $\nu(K) = v g^{(3)}(K) v$. From Eq. (22) one sees that $g^{(3)}$ is as function of the four-particle Green's function, which should be coupled to the five-body one, and so on. Therefore, one can construct a hierarchy of coupled equations for Green's functions with any number of particles.

4 Hierarchy Equations and Summary

The light-front Green's function of the two-body system expressing the covariant BS equation is written as:

$$\begin{aligned} g^{(2)}(K) &= g_0^{(2)}(K) + g_0^{(2)}(K) v g^{(3)}(K) v g^{(2)}(K) , \\ g^{(3)}(K) &= g_0^{(3)}(K) + g_0^{(3)}(K) v g^{(4)}(K) v g^{(3)}(K) , \\ g^{(4)}(K) &= g_0^{(4)}(K) + g_0^{(4)}(K) v g^{(5)}(K) v g^{(4)}(K) , \\ &\cdots \\ g^{(N)}(K) &= g_0^{(N)}(K) + g_0^{(N)}(K) v g^{(N+1)}(K) v g^{(N)}(K) , \\ &\ldots , \end{aligned} \tag{23}$$

This set of equations contains all two-body irreducible diagrams with exception of those including closed loops of bosons Φ_1 and Φ_2 and part of the cross-ladder diagrams. The truncation of the light-front Fock space allows only two bosons states with two particles Φ_1 and Φ_2 in the intermediate states, without any restriction on the number of bosons σ, which excludes the complete representation of the crossed ladder diagrams. It also resembles the iterated resolvent method of ref. [6]. To get the two-body propagator for light-front times in the covariant ladder approximation, the kernel of the hierarchy equations should be restricted.

A systematic expansion can be obtained by truncating the light-front Fock space up to N particles in the intermediate states (boson 1, boson 2 and $N-2$ σ's) in the set of Eqs. (23), which amounts to substituting $g^{(N)}(K) \to g_0^{(N)}(K)$. By restricting to up to four-particles in the intermediate state propagation, we get the following nonperturbative equation for the Green's function:

$$g^{(2)}(K) = g_0^{(2)}(K) + g_0^{(2)}(K) v g^{(3)}(K) v g^{(2)}(K) , \tag{24}$$

$$g^{(3)}(K) = g_0^{(3)}(K) + g_0^{(3)}(K) v g_0^{(4)}(K) v g^{(3)}(K) . \tag{25}$$

The kernel of Eq. (24) still contains an infinite sum of light-front diagrams, that are obtained by solving Eq. (25). To get the ladder approximation up to order g_S^4, or up to second order in the quasi-potential expansion, with the effective interaction of Eq. (23), only the free and first order terms are kept in Eq. (25), while restricting it only to the ladder.

In summary, we discussed the general framework for constructing the light-front two-body Green's function with quasi-potential reduction of the Bethe-Salpeter equation in the light-front. We displayed a set of coupled hierarchy equations which gives the two-body propagator in several cases, including the ladder approximation. Finally we pointed out that the truncation in the hierarchy can be performed consistently by approximating the N-body Green's function with the free operator, i.e., the light-front N-body intermediate state is accounted in lowest order in the two-body propagation.

Acknowledgement. We would like to thank Fundação de Amparo à Pesquisa do Estado de São Paulo (FAPESP) and the Conselho Nacional de Desenvolvimento Científico e Tecnológico (CNPq) of Brazil for partial support.

References

1. J.H.O. Sales, T. Frederico, B.V. Carlson and P.U. Sauer, Phys. Rev. **C61**, 044003 (2000); *ibid.*, Phys. Rev. **C63**, 064003 (2001)
2. P.A.M. Dirac, Rev. Mod. Phys. **21**, 392 (1949)
3. S. Weinberg, Phys. Rev. **150**, 1313 (1966)
4. B.C. Tiburzi and G.A. Miller, hep-ph/0205109
5. R.M. Woloshyn and A.D. Jackson, Nucl. Phys. **B64**, 269 (1973)
6. S.J. Brodsky, H.-C. Pauli and S.S. Pinsky, Phys. Rep. **301**, 299 (1998)

Few-Body Systems Suppl. 15, 237–245 (2003)

Energy Spectra of $^{9}_{\Lambda}$Be

E. Cravo[1], A.C. Fonseca[1,*], Y. Koike[2]

[1] Centro de Física Nuclear, University of Lisbon, Av. Prof. Gama Pinto 2, 1649-003 Lisbon, Portugal

[2] Science Research Center, Hosei University, 2-17-1, Fujimi, Chiyoda-ku, Tokyo 102, Japan

Abstract. The hypernucleus $^{9}_{\Lambda}$Be is investigated using the $\alpha + \Lambda + \alpha$ cluster model. The corresponding Faddeev equations are solved for different $\alpha - \alpha$ and $\Lambda - \alpha$ interactions that describe both ^{8}Be and $^{5}_{\Lambda}$He energy spectra. For the ground state $J^{\pi} = \frac{1}{2}^{+}$ and excited states $\frac{3}{2}^{+}$ and $\frac{5}{2}^{+}$ we include the Coulomb repulsion between the α's and calculate, in addition to the energy eigenvalues, the $\alpha-$ and $\Lambda-$particle mean radius, the rms charge radius, the electric quadrupole moment (Q), as well as the magnetic dipole. Structural differences between ^{9}Be and $^{9}_{\Lambda}$Be lead to values of Q and μ that have opposite sign. Unlike previous theoretical work we find only two degenerate negative parity resonances $\frac{1}{2}^{-}(\frac{3}{2}^{-})$ but, in addition, we get two degenerate positive parity resonances with $J^{\pi} = \frac{7}{2}^{+}(\frac{9}{2}^{+})$ at higher energy which, together with the bound states, complete the positive parity rotational band.

1 Introduction

As pointed out by Feshbach [1] a number of years ago, one of the main purposes of hypernucleon physics is to study new many-body spectroscopy through the implant of an "impurity" in the condensed matter sense to unveil dynamical symmetries that are forbidden in ordinary nuclei by the Pauli principle. One nucleus where this phenomenon plays an important role is $^{9}_{\Lambda}$Be which, vis-à-vis ^{9}Be, presents, not only a very different spectra of bound states and resonances, but also different dynamical structure, possibly leading to surprising results for observables such as the rms charge radius, the quadrupole moment Q and the magnetic moment μ. This investigation, together with a new search for resonances in the continuum, is the main goal of the present work whose extended version is published in ref. [2]. Our work follows the traditional three-body

* *E-mail address:* fonseca@cii.fc.ul.pt

$\alpha-$cluster approach which has been used before by some of us to study ^{9}Be [3, 4] and also by others [5, 6] for $^9_\Lambda$Be.

Unlike ^{9}Be, which is typically a Boromean nucleus due to neither ^{5}He nor ^{8}Be being stable nuclei, $^9_\Lambda$Be may be thought as $^5_\Lambda$He $+\alpha$ due to the existence of $^5_\Lambda$He as a stable hypernucleus with $B_\Lambda \simeq 3.12$ MeV. Given the small binding energy of $^5_\Lambda$He relative to the α-particle binding and the large energy gap between $\Lambda+^4$He and $^4_\Lambda$He $+ n$, the natural choice is a three-body cluster model based on two alpha particles and the hyperon Λ. The difficulty now lies on the choice of $\alpha - \alpha$ and $\Lambda - \alpha$ interactions that are required to map the low lying spectra of ^{8}Be and $^5_\Lambda$He and provide, in the case of $\alpha - \alpha$, a reasonable description of Pauli blocking between nucleons in separate clusters.

A number of $\alpha - \alpha$ interactions already exist; some treat Pauli blocking by means of a short range repulsion, others by including Pauli forbidden states which are subsequently removed. Therefore, we use, not only Ali-Bodmer (AB) and Chien-Brown (CB) potentials, but also one derived from resonating group theory (UIM) which is non local due to the exchange nature of the RGM core potential (see ref. [2] for further details).

For the $\Lambda - \alpha$ interaction we use a number of effective potentials that have been developed over the years using different folding techniques or phenomenology. They are the Tang-Herndon (TH) which has a simple Gaussian shape, the Dalitz- (DA) and the Deloff-type (DE) potentials which were derived by folding based on $\Lambda - N$ interactions with a hard-core proposed by Dalitz et al. and Deloff, and the Maeda-Schmid (MS) which uses two Wood-Saxon phenomenological potentials to fit the $\Lambda - \alpha$ interaction also calculated by folding. All potentials reproduce the $^5_\Lambda$He binding energy reasonably well (see ref. [2] for further details).

For the chosen $\alpha - \alpha$ and $\Lambda - \alpha$ interactions we solve the bound state Faddeev equations [7] including the Coulomb repulsion between the α's. The general form of the equations was first proposed by Lehman et al. [8] and subsequently generalized by Cravo [4] to study the structure of ^{9}Be. The $\alpha - \alpha$ interaction is taken as a local interaction in different $\alpha - \alpha$ partial waves (s-, d-, and g-waves) or as a finite rank operator in the case of the RGM-based potential. The $\Lambda - \alpha$ interaction is always expanded as a sum of separable terms using the method developed by Koike [9]. The convergence of the calculation in terms of rank "r" and number of partial waves in the $\Lambda - \alpha$ interaction is shown in ref. [2]; all s-, p-, and d-waves are included.

2 Structure of $^9_\Lambda$Be Bound States

In the absence of a $\Lambda - \alpha$ spin-orbit force, two of the three bound states we find are degenerate ($J^\pi = \frac{3}{2}^+$and $\frac{5}{2}^+$). The ground state, $J^\pi = \frac{1}{2}^+$, lies 2.3 - 3.0 MeV below the excited states. Our results are shown in Table 1 for $J^\pi = \frac{1}{2}^+$ and in Table 2 for the excited states $\frac{3}{2}^+(\frac{5}{2}^+)$. The Coulomb interaction has been included up to partial waves L = 6 for convergence. Experimental values are taken from refs. [10] and [11]. The results indicate that the calculated binding energies depend more on the choice of $\Lambda - \alpha$ interaction than on the $\alpha - \alpha$ potential one uses. This is expected given the lack of sufficient experimental information on ${}^{5}_{\Lambda}$He to constrain the $\Lambda - \alpha$ interaction. Although the calculated binding energies for the ground state and excited state span a wide range of energies, the excitation energy between the two states is reasonably independent of the combination of $\alpha - \alpha$ and $\Lambda - \alpha$ potentials one uses, and in close agreement with the experimental excitation energy (3.1 MeV). The best results, compared to data, are obtained with MS $\Lambda - \alpha$ potential in combination with AB (or CB) $\alpha - \alpha$ interaction.

Table 1. $\frac{1}{2}^+$ binding energies (MeV) for different $\alpha - \alpha$ and $\Lambda - \alpha$ interactions. The Coulomb repulsion has been included in all partial waves needed for convergence.

	UIM	AB	CB
TH	-5.96	-5.98	-6.02
MS	-6.50	-6.73	-6.75
DE	-7.08	-7.36	-7.37
DA	-7.74	-8.27	-8.19
Exp. [10]		-6.71	

Table 2. Same as in Table 1 for $\frac{3}{2}^+(\frac{5}{2}^+)$ binding energies (MeV).

	UIM	AB	CB
TH	-3.44	-3.25	-3.17
MS	-4.05	-3.97	-3.80
DE	-4.63	-4.59	-4.41
DA	-5.18	-5.35	-5.05
Exp. [11]		-3.67	

Next we proceed to characterize these two states in terms of $\alpha-$particle and $\Lambda-$particle point rms radius, charge radius, electric quadrupole moment and magnetic dipole. Although most of these observables have not been measured for ${}^{9}_{\Lambda}$Be, they provide a better understanding of the structure of this hypernucleus and allow a comparison with ^{9}Be. All calculations include the Coulomb repulsion between the α's.

Both the $\alpha-$particle and the $\Lambda-$particle r.m.s. radius are calculated in the point-particle approximation. Although both values are slightly bigger for the excited state $\frac{3}{2}^+(\frac{5}{2}^+)$ than for the ground state $\frac{1}{2}^+$, they are comparable and almost independent of the choice of potentials. Therefore we can claim that $\bar{r}_\alpha \simeq 1.83$ fm and $\bar{r}_\Lambda \simeq 2.15$ fm. Comparing with ^{9}Be [4] we find that $\bar{r}_\alpha$ is about the same, but that $\bar{r}_n \simeq 3.1$ fm. Therefore one may conclude that the neutron in ^{9}Be moves well outside the average location of the α−particles relative to the center of mass of the system, while the Λ, in $^9_\Lambda$Be, shares its position with the α's. This gives a certain credibility to hypernuclear models that take $^5_\Lambda$He as a cluster.

Next we calculate the values for the charge radius of $^9_\Lambda$Be including the contribution of the alpha particle form factor. The result is also almost independent of the potential choice or state, leading to a value for $r_c \simeq 2.48$ fm. This value compares with $r_c = 2.48$ fm for ^{9}Be calculated with AB potential. This clearly indicates that the size of both nuclei is about the same from the view point of the α subclusters but that the neutron in ^{9}Be sticks much further out than the Λ in $^9_\Lambda$Be. We find no evidence of contraction of the ^{8}Be core due to the substitution of the neutron by the Λ− particle as pointed out by others. This may indicate that the Pauli repulsion between the α's is strong enough to block any additional binding effect created by the Λ− particle.

Table 3. Electric Quadrupole moment Q in $e$fm^2 of the $\frac{3}{2}^+$ and $\frac{5}{2}^+$ excited states for all $\alpha-\alpha$ and $\Lambda-\alpha$ potentials. For comparison we also show the calculated quadrupole moment of ^{9}Be $\frac{3}{2}^-$ state [4].

	$\frac{3}{2}^+$			$\frac{5}{2}^+$		
	UIM	AB	CB	UIM	AB	CB
TH	-5.53	-5.98	-6.00	-7.89	-8.54	-8.57
MS	-5.06	-5.27	-5.26	-7.24	-7.54	-7.52
DE	-4.91	-5.08	-5.05	-7.02	-7.26	-7.21
DA	-5.07	-5.07	-5.12	-7.24	-7.25	-7.32
^{9}Be		4.79				

Further understanding is obtained by calculating the nonvanishing electric and magnetic moments. The electric quadrupole moment Q is zero for the $\frac{1}{2}^+$ state due to Wigner Echart theorem, but for $\frac{3}{2}^+$ and $\frac{5}{2}^+$ states we show in Table 3 the corresponding results for different potential models. The quadrupole moment, as expected, is large but of negative sign: on the average we get $Q \simeq -5.28$ $e$fm^2 for the $\frac{3}{2}^+$ state and $Q \simeq -7.55$ $e$fm^2 for the $\frac{5}{2}^+$ state. Here we observe an important structural difference relative to ^{9}Be that carries a positive quadrupole moment whose experimental value is 5.3 ± 0.3 $e$fm^2 [12]. This is a direct consequence of the underlying differences in the dominant partial waves that bind $^9_\Lambda$Be and ^{9}Be as three-body cluster nuclei. In the $\frac{3}{2}^+$ state of $^9_\Lambda$Be the dominant channel in the wave function involves the two α's in relative orbital angular momentum $L = 2$ while the Λ carries orbital angular momentum $\ell = 0$ relative to the center of mass of the α's. Since the matrix element for Q involves $J = J_z = \frac{3}{2}$, there are two possible contributions to Q coming from $(L = 2, M = 2)$ together with Λ spin projection $m_\Lambda = -\frac{1}{2}$ and $(L = 2, M = 1)$ with $m_\Lambda = \frac{1}{2}$. The well known angular distribution function $Y_{22}(\hat{R})$ for the $\boldsymbol{R}$ vector between the α's is an elongated torus revolving around the z axis but sitting on the xy plane, leading to a large negative contribution to the quadrupole moment which is not compensated by the smaller but positive contribution from $Y_{21}(\hat{R})$. Comparing with ^{9}Be we find instead that the dominant $\frac{3}{2}^-$ wave function components are $L = 0$ and $L = 2$ with $\ell = 1$ for the neutron orbital angular momentum relative to the center of mass of the α's. The $L = 0$ contribution to Q is zero and from $L = 2$ we have three terms: $(L = 2, M = 0)$ with $m_j = \frac{3}{2}$,$(L = 2, M = 1)$ with $m_j = \frac{1}{2}$ and $(L = 2, M = 2)$ with $m_j = -\frac{1}{2}$ where $m_j = m + m_n$ such that m is the z component of $\boldsymbol{\ell}$ and m_n is the neutron spin projection. The angular distribution function for $Y_{20}(\hat{R})$ which stretches along the z axis gives rise to a large positive contribution to Q that dominates the other two, leading to a calculated value of 4.79 $e$fm^2 [4] for the AB $\alpha - \alpha$ potential. Therefore one important conclusion is that $^9_\Lambda$Be is an oblate nucleus while ^{9}Be is prolate. This behaviour is further enhanced in $^9_\Lambda$Be $\frac{5}{2}^+$ state because $J_z = \frac{5}{2}$ implies the contribution of $(L = 2, M = 2)$ with $m_\Lambda = \frac{1}{2}$ alone, leading to a single large negative contribution from $Y_{22}(\hat{R})$.

Table 4. Magnetic moment μ (in μ_N) of the $\frac{3}{2}^+$ and $\frac{5}{2}^+$ excited states for all $\alpha - \alpha$ and $\Lambda - \alpha$ potentials. We also show, for comparison, the calculated magnetic moment of ^{9}Be [4] and, in parenthesis, the corresponding value one gets when the intrinsic magnetic moment of the neutron is the same as Λ and one neglects the contribution from the neutron form factor.

	$\frac{3}{2}^+$			$\frac{5}{2}^+$		
	UIM	AB	CB	UIM	AB	CB
TH	1.26	1.25	1.25	0.38	0.37	0.37
MS	1.26	1.26	1.26	0.38	0.38	0.38
DE	1.27	1.27	1.27	0.39	0.39	0.39
DA	1.27	1.27	1.27	0.39	0.39	0.39
^{9}Be		-1.15	(-0.15)			

Since the ground state of $^9_\Lambda$Be is $\frac{1}{2}^+$ the dominant channels involve $\ell = L = 0$. Therefore the magnetic moment $\mu = \mu_\Lambda$ for such state. Nevertheless the same sign change effect relative to ^{9}Be takes place in the magnetic moment μ of $^9_\Lambda$Be excited states which are shown in Table 4: again, on the average, we find $\mu \simeq 1.26\ \mu_N$ for the $\frac{3}{2}^+$ state and $\mu = 0.38\ \mu_N$ for the $\frac{5}{2}^+$ state. Although the $\Lambda-$particle and the neutron have magnetic moments of the same sign, once embeded in ^{8}Be, one gets magnetic moments of opposite sign for reasons similar to the ones explained for Q. Even if we make the intrinsic magnetic moment of the neutron $\mu_n = \mu_\Lambda$ in ^{9}Be we get a smaller but negative value as shown in Table 4 in parenthesis. The difference between this new result and the value obtained for $^9_\Lambda$Be is solely due to structural differences between the two nuclei.

3 Resonances in ${}^{9}_{\Lambda}$Be

In the present work we find two pairs of resonances: one below breakup threshold and one above breakup threshold. The first one is $\frac{1}{2}^-(\frac{3}{2}^-)$ while the second one is $\frac{7}{2}^+(\frac{9}{2}^+)$. The results are shown in Table 5 and 6 respectively. The $\frac{1}{2}^-(\frac{3}{2}^-)$ resonance is located just above the $\Lambda-\alpha$ threshold and has a large decay width. Nevertheless, unlike the work in ref. [5], we find no negative parity states above breakup threshold, but unveil instead a positive parity state $\frac{7}{2}^+(\frac{9}{2}^+)$ that relies on the $\alpha-\alpha$ g-wave interaction. This last resonance has not been found before by other continuum calculations but, together with the $\frac{1}{2}^+$ bound state and the $\frac{3}{2}^+(\frac{5}{2}^+)$ excited state, complete the even parity rotational band. On the contrary, the absence of a large number of negative parity states in ${}^{9}_{\Lambda}$Be is two fold: **a)** a direct result of the Pauli principle that allows the Λ to stay predominantly in s-state while keeping the fifth neutron in ^{9}Be in a p-state; **b)** proper coupling to the decay channel $\alpha+\Lambda+\alpha$ and correct location of all underlying branch cuts of the $\alpha-\alpha$ system in the complex energy plane. We have tried to search for $\frac{5}{2}^-(\frac{7}{2}^-)$ resonances corresponding to total orbital angular momentum $L=3$ but could not find any. In Table 7 we show the excitation energy E_x relative to the $\frac{1}{2}^+$ ground state for both $\frac{1}{2}^-(\frac{3}{2}^-)$ and $\frac{7}{2}^+(\frac{9}{2}^+)$ resonances. We find that $5.7 < E_x < 8.3$ for the negative parity state, while $8.3 \leq E_x \leq 10.3$ for the positive parity resonance. If we consider, for example, the MS $\Lambda-\alpha$ potential together with CB $\alpha-\alpha$ interaction the $\frac{1}{2}^-(\frac{3}{2}^-)$ resonance corresponds to $\simeq 6.9$ MeV excitation and the $\frac{7}{2}^+(\frac{9}{2}^+)$ to $\simeq 9.7$ MeV.

Table 5. Position and width of the $\frac{1}{2}^-(\frac{3}{2}^-)$ resonance (MeV) for different $\alpha-\alpha$ and $\Lambda-\alpha$ potentials in the absence of the Coulomb repulsion.

	UIM		AB		CB	
	E	$\Gamma/2$	E	$\Gamma/2$	E	$\Gamma/2$
TH	-2.00	0.62	-1.98	0.59	-2.03	0.68
MS	-1.69	0.56	-1.65	0.55	-1.67	0.64
DE	-1.89	0.58	-1.84	0.57	-1.86	0.66
DA	-1.92	0.41	-1.89	0.40	-1.87	0.47

So far the most accepted understanding of the level structure of ${}^{9}_{\Lambda}$Be is the work of Yamada et al. [13] where a variational $(\alpha + 3\text{N} + \text{N}) + \Lambda$ cluster model that includes $\alpha-$ particle breakup modes is applied to achieve a unified understanding of this hypernucleus in the energy region up to 20 MeV excitation. They find a very rich spectra of rotational bands with different symmetries [14] of which we can only find the lowest one made up of $\frac{1}{2}^+$, $\frac{3}{2}^+(\frac{5}{2}^+)$, and $\frac{7}{2}^+(\frac{9}{2}^+)$ states. The two negative parity bands are absent from our calculation except for the lowest state $\frac{1}{2}^-(\frac{3}{2}^-)$ in the first band which carries total orbital angular

Table 6. Position and width of the $\frac{7}{2}^+(\frac{9}{2}^+)$ resonance (MeV) for different $\alpha - \alpha$ and $\Lambda - \alpha$ potentials in the absence of the Coulomb repulsion.

	UIM		AB		CB	
	E	$\Gamma/2$	E	$\Gamma/2$	E	$\Gamma/2$
TH	1.29	0.14	0.59	0.06	1.66	0.20
MS	0.51	0.06	0.06	0.03	1.09	0.12
DE	-0.06	0.04	-0.51	0.02	0.51	0.09
DA	-0.30	0.04	-0.64	0.02	0.29	0.09

Table 7. $\frac{1}{2}^-(\frac{3}{2}^-)$ and $\frac{7}{2}^+(\frac{9}{2}^+)$ excitation energies (MeV) for different $\Lambda-\alpha$ and $\alpha-\alpha$ potentials.

	$\frac{1}{2}^-(\frac{3}{2}^-)$			$\frac{7}{2}^+(\frac{9}{2}^+)$		
	UIM	AB	CB	UIM	AB	CB
TH	6.57	5.74	5.79	9.86	8.31	9.48
MS	7.38	6.84	6.90	9.58	8.55	9.66
DE	7.77	7.30	7.35	9.60	8.63	9.72
DA	8.26	8.15	8.13	9.88	9.40	10.29

momentum $L = 1$. Two reasons may explain such disagreement: **a)** lack of internal structure of the α–particle in our calculation; **b)** lack of appropriate coupling to the continuum in their calculation.

As far as we know, the experimental evidence for the spectra of ${}^9_\Lambda$Be comes from ref. [15, 16]. In ref. [15] only two peaks are found above the $\alpha + \Lambda + \alpha$ threshold. The second peak, at about 20 MeV above breakup threshold, is clearly out of the domain of applicability of our calculation due to lack of structure in the $\alpha-$ particle. The first peak, also centered at about 5 MeV above breakup, may not correspond to any of the states we have found because of both location and width; the $\frac{1}{2}^-(\frac{3}{2}^-)$ has the right width but is located very close to the breakup threshold once we add about 1.8 MeV for the Coulomb energy, while the $\frac{7}{2}^+(\frac{9}{2}^+)$ has the correct position but is too narrow to correspond to the experimental peak. Instead the experimental work of ref. [16] finds two resonance peaks corresponding to excitation energies of $E_x = 6$ and 10 MeV above the ground state, which may correspond to the states we find. The first one corresponds to the $\frac{1}{2}^-(\frac{3}{2}^-)$ state (1^-), while the second to the $\frac{7}{2}^+(\frac{9}{2}^+)$ state (4^+). In ref. [16] this last state has been denoted as 3^- in accordance with the theoretical predictions of ref. [13] but, as we mentioned above, we find no additional negative parity resonance to justify such assignment.

4 Conclusions

In the present work we study ${}^{9}_{\Lambda}$Be as a cluster of two α−particles and a Λ and calculate the structure of bound states and resonances. As already found by others [5] we confirm the existence of three even parity bound state $\frac{1}{2}^+, \frac{3}{2}^+, \frac{5}{2}^+$ strongly associated with the $L = 0$ ($\frac{1}{2}^+$) and $L = 2$ ($\frac{3}{2}^+, \frac{5}{2}^+$) rotational states in ^{8}Be. In the absence of $\Lambda - \alpha$ spin-orbit force, the ($\frac{3}{2}^+$) and ($\frac{5}{2}^+$) states are degenerate. Depending on the combination of $\Lambda - \alpha$ and $\alpha - \alpha$ interactions one uses, the energy difference between the two states lies between 2.3 MeV and 3.0 MeV. This compares with the experimental excitation energy of $\simeq$ 3.1 MeV.

In the framework of this model we predict that the electric quadrupole moments of the $\frac{3}{2}^+$ and $\frac{5}{2}^+$ states are negative. The calculated $Q(\frac{3}{2}^+) \simeq -5.28$ $e$fm^2 while $Q(\frac{5}{2}^+) \simeq -7.55$ $e$fm^2. Therefore ${}^{9}_{\Lambda}$Be is an oblate nucleus much like ^{8}Be, while ^{9}Be is prolate with an experimental $Q = 5.3 \pm 0.3$ $e$fm^2. The same sign change takes place if one compares the magnetic moments of both ${}^{9}_{\Lambda}$Be and ^{9}Be nuclei. While ^{9}Be $\frac{3}{2}^-$ ground state has a negative $\mu = -1.18\,\mu_N$, we predict that $\mu(\frac{3}{2}^+) \simeq 1.26\,\mu_N$ and $\mu(\frac{5}{2}^+) \simeq 0.38\,\mu_N$. The $\frac{1}{2}^+$ ground state has $\mu = \mu_\Lambda$. It should be extremely interesting to find out if shell-model calculations of ${}^{9}_{\Lambda}$Be and ^{9}Be would reproduce this sign change for both Q and μ.

In addition, we calculate the charge radius for all ${}^{9}_{\Lambda}$Be states and find it to be the same as in ^{9}Be. Therefore we do not find any contraction of the system due to the strong $\Lambda - \alpha$ s-wave attraction, but find instead that the Λ shares its location with the α's ($\overline{r}_\Lambda \simeq \overline{r}_\alpha \simeq 2$ fm) while the fifth neutron in ^{9}Be sticks much further out ($\overline{r}_n \simeq 3$ fm).

In the resonance region we locate the even parity states ($\frac{7}{2}^+, \frac{9}{2}^+$) corresponding to an excitation energy $E_x \simeq 9.6$ MeV above the ground state, that complete the ^{8}Be like rotational band. Given the similarity between the even parity states in ${}^{9}_{\Lambda}$Be and ^{8}Be we find it plausible to assume these to be ^{8}Be-like states as discussed in Refs. [13, 17].

Close to breakup threshold we find a single pair of negative parity states ($\frac{1}{2}^-, \frac{3}{2}^-$) which depend strongly on the strengh of the $\Lambda - \alpha$ s-wave interaction. Therefore we cannot denote this state as a ^{9}Be like resonance but instead as part of a genuine hypernuclear band [14].

Given the strengh of the $\Lambda - \alpha$ effective p-wave interaction we cannot find ^{9}Be like states independently of the $\Lambda - \alpha$ interaction we choose. If they exist they have such a large width that we cannot find them numerically in our three-body cluster model. Presumably they were found by Yamada et al [13] due to lack of coupling to the continuum in their calculation.

References

1. H. Feshbach: In: *Proc. of the Summer Study Meeting on Kaon Physics and Facilities*, ed. by H. Palevsky, p. 185 (BNL 18335) 1973
2. E. Cravo, A.C. Fonseca, Y. Koike: Phys. Rev. **C66** 014001 (2002)

3. A.C. Fonseca, M.T. Peña: Nucl. Phys. **A487**, 92 (1988)
4. E. Cravo: Phys. Rev. **C54**, 523 (1996)
5. S. Oryu et al.: Few-Body Systems **28**, 103 (2000)
6. Y. Sunami, H. Narumi: Prog. Theor. Phys. **66**, 355 (1981)
7. L.D. Faddeev: JETP **12**, 1014 (1961)
8. D.R. Lehman et al.: Phys. Rev. **C29**, 1450 (1984)
9. Y. Koike: Phys. Rev. **C42**, R2286 (1990)
10. M. Jurič et al.: Nucl. Phys. **B52**, 1 (1973)
11. H. Akikawa et al.: Phys. Rev. Lett. **88**, 082501 (2002)
12. F. Ajzenberg-Selove: Nucl. Phys. **A490**, 1 (1988)
13. T. Yamada et al.: Phys. Rev. **C38**, 854 (1988) and references therein
14. R.H. Dalitz, A. Gal: Phys. Rev. Lett. **36**, 362 (1976)
15. W. Brüeckner et al.: Phys. Lett. **55B**, 107 (1975); **79B**, 157 (1978)
16. S. Ajimura et al.: Nucl. Phys. **A639**, 93C (1998)
17. B.F. Gibson, E.V. Hungerford III: Phys. Rep. **257**, 349 (1995)

Few-Body Systems Suppl. 15, 247–252 (2003)

Analysis of Light Nuclei by the AMD Method with Realistic NN Potentials

Y. Taniguchi, T. Watanabe, N. Sawado, S. Oryu*

Department of Physics, Faculty of Science and Technology, Frontier Research Center for Computational Science, Tokyo University of Science, Noda, Chiba 278-8510, Japan

Abstract. The AMD simulation for light nuclei with the realistic Paris potential and with the Coulomb interaction is investigated. The results for the binding energies are in qualitatively good agreement with ground state energies of nuclei having A=2 to 8. The framework shows a lot of promise for studying the structure of nuclei. Nuclear density distributions are also illustrated.

1 Introduction

The motivation for this work is a desire to formulate a (new) microscopic method with which one can investigate a system composed of a finite number of fermions. We would like to insure that the following properties hold. First of all, the method should handle not only realistic NN forces but also long range forces, and it should include higher partial waves. Secondly, the method should handle cases from low energy to higher energy and permit the treatment of relativistic systems. From the applications point of view, however, the method should be able to handle two-body to A-body systems. One such candidate method is the antisymmetrized molecular dynamics, or AMD, method. The AMD method provides a simulation determined by a time-dependent variational principle which does not require a partial-wave decomposition [1, 2]. It can handle the Coulomb potential as well as three-body forces. Furthermore, it is able to handle heavy ion collisions at medium and higher energies. Unfortunately, the AMD method has not been applied to few-body systems, or the results were rather poor because of lack of precision in the calculations. In this paper we wish to explore the use of the AMD method with realistic NN potentials to calculate the binding energy of light nuclei which could be appear in primordial nucleosynthesis.

* *E-mail address:* oryu@ph.noda.tus.ac.jp

2 Theory

We would like to introduce a time evolution equation for the nuclear system starting from the time dependent variational principle

$$\delta \int_{t_1}^{t_2} dt \frac{\langle \Phi(\{Z\})|i\hbar \frac{d}{dt} - H|\Phi(\{Z\})\rangle}{\langle \Phi(\{Z\})|\Phi(\{Z\})\rangle} = 0\,, \tag{1}$$

where H is the total Hamiltonian of the system, and $|\Phi(\{Z\})\rangle$ is the anti-symmetrized wave function for A nucleons with respect to a set of complex vector variational parameters for each nucleon $\{Z\}=(\mathbf{Z}_1, \mathbf{Z}_2, \mathbf{Z}_3, ..., \mathbf{Z}_A)$. The function is given by a Slater determinant where the i-th nucleon wave function is $\varphi_{\mathbf{Z}_i}(\mathbf{r}_j)$. The individual nucleon wave-packet is defined by the spatial part and the spin-isospin wave functions χ_i, τ_i; i.e.,

$$\varphi_{\mathbf{Z}_i}(\mathbf{r}_j) = \langle \mathbf{r}_j|\varphi_{\mathbf{Z}_i}\rangle = (\frac{2\nu}{\pi})^{3/4} \exp[-\nu(\mathbf{r}_j - \frac{\mathbf{Z}_i}{\sqrt{\nu}})^2 + \frac{1}{2}Z_i^2]\chi_i\tau_i\,, \tag{2}$$

where ν is a width parameter taken as a constant in this work.

The AMD equation is introduced from the usual time dependent variational equation[1, 2]; i.e.,

$$\begin{aligned} \frac{dZ_{j\tau}}{dt} &= \frac{1}{i\hbar}\sum_{i\sigma}(C^{-1})_{i\sigma j\tau}\frac{\partial \mathcal{H}}{\partial Z^*_{i\sigma}} \qquad (\sigma, \tau = x, y, z) \\ C_{i\sigma j\tau} &= \frac{\partial^2}{\partial Z^*_{i\sigma}\partial Z_{j\tau}} \log \mathcal{N}\,, \end{aligned} \tag{3}$$

where $\mathcal{N}$ is the norm of the wave function $|\Phi\rangle$, and $\mathcal{H}$ is the eigenvalue of the Hamiltonian H. In order to obtain the binding energies of nuclei, we use the friction cooling technique putting the complex factor $(\lambda - i\mu)$ on the equation (3), with positive parameters λ and μ, i.e.,

$$i\hbar \sum_{j\tau} C_{i\sigma j\tau}\frac{dZ_{j\tau}}{dt} = (\lambda - i\mu)\frac{\partial \mathcal{H}}{\partial Z^*_{i\sigma}} \qquad (\sigma, \tau = x, y, z)\,. \tag{4}$$

It should be noted that the assumed wave function with a Gaussian shape causes zero point oscillations for the constituent nucleons (T_0). Therefore, the spurious kinetic energy should be removed from the expectation value of Hamiltonian $\mathcal{H}=\langle\Phi|H|\Phi\rangle/\ \mathcal{N} \equiv \langle H\rangle/\mathcal{N}$. Then the genuine eigenvalue can be expressed as

$$\mathcal{E} = \mathcal{H} - \frac{3\hbar^2\nu}{2M}A + T_0(A - N_F)\,, \tag{5}$$

where A is the total mass number, M is the nucleon mass, and N_F is the fragment number. The second term of the r.h.s. is the sum of the zero-point oscillations of A nucleons, and the third term is the correction for the zero-point oscillations of the fragments. Here, T_0 should be $3\hbar^2\nu/2M$ in principle, but T_0 in the third term is usually treated as an adjustable parameter fitted to a binding energy. In the beginning, for the Volkov No. 1 potential [3] T_0=7.7 MeV was used extensively to fit massive nuclei.

3 Calculation and Discussion

3.1 *NN-Potential*

In order to modify our realistic Paris NN potential [4] to fit the AMD method, the potential was transformed into a form with Wigner, Bartlett, Heisenberg, and Majorana terms in which P_σ and P_τ are the spin and isospin exchange operators, while Ω_{LS}, Ω_T, Ω_{SO_2} are the LS, the tensor, and the quadratic spin-orbit operators, respectively. That is, the Paris potential is expressed as

$$\begin{aligned} V(\mathbf{r},p^2) &= V_W(\mathbf{r},p^2) + V_B(\mathbf{r},p^2)P_\sigma - V_H(\mathbf{r},p^2)P_\tau - V_M(\mathbf{r},p^2)P_\sigma P_\tau \\ &\quad + V_{LS}(r)\Omega_{LS} + V_T(r)\Omega_T + V_{SO_2}(r)\Omega_{SO_2} \, . \end{aligned} \tag{6}$$

Furthermore, we converted each spatial term to an equivalent Gaussian-basis function. The merit of the expansion is in the allowed analytic treatment of the potential. Unfortunately, it was found that the original Paris potential did not converge to a stable solution in the AMD calculation, although the LS and the tensor potential terms are omitted in this work. As a consequence, we modified the Paris potential's shape for the triplet central potential ($T = 0$, $S = 1$) (see ref. [5]). Since the modification was made only in the short-range region, the phase shifts differ from the original ones only in the $^3D_1$, $^3D_2$ and $^3D_3$ partial waves at energies higher than 100 MeV, which was discussed in ref. 5.

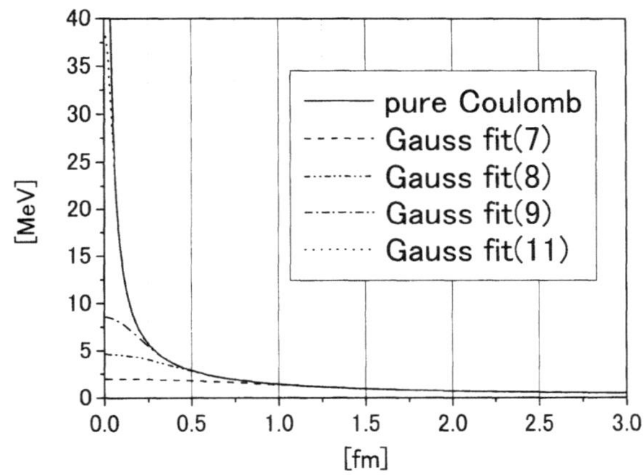

Figure 1. The Coulomb potential fit is illustrated. The Gaussian basis expansions are shown for N=7,8,9 and 11 terms, respectively.

3.2 *Coulomb Potential*

We take into account the Coulomb interaction by expanding the Coulomb potential function e^2/r in a Gaussian basis,

$$\frac{e^2}{r} \sim e^2\sqrt{\nu}\sum_{m=0}^{N} c_n \exp[-(\sqrt{\nu}r/\mu_n)^2]\,. \tag{7}$$

As a result we obtain a very good function in the long range region (up to 20 fm) by $N = 7$ [5]. In order to see the convergence of the Coulomb effect, $N = 7, 8, 9$ and $N = 11$ are chosen (see Fig. 1). Finally, the eleven-term function can fit the pure Coulomb potential in the region from 0.1 fm to 20 fm. The results are shown in Table 1. The Paris potential fits well the experimental data with $T_0 = 7.97$ MeV, except for the three-nucleon system. T_0 was chosen to fit the deuteron binding energy. It is found that the convergence of the Coulomb effect is very good in the binding energy. The error is less than 1 keV with the eleven-term function.

The results agree reasonably well with the experimental data for the binding energies of the massive nuclei, but some discrepancies arise in the very light nuclei, especially for the three-nucleon system (Fig. 2). Now, the energy difference $\Delta E_{\rm cal}$ = BE(^{3}H)-BE(^{3}He) = 0.59 MeV in our calculation comes from the Coulomb energy.

Finally, we illustrate the density distribution in Fig. 3. It is seen that the cluster structure is well developed in the nuclei $5 \leq A \leq 8$ in which the fragment number $N_F = 2$ is adopted, while $N_F = 1$ for $A \leq 4$ nuclei. Fortunately, our formalism already includes the Coulomb force and all higher partial waves of the NN interaction automatically. Furthermore, the three-body force can easily be taken into account in this formalism [6]. These facts suggest the merit of the present method, although many issues still exist in the Faddeev formalism. The present calculation should be improved by adding the LS and tensor forces, as well as three-body forces, but it is a promising method for various applications.

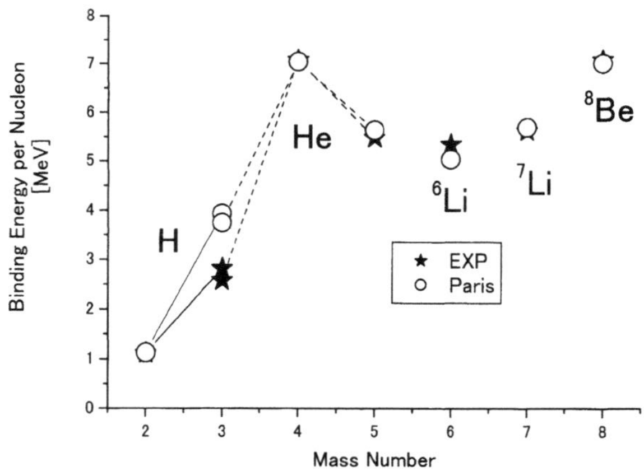

Figure 2. The ground-state binding energies for the light nuclei: ^{2}H, ^{3}H, ^{3}He, ^{4}He, ^{5}He, ^{6}Li, ^{7}Li, ^{7}Be, and ^{8}Be.

Table 1. The binding energies of the ground states for the light nuclei. N stands for the maximum number of the Gaussian basis for the Coulomb potential. ΔE(exp.-cal.) is the energy difference in MeV between the experimental binding energy and the calculated one. "Coulomb" means the Coulomb energy contribution to the binding energy.

	exp.[MeV]	cal.[MeV]	ΔE(exp.-cal.)	Coulomb[MeV]	
^{2}H	2.225	2.225	0	0	
^{3}H	8.452	11.77	-3.32	0	
^{3}He	7.718	11.21	-3.49	0.563	
^{4}He	28.29	28.07	0.22	0.563	N=7
^{5}He*	27.33	28.07	-0.74	0.563	
^{6}Li	31.99	30.12	1.78	0.747	
^{7}Li	39.25	39.67	-0.42	0.741	
^{8}Be*	56.50	55.82	0.68	1.460	
^{2}H	2.225	2.225	0	0	
^{3}H	8.452	11.77	-3.32	0	
^{3}He	7.718	11.19	-3.47	0.583	
^{4}He	28.29	28.05	0.24	0.583	N=8
^{5}He*	27.33	28.05	-0.72	0.583	
^{6}Li	31.99	30.10	1.89	0.767	
^{7}Li	39.25	39.65	-0.40	0.760	
^{8}Be*	56.50	55.77	0.73	1.499	
^{2}H	2.225	2.225	0	0	
^{3}H	8.452	11.77	-3.32	0	
^{3}He	7.718	11.18	-3.46	0.586	
^{4}He	28.29	28.05	0.24	0.586	N=9
^{5}He*	27.33	28.05	-0.72	0.586	
^{6}Li	31.99	30.09	1.90	0.770	
^{7}Li	39.25	39.65	-0.40	0.764	
^{8}Be*	56.50	55.77	0.73	1.506	
^{2}H	2.225	2.225	0	0	
^{3}H	8.452	11.77	-3.32	0	
^{3}He	7.718	11.18	-3.46	0.587	
^{4}He	28.29	28.05	0.24	0.587	N=11
^{5}He*	27.33	28.05	-0.72	0.587	
^{6}Li	31.99	30.09	1.90	0.772	
^{7}Li	39.25	39.64	-0.39	0.765	
^{8}Be*	56.50	55.77	0.73	1.509	

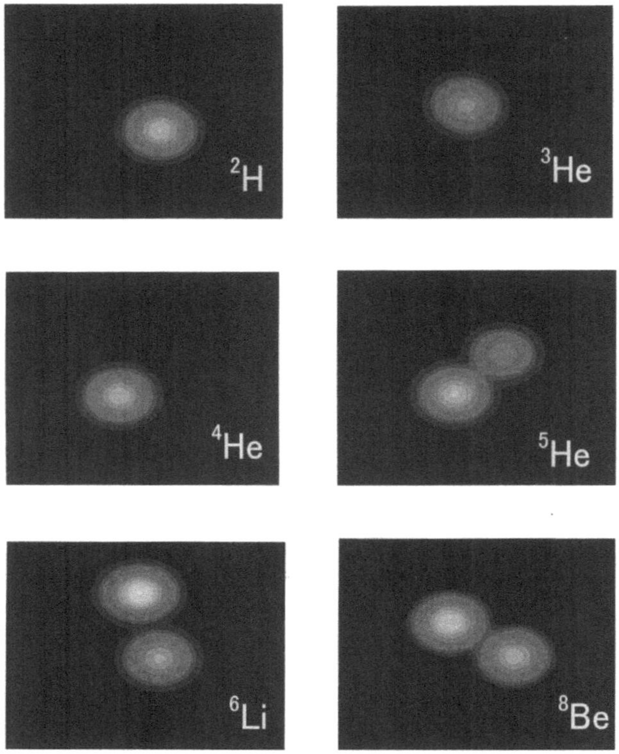

Figure 3. The density of various nuclei with the modified Paris potential. The cluster structure is well developed in the nuclei $5 \leq A \leq 8$.

References

1. A. Ono, H. Horiuchi, T. Maruyama, A. Ohnishi: Prog. Theor. Phys. **87** 1185 (1992)
2. A. Ono, H. Horiuchi, T. Maruyama, A. Ohnishi: Phys. Rev. Lett. **68** 2898 (1992)
3. A.V. Volkov: Nucl. Phys. **74**, 33 (1965)
4. M. Lacombe, B. Loiseau, J.M. Richard, R. Vinh Mau: Phys. Rev. **C 21** 861 (1980)
5. T. Watanabe, Y. Taniguchi, N. Sawado, S. Oryu: In: *Proceedings of the Second Asia Pacific Conference on Few-Body Problems in Physics*, August 27-30, 2002 Shanghai, China. Singapore: World Scientific (to be published)
6. A. Ando, K. Ikeda, A. Tohsaka: Prog. Theor. Phys. **64** 1608 (1980)

Few-Body Systems Suppl. 15, 253–258 (2003)

Few-Body Systems

Evidence of Tensor Correlations in the Nuclear Many-Body System Using a Modern NN Potential

J.O. Fiase,[1,*] J.S. Nkoma[2], L.K. Sharmaand[2], A. Hosaka[2]

[1] Department of Physics, University of Botswana, P.B. 0022, Gaborone, Botswana

[2] Research Center for Nuclear Physics (RCNP), Osaka University, Ibaraki 567-0047, Japan

Abstract. In this paper we show evidence of the importance of tensor correlations in the nuclear many-body system by calculating the effective two-body nuclear matrix elements in the frame work of the Lowest-Order Constrained Variational (LOCV) technique with two-body correlation functions using the *Reid93* potential. We have achieved this by switching on and off the strength of the tensor correlations, α^k. We have found that in order to obtain reasonable agreement with earlier calculations based on the G-matrix theory, we must turn on the strength of the tensor correlations especially in the triplet even (TE) and tensor even (TNE) channels to take the value of approximately, 0.05. As an application, we have estimated the value of the Landau - Migdal parameter, $g'_{\rm NN}$ which we found to be $g'_{\rm NN} = 0.65$. This compares favorably with the G-matrix calculated value of $g'_{\rm NN} = 0.54$.

1 Introduction

The microscopic description of nuclei from the bare nucleon-nucleon force has been a major topic in recent years [1].

Within the frame-work of the cluster expansion technique, we have been studying the structure of nuclei via the lowest order constrained variational (LOCV) approach using two-body correlation functions [2]. Our findings suggest that one can in fact use very simple two-body correlation functions to obtain very accurate results comparable with the best available methods provided the effect of tensor correlations are included. In an earlier approach, we had used the Reid [3] N-N potential and had obtained very reasonable results [4]. The improvement in N-N phase-shift data, leading to more accurate N-N

* *E-mail address:* fiasejo@hotamail.com

potentials calls for an update of our earlier calculations. One such high quality N-N potential is the Nijgemen Reid-like potential hereafter designated *Reid93* [5] which is used in this work.

As an application of our potential model we have estimated a value for the Landau-Migdal parameter, $g'_{\rm NN}$. Recently [6] there has been a renewed interest in spin and isospin response functions. The new experimental data suggests the breaking of universality and smaller values in the delta (Δ) sector, such that $g'_{\rm NN}= g'_{{\rm NN}\Delta}=g'_{\Delta\Delta}$ [7] no longer holds. However, before extending to the delta sector, we will test our model in the N-N sector by calculating $g'_{\rm NN}$.

In what follows we shall describe briefly how this estimate can be achieved.

2 The Method

Here we briefly discuss the approach developed in Refs. [2,8] for the determination of the two-body effective interactions. We start with the non-relativistic Hamiltonian:

$$H_0 = \sum_i p^2/2m + \sum_{i>j} V_{\rm NN}(\boldsymbol{r}_i - \boldsymbol{r}_j), \tag{1}$$

defined in the rest-frame of the nucleus as:

$$H_0 \to \bar{H} = H_0 - E_{CM} = \sum_{i>j}(p_{ij}^2/M + V_{\rm NN}(\boldsymbol{r}_i - \boldsymbol{r}_j)), \tag{2}$$

where $M \sim m_N A$ is the total mass of the nucleus, E_{CM} is the translational kinetic energy of the centre of mass of the nucleus, $\boldsymbol{p}_{ij} = \frac{1}{2}(\boldsymbol{p}_i - \boldsymbol{p}_j)$ is the relative momentum of the interacting pairs while $V_{\rm NN}(\boldsymbol{r}_i - \boldsymbol{r}_j) = V(ij)$ is taken to be the *Reid93* [5] potential.

The Hamiltonian of Eq. (2) is a many-body problem and is difficult to determine exactly and one has to make approximations. The usual approach is to approximate the interaction by two-body effective interactions. We define then an effective two-body interaction in the form (see ref. [8]):

$$H_{\rm eff}^{(2)} = \sum_{i>j} f_{\rm NN}^{(2)}(ij)(p_{ij}^2/M + V_{ij})f_{\rm NN}^{(2)}(ij) \tag{3}$$

where $f_{\rm NN}^{(2)}(ij)$ are the two-body correlation operators. We choose the two-body correlation functions to be consistent with the form of the inter-nucleon potential. In the case of the *Reid93* potential [5], they may be expanded as [8]:

$$f_{\rm NN}^{(2)}(ij) = \sum_k f^{(k)}(r_{ij})O_k(ij), \tag{4}$$

with $\{O_k\} = \{1, S_{ij}, P_{ls}\}$. This form of two-body correlations functions have been studied by many authors. In particular the studies of Irvine and collaborators [2,8] have shown that tensor correlations have the most significant contribution. This has led to a parametrization of the two-body correlation functions as [2,8]:

$$
\begin{aligned}
r_{ij} < r_c \quad &: \quad f^{(k)}(r_{ij})O_k(ij) = 0\,, \\
r_{ij} \geq r_c \quad &: \quad f^{(k)}(r_{ij})O_k(ij) = f(r_{ij})(1 + \alpha^k(A)S_{ij})\,,
\end{aligned}
\tag{5}
$$

where

$$
f(r_{ij}) = 1 - e^{-\beta(r_{ij}-r_c)^2}\,, \tag{6}
$$

with $r_c = 0.25$fm and $\beta = 25\text{fm}^{-2}$. The parameter, $\alpha^k(A)$ represents the strength of the tensor correlations and is non-zero only in the lowest coupled $^3S_1$ - $^3D_1$ channel.

For the purpose of this work we have defined an effective two-body potential energy term as:

$$
H_{\text{eff}}^{(2)'} = \sum_{i>j} f_{\text{NN}}^{(2)}(ij)V_{ij}f_{\text{NN}}^{(2)}(ij) \tag{7}
$$

3 An Approximate Effective NN Interaction and the Landau-Migdal Parameter, g'_{NN}

We first fit our oscillator SO, TO, SE and TE (respectively the singlet-odd, triplet-odd, singlet-even and triplet-even channels) two-body matrix elements to those of the oscillator sum of Yukawa functions with different ranges. The central Yukawa V_c, the spin-orbit V_{ls} and the tensor V_t potential functions are given by:

$$
\begin{aligned}
V_c &= \sum_m A_m F(r_{ij}/R_m), \\
V_{ls} &= \sum_m A_m F(r_{ij}/R_m)\boldsymbol{L}\cdot\boldsymbol{S}, \\
V_t &= \sum_m A_m F(r_{ij}/R_m),
\end{aligned}
\tag{8}
$$

where $F(y) = \frac{e^{-y}}{y}$. The R_m are the ranges while the A_m are the strengths of the interaction which are determined by fitting the oscillator matrix elements of Eq. (7) to the oscillator matrix elements of Eq. (8). In this way we obtain a potential in each of these channels that will fit approximately the *Reid93* NN potential [5] two-body matrix elements of Eq. (7). We chose the ranges with $m \leq 4$ to be 0.25, 0.4, 0.7 and 1.414 fm, which are consistent with the one-boson exchanges. The shorter ranges correspond to heavier meson exchanges such as σ, ρ and ω mesons, while the longest range 1.414 fm corresponds to the one-pion exchange.

It is useful to recast our potential model in the spin-isospin formalism as:

$$
V_{\sigma\tau} = \frac{1}{16}(V_{\text{SO}}(r) + V_{\text{TO}}(r) - V_{\text{SE}}(r) - V_{\text{TE}}(r)). \tag{9}
$$

For completeness in subsequent discussions we shall also include the tensor-even (TNE), tensor-odd (TNO), the spin-orbit-odd (LSO) and the spin-orbit-even (LSE) components. At this stage we make a Fourier transform of the potential function and write the direct and the exchange terms of the Landau-Migdal parameter, $g'_{\rm NN}$ respectively as [9]:

$$g'_{\rm NN}(\rm dir) = \lim_{q->0}(\frac{m_\pi}{f_\pi})^2\frac{1}{16}(V_{SO}(q) + V_{TO}(q) - V_{SE}(q) - V_{TE}(q)), \tag{10}$$

and

$$g'_{\rm NN}(\rm exch) = -(\frac{m_\pi}{f_\pi})^2\frac{1}{16}(V_{SO}(q) + V_{TO}(q) + V_{SE}(q) + V_{TE}(q)), \tag{11}$$

where for Eq. (11) $q = 2p_F$ with p_F being the Fermi momentum of nucleons.

4 Results

In Table 1 we present the set of relative two-body matrix elements of our effective interaction for the $A = 16$ system for the SE, SO, TE, TO, TNE, TNO, LSO and LSE channels using the *Reid93* [5] potential. We calculated the matrix elements in a harmonic oscillator basis with $\hbar\omega = 14.0$MeV suitable for $A = 16$ nuclei. In this work only the TE and TNE channels are affected by the tensor correlations. Thus, by setting $\alpha^k(A) = 0.0$, we have no tensor correlations.

In the present case, we have found the set $\{\hbar\omega, \alpha^k(A)\} = \{14.0\text{MeV}, 0.050\}$ to give the best agreement with the G-matrix calculation of of Hosaka et al. [9].

From Table 1 we note the excellent agreement with the G-matrix results of Hosaka et al. [9] in most angular momentum channels. The effects of tensor correlations are seen especially in the TE channels, which gave excellent agreement with the G-matrix results of Hosaka et al. [9], by switching on the strength of the tensor correlations, compared to when they are switched off.

4.1 *Strength of Interactions*

Table 2 gives the strengths of our fitted potential in the relevant channels used in this work which should be compared with those of Anantaraman et al. [10].

In Table 3 we give our calculated value of the Landau-Migdal parameter which we found to be 0.64 with tensor correlations and 0.52 without tensor correlations.

5 Conclusions

We have performed an LOCV calculation with the *Reid93* [5] potential folded with two-body correlations functions. Our findings indicate that tensor correlations are very important especially in TE channels. Finally, we have used our

Table 1. Calculated relative matrix elements for $A = 16$ using $\hbar\omega = 14.0$MeV. The first entry of a column is the present results with the *Reid 93* potential and $\alpha = 0.05$. The second entry of a column is the G-matrix results of Hosaka et al. [9]. The third entry of a column for the TE and TNE channels are the present results with the *Reid 93* potential and $\alpha = 0.0$.

SE				TE			
(S/S)	$n = 0$	1	2	(S/S)	$n = 0$	1	2
$n' = 0$	−7.6869	−6.7809	−5.4889	$n' = 0$	−10.5007	−8.5904	−6.4883
	(-6.5801)	(-5.2924)	(-3.7431)		(−10.049)	(−8.3732)	(−6.6530)
					−3.1697	−2.0698	−0.9538
1		−6.6348	−5.5981	1		−8.3187	−6.7631
		(-4.6438)	(-3.3222)			(−7.4208)	(−6.2239)
						−1.5311	−0.5847
2			−4.9555	2			−6.0328
			(-2.3858)				(−5.4977)
							0.0449

SO				TO			
(P/P)	$n = 0$	1	2	(P/P)	$n = 0$	1	2
$n' = 0$	2.5449	2.6564	2.6122	$n' = 0$	0.1311	0.2193	0.3441
	(2.5310)	(2.4564)	(2.2570)		(−0.0803)	(−0.1174)	(−0.0972)
1		3.5558	3.7380	1		0.4295	0.6174
		(3.1844)	(3.2178)			(−0.0937)	(−0.0535)
2			4.3010	2			0.8674
			(3.6677)				(0.0254)

TNE				TNO			
(S/D)	$n = 0$	1	2	(P/P)	$n = 0$	1	2
$n' = 0$	−5.0671	−7.0551	−8.2243	$n' = 0$	0.7624	0.7456	0.6522
	(−5.6927)	(−7.8798)	(−9.0517)		(0.7787)	(0.7333)	(0.6274)
	−5.6687	−7.8539	−9.115				
1	−2.5023	−4.8436	−6.6104	1		0.9106	0.8793
	(−2.7039)	(−5.3284)	(−7.2073)			(0.8997)	(0.8643)
	−2.7400	−5.3271	−7.2675				
2	−1.2770	−2.7659	−4.4539	2			0.9455
	(−1.3263)	(−2.9611)	(−4.7745)				(0.9456)
	−1.3520	−2.9778	−4.8382				

LSE				LSO			
(D/D)	$n = 0$	1	2	(P/P)	$n = 0$	1	2
$n' = 0$	0.0020	0.0074	0.0139	$n' = 0$	−0.7258	−1.0436	−1.2601
	(-0.0241)	(-0.0388)	(-0.0508)		(−0.5685)	(−0.8109)	(−0.9704)
1		0.0164	0.0268	1		−1.5143	−1.8406
		(−0.0747)	(−0.0949)			(−1.1678)	(−1.4062)
2			0.0411	2			−2.2501
			(−0.1361)				(−1.7038)

Table 2. Best-fit interaction strengths (MeV) for $A = 16$, $\hbar\omega = 14.0$MeV, $\alpha = 0.05$. The values in brackets are taken from the work of Anantaraman et al. [10]. Here we have presented only the strengths used in calculating the Landau-Migdal parameter.

	Channel	$R_1 = 0.25$fm	$R_2 = 0.40$fm	$R_3 = 0.70$fm	$R_4 = 1.414$fm
1	SE	9758	-3232		-10.463
		(11466)	(-3556)		(-10.463)
2	TE	15203	-4947		-10.463
		(13967)	(-4594)		(-10.463)
3	SO	32.35	1159.93		31.389
		(-1418)	(950)		(31.389)
4	TO	20.3	572.7		3.488
		(11345)	(-1900)		(3.488)

Table 3. Estimates of g'_{NN} of the present calculation compared with estimates of other works.

Source	g'_{dir}	g'_{exch}	g_{total}
Present (a)tensor	0.81	-0.17	0.65
(b)no tensor	0.69	-0.17	0.52
G-Matrix [9]	0.94	-0.40	0.54
$\pi + \sigma$ [9]	0.97	-0.11	0.86
phenomenological [7]	-	-	0.6 - 0.7

potential to calculate a value for the Landau-Migdal parameter. This is found to agree reasonably with those of the other workers shown in Table 3.

Acknowledgement. (J.O.F) wishes to thank Dr. M.C.M. Rentmeester for his help on the use of *Reid 93* potential.

References

1. M. Hjorth-Jensen, T.T.S. Kuo, E. Osnes: Phys. Repts **261**, 125 (1995)
2. J. Fiase, A. Hamoudi, J.M. Irvine, F. Yazici: J. Phys. G. Nucl. Phys. **14**, 27 (1988)
3. R.V. Reid: Ann. Phys. (N.Y) **50**, 411 (1968)
4. J.O. Fiase, K.R.S. Devan, A. Hosaka: Phys. Rev **C66**, 014004(2002)
5. V.G.J. Stoks, R.A.M. Klomp, C.P.F. Terheggen, J.J. De Swart: Phys. Rev **C49**, 2950 (1994)
6. T. Wakasa et al.: Phys. Rev **C55**, 2909(1997)
7. T. Suzuki, H. Sakai, T. Tastsumi: nucl-th/9901097
8. J.M. Irvine: Prog. in Part. and Nucl. Phys. **5**, 1 (1980)
9. A. Hosaka, K. I. Kubo and H. Toki: Nucl. Phys. **A444**, 76 (1985)
10. N. Anantaraman, H. Toki and G.F. Bertsch: Nucl. Phys. **A398**, 269 (1983)

Few-Body Systems Suppl. 15, 259–266 (2003)

Few-Body Systems

Experimental Study of Nuclear Few-Body Systems at Jefferson Lab

A. Saha* for the Jefferson Lab Hall A Collaboration

Jefferson Lab, 12000 Jefferson Avenue, Newport News, VA 23606, USA

Abstract. In a new and detailed experimental program carried out at Jefferson Laboratory, the few nucleon systems has been studied via electrodisintegration of deuterium and the helium nuclei. Interesting new results probing the high momentum structure have been studied via the $(e, e'p)$ reaction using the two high resolution spectrometers in Hall A and various kinematic regions have been investigated. Detailed and precise information on the effective bound state momentum distributions have been measured for the first time up to missing momenta of 1 Gev/c in ^{3}He, exhibiting significant strength at these high momenta which standard nuclear calculations fail to explain and could be an indication of the onset of non-nucleonic degrees of freedom. The continuum region in ^{3}He has also been investigated in great detail looking for nuclear correlations and quasi-deuteron strength. Response function separations have been carried out to extract details of the few body structure and relativistic dynamics. Measurements made so far on D and ^{4}He have focussed on specific kinematic regions and there are plans to extend them over kinematic ranges similar to the ^{3}He investigation.

1 Introduction

Electron quasi-elastic scattering is a powerful tool for the study of nuclear structure. In particular, $(e, e'p)$ reactions have provided a wealth of information such as the effective nucleon momentum distributions in nuclei as well as separated response function distributions. In addition, these measurements provide information on ground state short-range correlations, final state interactions, meson exchange contributions and relativistic effects in the few body wave function. At JLab, this is being accomplished by extending the domain of momentum transfers towards higher values where short-range effects and possibly the internal structure of the nucleons are manifested, by exploring nuclear structure in its extreme conditions, and by investigating the high momentum part of the

* *E-mail address:* saha@jlab.org

wave function. We will also increase the knowledge of the nuclear structure by separating the response functions associated with different polarization states of the virtual photon.

At Jefferson Lab a series of $(e, e'p)$ measurements have been performed on the Deuterium [1, 2] and Helium [3, 4] isotopes designed to exploit these new possibilities. Deuteron is the simplest nuclear system where exact calculations can be performed and its structure can be investigated in great detail. Next to the deuteron, the A=3 and A=4 nuclei are the simplest systems in which all basic ingredients of a complex nucleus exist. Sophisticated methods to solve the many-body Schrödinger equation almost exactly have been applied to the A=3 nuclei and have been extended to ^{4}He. Microscopic calculations of FSI and MEC contributions have been developed and applied to reactions on few-nucleon systems. The data provided by these experiments test the validity of these models in the high Q^2 and high missing momentum regime. In this paper the preliminary results of the experiments will be presented and discussed with special emphasis on the e89044 experiment.

2 The Experiments

In the e89044 experiment [3], different kinematics regions were studied to exploit various aspects of the nuclear structure and reaction mechanisms. In constant (q, ω) kinematics, also known as perpendicular kinematics, a constant $q = 1.5$ GeV/c and $\omega = 0.845$ GeV was maintained. In this kinematics the single nucleon structure of ^{3}He was studied with special emphasis on high momenta up to $p_{\text{miss}} = 1$ GeV/c. We also did a complete in-plane separation of the response functions R_{TL}, R_T, and the combination of R_{L+TT} up to missing momenta of 0.55 GeV/c.

In parallel kinematics, the q dependence of the reaction was determined by performing an R_L/R_T (longitudinal/transverse) Rosenbluth separation for protons emitted along $\mathbf{q}$ (in parallel kinematics), up to $q = 3$ GeV/c. This was performed in both quasi free kinematics ($p_{\text{miss}} = 0$) and for $q = 1$ and 2 GeV/c at $p_{\text{miss}} \pm 0.3$ GeV/c.

Also, the continuum region was studied in order to search for correlated nucleon pairs. This was done in both parallel and perpendicular kinematics with full in-plane separation of the response functions.

In the D$(e, e'p)$ experiment [1], the effective momentum distribution was mapped out to 600 MeV/c in missing momentum in perpendicular kinematics. In a presently run experiment on the Deuterium [2], the Bjorken x distribution of the cross section at various values of missing momenta (200 Mev/c, 400 MeV/c and 500 MeV/c) are being mapped out to look for Color Transparency effects.

In a recently approved experiment on ^{4}He [4], a detailed study will be made through $(e, e'p)$ measurements at high momentum transfers in kinematics similar in range and scope to the ^{3}He$(e, e'p)d$ [3] study.

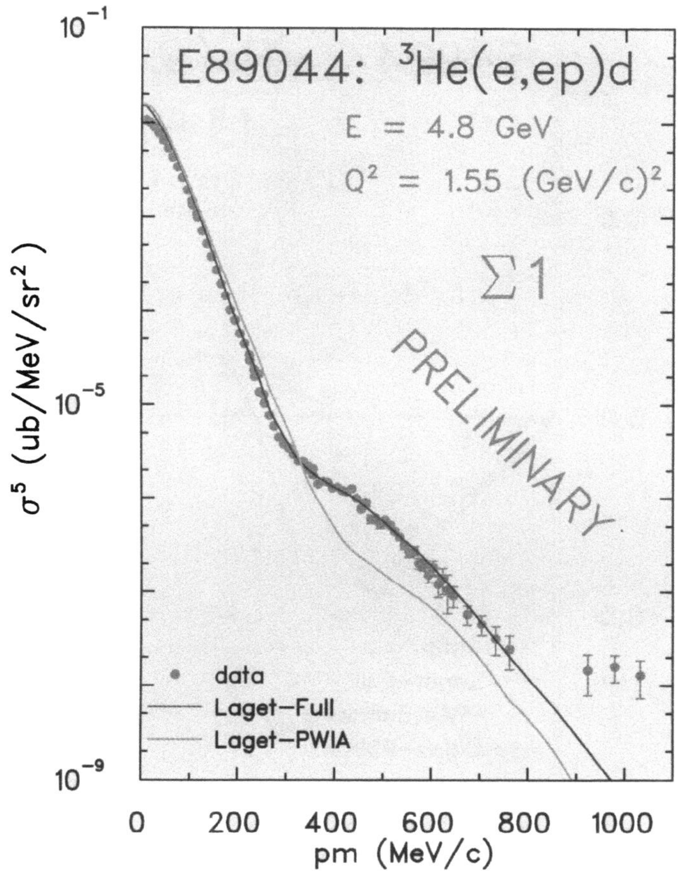

Figure 1. Shown are the preliminary cross section results for the reaction ^{3}He$(e, e'p)d$ as a function of missing momentum with a beam energy of 4807 MeV and with a fixed $q = 1500$ MeV/c and $\omega = 840$ MeV with $\phi = 180°$. The theory curves show a PWIA calculation using a spectral function of Salme along with the latest calculations of Laget.

3 Results

The preliminary cross section results of the in ^{3}He$(e, e'p)d$ reaction [3] in perpendicular kinematics are shown in Fig. 1 for the forward kinematics. It has generated considerable theoretical interest. The curves show the most recent calculations of Laget [5]. There is good agreement of Laget's full calculation with the data up to 750 MeV/c. One sees that in the region between 0 and 200 MeV/c in $p_{\rm miss}$ there is good agreement with the PWIA calculation except at the very low values of $p_{\rm miss}$ ($\leq$50 MeV/c) and this is presently being investi-

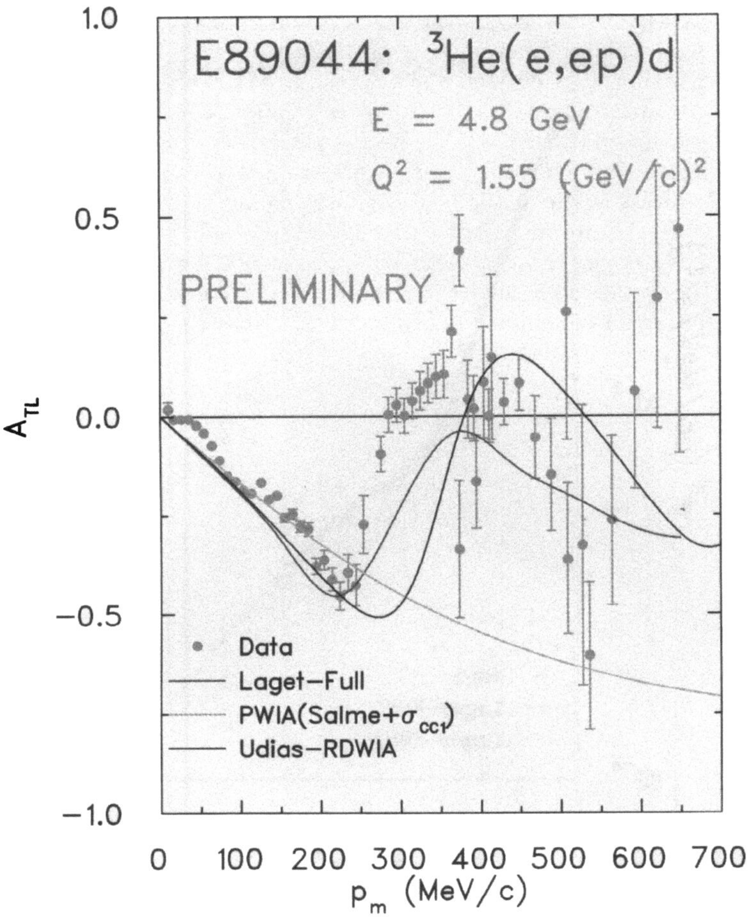

Figure 2. Shown is the preliminary A_{TL} data. This result is obtained by taking the sum over the difference of the two forward kinematics. The curves show the latest calculation of Laget [5] along with a preliminary results of Udias [6]. At lower missing momentum the theories are most sensitive to relativistic effects, while at the larger missing momentum the theories become sensitive to final state interaction effects.

gated more thoroughly. In the region between 200 MeV/c and 400 MeV/c, the data lies below the PWIA calculation and this is due to the interference between the PWIA and the rescattering amplitudes - also referred to as Glauber screening. In the region between 400 MeV/c and 750 Mev/c most of the enhancement in the strength of the data over the PWIA calculation is due to final state interactions (FSI). There is as yet no clear indication as to what is causing the cross section at the largest missing momentum at around 1 GeV/c to be nearly an order of magnitude greater than predicted.

The A_{TL} response function is a sensitive tool to study the details of the theoretical inputs. It is obtained by taking the sum over the difference of the two forward kinematics: $\sigma(\phi = 0°)$ and $\sigma(\phi = 180°)$. The curves in Fig. 2 show the latest calculation of Laget [5] as well as preliminary and rough calculations of Udias [6]. At lower missing momentum the theories are most sensitive to relativistic effects, while at the larger missing momentum the theories become sensitive to final state interaction effects. Udias is now working on making a full calculation of the cross sections and A_{TL} using a realistic potential.

With the cross section analysis of the perpendicular kinematics now nearly finished, the process of making the response function separations is now underway and should be available soon.

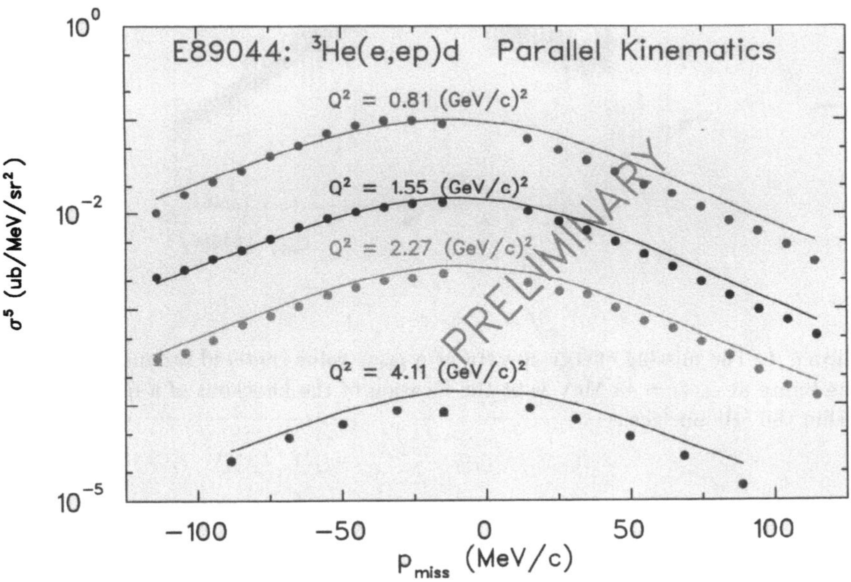

Figure 3. The forward (high epsilon) kinematics for Q^2 values of 0.8, 1.5, 2.2 and 4.1 $(\mathrm{GeV/c})^2$ in parallel kinematics. The PWIA results of Laget are also shown.

In parallel kinematics, in Fig. 3 we show the results of the forward (high epsilon) kinematics for Q^2 values of 0.8, 1.5, 2.3 and 4.1 $(\mathrm{GeV/c})^2$. The PWIA results of Laget are also shown and they agree with the data at the low Q^2 values. Once the backward angle kinematics (low epsilon) data are analysed we shall have the R_L/R_T (longitudinal/transverse) Rosenbluth separation done. This will enable us to probe the electromagnetic properties of the proton embedded in the nucleus and see if they are modified from those of the free nucleon.

We have made good progress in the analysis of the continuum region. Fig. 4 shows the missing energy spectra at a p_{miss} value centered around 425 MeV/c.

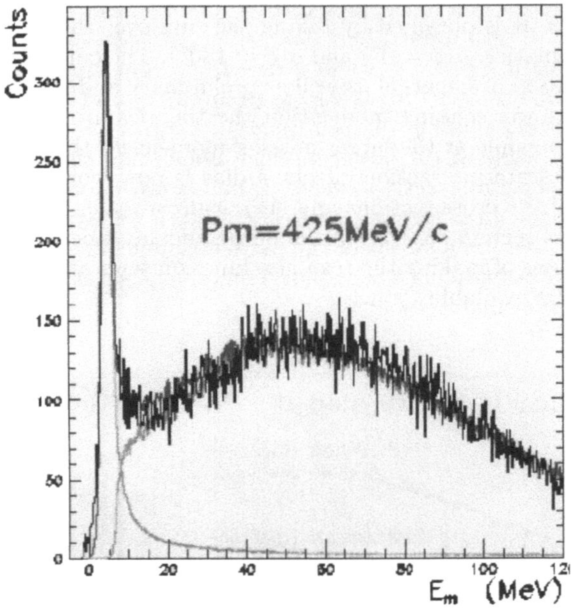

Figure 4. The missing energy spectra at a p_{miss} value centered around 425 MeV/c. The bump at $e_{\text{miss}} = 45$ MeV is at the location of the knockout of a quasi-deuteron within the ^{3}He nucleus..

The bump at $e_{\text{miss}} = 45$ MeV is at the location of the knockout of a quasi-deuteron within the ^{3}He nucleus.

Experiment E94-004 [1] performed the $d(e, e'p)n$ reaction and the results are presented in Fig. 5. The data was taken in perpendiclular kinematics analogous to the $^3\text{He}(e, e'p)$ kinematics but at a lower value of $Q^2 = 0.665$ (GeV/c)^2. The data show the same characteristics as the ^{3}He data described above with lowering of the cross section with respect to the PWIA calculation in the p_{miss} region around 200 MeV/c and enhancement over the PWIA calculation in the region around $p_{\text{miss}} = 400$ MeV/c due to final state interactions. The full calculations are from Arenhovel [7].

4 Conclusions

The Jefferson Lab E89-044 $^3\text{He}(e, e'p)$ experiment has been successfully completed and preliminary results are starting to become available. From the preliminary cross section results along with the A_{TL} asymmetry we can already

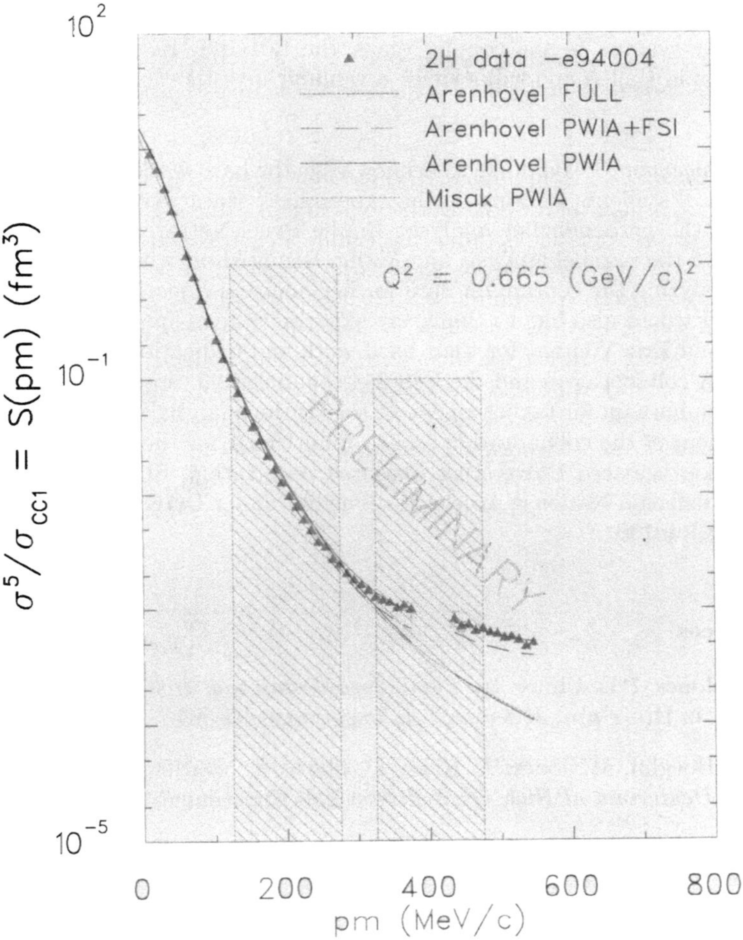

Figure 5. The $d(e,e'p)$data was taken in perpendiclular kinematics at $Q^2 = 0.665$ $(\mathrm{GeV/c})^2$. The data show the same characteristics as the ^{3}He data described earlier with lowering of the cross section with respect to the PWIA calculation in the p_{miss} region around 200 MeV/c and enhancement over the PWIA calculation in the region around $p_{\mathrm{miss}} = 400$ MeV/c due to final state interactions. The full calculations are from Arenhovel [7].

see regions where the theoretical models clearly predicted the experimental results, such as the rise in the cross section at 300 MeV/c in Fig. 1. Also seen are the deficiencies in the theories, as indicated by much larger then expected cross section at extremely large missing momentum in Fig. 1 and by the A_{TL}. As the analysis now continue with the response function separations, it should become clear what effects are causing the discrepancies between theory and experiment, and allow a significant improvement in our understanding of the three-body system. In the coming years, the E01-108 ^{4}He$(e, e'p)$ experiment will be run in Hall A and will provide a valuable investigation of the four-body system.

Acknowledgement. I would like to acknowledge the hard work and dedication of the doctoral students working on the experiment. Marat Rvachev with M.I.T. has done the perpendicular analysis. Emilie Penel-Nottaris with Grenoble is working on the parallel analysis and Fatiha Benmokhtar with Rutgers University is analyzing the continuum data for her doctoral research.

Also, I would also like to thank my experimental co-spokespersons, Marty Epstein and Eric Voutier, for their hard work and dedication along with both the Hall A collaboration and the E89-044 collaboration. Special thanks also to Doug Higinbotham for taking an active role in the analysis of the data and the management of the collaboration and to Paul Ulmer for the $d(e, e'p)$ results.

The Southeastern Universities Research Association (SURA) operates the Thomas Jefferson National Accelerator Facility under DOE Contract No. DE-AC05-84ER40150.

References

1. M. Jones, P.E. Ulmer: In: *Plane Separations and High momentum Structure in* D$(e, e'p)n$, Jefferson Lab Experiment 94-004
2. W. Boeglin, M. Jones, A. Klein, P. Ulmer, E. Voutier: *$(e, e'p)$ Studies of the Deuterium at High Q^2*, Jefferson Lab Experiment 01-020
3. M. Epstein, A. Saha, E. Voutier: *Selected Studies of the ^{3}He and ^{4}He Nuclei through Electrodisintegration at High Momentum Transfer*, Jefferson Lab Experiment 89-044
4. K. Aniol, D.W. Higinbotham, S. Gilad, A. Saha: *Detailed Study of the ^{4}He Nuclei Through Response Function Separations at High Momentum Transfers*, Jefferson Lab Experiment 01-108
5. J.M. Laget: Phys. Rev. **D579**, 333 (1994)
6. J. Udias: private communication
7. W. Fabian, H. Arenhovel: Nucl. Phys. **A314**, 253 (1979)

Few-Body Systems Suppl. 15, 267–270 (2003)

Few-Body Systems

From $\gamma - d$ to $\gamma-$Tumor: Peter Sauer's Influence as a Teacher

U. Oelfke *

Department of Medical Physics, Deutsches Krebsforschungszentrum (DKFZ), DKFZ, E0401, Im Neuenheimer Feld 280, 69120 Heidelberg, Germany

Abstract. As indicated by its title this short contribution to a rigorous physics journal does not contain any genuine scientific contribution. It's main purpose is to honor one of the most respected members of the "Few Body Systems" community - Peter Ulrich Sauer (PUS) - for his life long commitment to educate students.

1 Introduction

Although being a student of PUS a long time ago, i.e., for almost 10 years from the early 1980's to 1990 when leaving from Hannover to TRIUMF, it was not difficult at all to recall his enormous influence he had on his students. I appreciate the opportunity given to me by my old friends - Teresa Penã, Jiří Adam and Alfred Stadler - to present my respective experience with PUS at their excellently organized workshop in Lisbon.

The aim of this note is two-fold. First, to show that a profound education in theoretical nuclear physics can be applied also to more practical areas of life, e.g., the development of more efficient cancer therapies through the clinical application of high energy photon, proton and heavy ion beams. Second, to emphasize and remember the role of PUS as a teacher, whose personal commitment to his students, in my opinion, was crucial for their success in their professional careers.

* *E-mail address:* u.oelfke@dkfz-heidelberg.de

2 Material and Methods

2.1 Nuclear Physics and its Application in Radiation Therapy

About ten to twelve years ago nuclear physics of few-body systems was mostly related to the canonical topics, i.e., form-factors, relativistic effects and wave functions, polarization observables, quark- or nucleonic degrees of freedom etc.. According to my impression from the scientific part of the workshop no revolutionary discovery has been made since; slightly new phenomena were discussed in a similar context and analyzed with comparable intellectual rigor. In the following I will shortly describe a different, more applied area of physics, whose direct benefit to society is immediately evident: the application of nuclear physics techniques to tumor therapy.

Proton Therapy of Inner-Ocular Tumors
Over the world, roughly 10 proton accelerators, originally built for basic research in nuclear physics – mostly cyclotrons with maximum proton energies between 60 and 70 MeV — were adapted for the treatment of inner-ocular tumors. Protons at these energies with a maximum range of 3 - 4 cm in water, can be delivered with a spatial accuracy and precision of less than a millimeter into tumor tissues, while other radio-sensitive structures of the eye, e.g., the optic nerve, can be highly spared from being irradiated. Almost 10.000 patients have had this extremely successful therapy, e.g., the achieved local tumor control rate ranges between 96% and 98%. Especially for 'Uveal Melanomas' this tumor therapy, pioneered at the Harvard Cyclotron in Boston and at the Paul Scherrer Institute in Villigen, has saved a considerable number of lives of cancer patients.

Intensity Modulated Radiation Therapy with High Energy Photons
The technique of modern linear accelerators applied in standard radiation therapy are based on the development of electron accelerators, applied for nuclear physics research. Furthermore, recent developments, have lead to a new form of radiation therapy, called 'intensity modulated radiation therapy'(IMRT). The idea is to divide a radiation beam into hundreds to thousands of small radiation sources, whose contribution to the overall tumor dose can be controlled independently. The construction of the respective dose delivery devices, so called electronically driven multi-leaf collimators, and the development of new therapy optimization methods ('inverse planning') was initiated by physicists. The IMRT concept allows the treatment of tumors which are in very close proximity to organs at risk with very high doses, e.g., it is expected that the clinical results for the irradiation of tumors adjacent to radio-sensitive tissues will substantially improve through the introduction of IMRT.

Intensity Modulated Proton and Heavy Ion Beams
The most advanced radiation therapy technology in the battle against cancer, again, originates from nuclear physics laboratories. The concept of intensity

modulated charged particle therapy (IMPT) utilizes high energy proton and heavy ion beams, which are magnetically scanned over the considered tumor volumes. In addition to the lateral position of the beam spots and the particle fluence one can also control the particle energy, which determines the penetration depth of these particles in tissue. Research on heavy ion therapy was initiated already in the 1950's in Berkeley and was perfected to its modern form, e.g., by pilot projects at research labs like the Harvard Cyclotron, PSI (Switzerland) or GSI (Germany). Several clinical centers were recently opened or are planned in the US, Japan and Europe. The potential of this new therapy form originates from it's significantly improved dose conformity around the tumor volume, e.g., for prostate tumors radiation induced complications in rectum and bladder can be reduced considerably.

2.2 Education by a "Center of Excellence": P. U. Sauer

The conversion from theoretical physics into quite a different field – an area where the knowledge of basic science is merged with medical experience – is naturally accompanied by a phase of learning completely new methods and strategies. The respective process requires a solid and profound education as a scientist, which is strongly correlated to the 'teacher' one is exposed to. In the following I briefly describe my personal experience with PUS as a teacher in order pay tribute to his efforts in guiding and helping students to become independent scientists.

Learning through Lectures and Visitors
The first outstanding fact about PUS' teaching are his lectures. They were always extremely well prepared, contained a very clear and concise presentation of the ideas which often were followed by well selected, explicit examples. Furthermore, PUS made the effort to teach all the required mathematical tools and foundations, so that students actually could 'do' calculations by themselves. Notes taken in classes taught by PUS could serve as reference of the respective topic for the rest of one's professional life.

The next cornerstone of PUS' education concept was that he provided opportunities for his students to interact with exceptional scientists from other institutions. While studying with PUS in the 80's we could enjoy long term visits and lectures from J. Koch, F. Coester, F. Gross, R. Rosenfelder, E. Truhlik, J. Adam, R. Mach, T. Peña, H. Garcilazo and S. N.Yang. The diversity of this group and the opportunity to observe and independently interact with these visitors provided his students with an invaluable experience for their future professional development.

Organizing Physics and Benevolent Support
Another specific education tool of PUS was his system of organizing lectures and projects in hand written notes, each labeled by a three letter abbreviation of the respective topic, e.g. PIP, $\gamma N\Delta$, RSS .... These treasures, well stacked in

one (maybe more) of his metal cabinets, provided an extremely helpful source of information for many students and collaborators of his.

Besides PUS's qualities as a scientist and teacher of physics I also want to mention his fair judgement and personal support of his students. From my experience PUS spent significant time and energy on a fair evaluation of his student's accomplishments. Although never obviously close to his students PUS has always shown considerable patience and respect to those working with him.

Papers and Pain

Writing papers with PUS naturally led to different stages of pain for his collaborators, mostly related to N+1 versions of a manuscript ($N \geq 8$), whose flood of 'red' comments on its margins simply never vanished ... However, we have to admit his quest for the highest standards in a publication, often paid off and at the end, there was a beneficial educational component for us too...

3 Conclusions

At present there are evolving exciting new applications of nuclear physics in medicine, e.g., the briefly sketched developments in radiation therapy. The ability to work as a scientist in basic physics research or one of these more applied professional areas requires a solid education mediated by teachers. The workshop in Lisbon provided us with the opportunity to honor PUS not only for his scientific achievements but also for his outstanding efforts in educating his students. His generous sharing of ideas, his excellent lectures, his personal support and his conduct as a scientist inspired us and was therefore crucial for the successful development of our professional lives.

Few-Body Systems Suppl. 15, 271–272 (2003)

Few-Body Systems

From Physics to Consulting

Ralf Schulze-Riegert*

Scandpower GmbH, Weidestr. 126, 22083 Hamburg, Germany

After almost a decade out of physics it was a pleasure to join this year's Fall Workshop on Nuclear Dynamics in Lisbon. The fact that the organisers have dedicated the workshop to my former teacher and supervisor Peter Sauer has made it a special event for myself. For one thing, this event gave me an opportunity to reflect on the status of research of few-nucleon systems I was actively involved in. For another, the organisers have prepared a pleasant atmosphere to meet and to talk to many of my former colleagues and collaborators.

Ways into and out of physics was not a major topic of this conference, but has still created some interest. For many graduate students in physics, a position in a research institution has an appealing perspective. A liberal environment and work dedicated to technical and intellectually demanding problems with few constraints along with publishing papers. However, perspectives in physics and nuclear physics in particular have changed with periodic ups and downs. Questions on future perspectives and working opportunities are therefore not just a matter of a particular time but keep recurring.

The consulting business has gained a major role in our world which requires a multitude of specialized services. It has grown significantly, contributing to economic growth in the nineties and still absorbs numerous graduate students. The consulting business for physicists is diverse, starting from the IT business, Research-and-Development consulting (which merges research and industry needs), business process consulting, financial risk management consulting up to scientific advisory board level, which creates a link to politics. Physicists are active in all of these fields, and the opportunities are as diverse as the term Consultant is used for different services. Outside of physics, graduates with a degree in physics are competing with other qualifications under continuously changing market requirements. Nevertheless, a technical and mathematical background has proven to be a valuable foundation. Consulting services in

* *E-mail address:* riegert@scandpower.de

industry require technical insights, the ability to share knowledge and social competence.

Working with Peter Sauer has combined these requirements. Graduate level work was often embedded in an international environment inasmuch as Peter Sauer worked together with his students and external collaborators to write scientific papers. Personally, I particularly wish to mention our collaborations with Tobias Frederico (ITA, Brazil) and Fritz Coester (Argonne, USA). These collaborations triggered many controversial as well as productive discussions in our group in Hannover, of which I have found memories. However, our goal to formulate physics problems was also very much linked to experiments, such as the calculations we performed in collaboration with MIT-Bates and others. I regard this as a valuable feature of few-nucleon physics, which defines a close link between theoretical and experimental physics.

Peter Sauer liked to travel but did not hesitate to send his students abroad to work and to present their work at conferences. In his summary to the Heisenberg Conference in 1991, C.D. v. Weizsäcker concluded, "physics conferences are partly organised to chat, perhaps with an interest to support one's career. But truly, they are also organised because we appreciate talking about things which we hope – if we continue to work on them for a few more decades – will allow us to make a really big step forward". Today, decades as a time frame for our work are out of sight, and we accept that big steps are rare and largely driven by efforts to put together pieces of a larger puzzle. This work can be organised with all the discipline and creativity it requires. Peter Sauer as a researcher and lecturer – or in business terms, as an auditor and research manager – concentrated a great deal on advising students and organising a culture of international collaborations. This created both challenges and opportunities to his students. In retrospect, it was a most valuable preparation for working both in research and industry.

Few-Body Systems Suppl. 15, 273–281 (2003)

Few-Body Systems

Celebrating Peter Sauer: an Occasion to Look at Nuclear Physics in Perspective

B. Frois*

Ministère de la Recherche, 1 rue Descartes, 75231 Paris, France

Abstract. This talk was presented at the symposium organized by Peter Sauer's friends in Lisbon. This symposium was a wonderful occasion to celebrate the importance of his achievements and examine the trends and perspectives in nuclear physics in 2003.

1 A Tribute to Peter Sauer

I would like first to give my warmest thanks to the organizers of this meeting celebrating Peter Sauer and his achievements. This is a very thoughtful initiative of his friends in Lisbon. I am very pleased to be able to share this precious moment with Peter Sauer and his friends.

Peter Sauer and his group have written an important page of the history of few-body physics. I would like to mention in particular their contribution to the understanding of the three-nucleon system. At the beginning of the 70's, the mathematical tools to solve the three-body systems were known, but calculations were showing some striking disagreements between experiments and theory. A discrepancy of 1 MeV was observed in the difference between the ^{3}H and ^{3}He binding energies. The charge distribution of ^{3}He observed by electron scattering seemed to have a hole at the center of the nucleus that no calculation could explain. One was beginning to believe that the role of non-nucleonic degrees of freedom such as π's and Δ's was not sufficiently well understood. There was a great need for consistent and reliable calculations of few nucleon sytems with realistic potentials.

Although nucleon-nucleon interactions are traditionally described in terms of meson-exchange processes, non-nucleonic degrees of freedom are omitted from calculations and nuclei are described as "dressed" nucleons interacting via a two-body potential. The first convincing observation of non nucleonic degrees of freedom was the precise measurement of the magnetization distribution

* *E-mail address:* bernard.frois@recherche.gouv.fr

involved in the electrodisintegration of the deuteron in 1985 at Saclay. When the energy transferred to the deuteron by the incoming electron corresponds to its break-up energy (2.2 MeV), the neutron-proton pair has a very small relative energy $E \simeq 0$. If, in addition, the electron is scattered at an angle $\theta \simeq 180°$, the electrodisintegration of the deuteron is a magnetic isovector M1 transition that leaves the two-nucleon system in a quasi-bound $^1S_0$ (T = 1) state. Theoretical calculations of electron scattering which take into account only nucleon degrees of freedom predict a minimum at $Q^2 = 0.5$ (GeV/c)^2; this is not observed experimentally. The filling of this minimum to higher Q^2 values is due entirely to the presence of meson-exchange currents (Fig. 1). Present calculations which include both nucleonic and mesonic degrees of freedom provide an excellent description of the data up to $Q^2 = 1$ (GeV/c)^2.

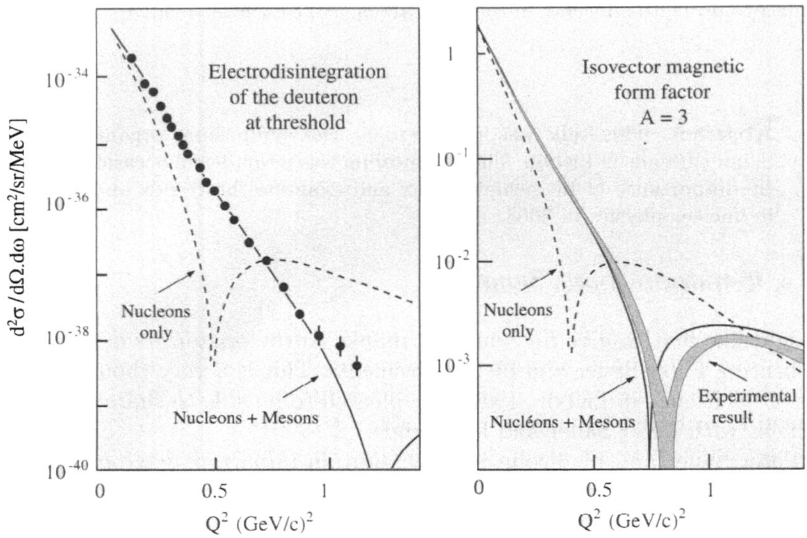

Figure 1. Comparison of the isovector magnetic form factors of the $A = 2$ and $A = 3$ systems.

The confirmation of the importance of non-nucleonic degrees of freedom came a few years later with the comparison between the measurements of the ^{3}He and ^{3}H charge and magnetic form factors and calculations performed by Peter Sauer and his group. They were the first to develop a new generation of consistent calculations for the $A = 3$ nuclei in terms of nucleons, pions and isobars based on a realistic potential. Their calculations also take into account a significant component of the three-body force. This work was a tremendous help to prepare and interpret the measurements of the form-factors of the ^{3}H and ^{3}H at Saclay. Tritium is radioactive nucleus. Building a tritium liquid target was a "tour de force" at Saclay that lead to very precise results and triggered considerable interest. The trinucleon charge and magnetic form factors were determined up to 1 $(\mathrm{GeV/c})^2$ by elastic electron scattering. These data allowed one to separate their isospin components up to this momentum transfer value. Thus a direct comparison of the isovector magnetic form factors of the two- and three-nucleon systems became possible. In particular, the isovector magnetic form factor can be compared with the cross section for the electrodisintegration of the deuteron at threshold.

Fig. 1 show that the isovector transitions in two- and three-nucleon systems has essentially the same behavior. The nucleonic contribution to the cross section vanishes at a relatively small momentum transfer due to the destructive interference between one-body amplitudes. When the same two-body currents that explain the isovector magnetic form factor for the electrodisintegration of the deuteron, are included in the theoretical description of the three-nucleon system, there is an excellent agreement between experiment and theory. Again, pion-exchange currents are the dominant contribution to the cross section up to $Q^2 \simeq 0.6$ $(\mathrm{GeV/c})^2$.

Thanks to Peter Sauer and his group, major theoretical progress has been achieved in the understanding of the three- and four-nucleon systems. We believe that the ground state wave function of the three-nucleon sytem is now well understood. Considerable progress was also achieved by Vijay Pandharipande and his group in Urbana, using Monte-Carlo techniques; they have shown that variational methods confirm these results and allow one to extend these calculations to the $A = 4$ system and heavier nuclei. The binding energy of the triton, the charge distributions and the charge radii of ^{3}He and ^{3}H are now well described by theory.

2 Trends and Perspectives in Nuclear Physics

2.1 Hadrons and Nuclei

At the end of the 80's, building a 4 GeV continuous and intense beam electron facility became the focus of discussion in Europe and in the United States. Unfortunately, endless discussions in Europe did not lead anywhere while in the United States a well organized community decided the construction of such a machine at Newport News, in Virginia. This machine (CEBAF) is today one of the major US facility in nuclear physics. CEBAF has now become Jefferson Lab, the Thomas Jefferson National Accelerator Facility. It is the flagship of a highly active community; in its recent long range plan, NSAC has recommended an evolutionary upgrade of Jefferson Lab to 12 GeV, as a priority of the US scientific community.

The most impressive success of CEBAF is the availability of three simultaneous 6 GeV electron beams with both high intensity and high polarization (70%) that are now routinely available. Advances in techniques using polarized electrons has yielded a wealth of data on the nucleon form factors, achieving an unprecedented level of precision. Until recently, the charge and magnetization distributions in the proton were assumed to be proportional to one another (corresponding to $\mu G_e/G_m$=1). New data show that this is not true, and is leading to a re-examination of the dynamics governing the proton's quark wavefunctions. An extensive research program on the electromagnetic structure of the proton and neutron form factors using polarization techniques is in progress.

Strange quark-antiquark pairs are constantly bubbling up in the proton. One would expect a substantial probability of strange quarks and a corresponding contribution to the charge and magnetization distributions of the proton. To disentangle this contribution from the dominant effects of the u and d quarks requires the use of the weak interaction (Z exchange) as a probe of the strange proton form factor. The HAPPEX experiment has shown that the actual contribution is considerably smaller than expected in most models.

After many years Japan has decided to build a major 50 GeV and 1 MW multipurpose proton facility. Several beams, including intense kaon beams, will be available for particle and nuclear physics. This should allow the study of strange hypernuclei to be pushed into new regions. Intense neutrino beams directed towards the Super-Kamiokande detector will allow one to study new features of neutrino oscillation. Finally, neutron physics and transmutation of nuclear wastes will be also important parts of the research program.

A major development is the considerable progress in lattice QCD that has been made recently and which is promised for the next few years. Developments in improved quark and gluon actions have increased to lattice spacing that one can use while still obtaining accurate continuum results. Funding agencies in Europe and the USA have agreed to support a number of dedicated High Performance Computers, all at the level of 10 Teraflops or thereabouts. These machines should permit full QCD simulations at quark masses only a factor of 3 or 4 above the physical light quark masses. The combination of improved chiral extrapolation and the new generation of supercomputers means that we can look to lattice QCD producing accurate hadron properties at the physical quark masses within the next five years.

2.2 The Limits of Nuclear Stability

One of the questions of current interest is whether or not the role of the magic numbers, well established along the valley of stability, remains important when an extreme excess of protons or neutrons is present in a nucleus.

The doubly-magic nucleus ^{48}Ni was recently observed for the first time at the National Heavy Ion Facility (GANIL) in France using a high-intensity ^{58}Ni beam at 74.5 MeV/A on a nickel target. It is the only case of a doubly-magic nucleus for which the mirror nucleus, ^{48}Ca, is bound. A lower limit of its half-life of about 0.5 μs and an estimate of the production cross-section around 0.05 pb were deduced.

In the domain of super-heavy element research, the discovery of the element 112 at GSI is confirmed while the existence of elements 114 and 118 remains a question.

Another important experimental result has permitted to confirm the existence of a decay mode that has been actively sought for 40 years by the nuclear physics community. In two very recent experiments, one at GANIL in France and one at GSI in Germany, an international team of physicists has demonstrated that the ground state of the atomic nucleus Iron-45 decays directly by the emission of two protons. For the first time, the results from the experiments at GSI and at GANIL demonstrate that a nucleus with very large proton excess can spontaneously disintegrate by double proton emission from its ground state with a comparatively long half-life, which can therefore be directly measured. This should allow one to study the mechanism of two-proton emission, thus opening up a new way for observing the internal forces governing the atomic nucleus.

2.3 Rare Isotope Beams

First generation Radioactive Nuclear Beam (RNB) facilities are operating or under construction in the three regions of the world where nuclear physics is most actively pursued, Europe, North America and Asia/Pacific. These facilities continue to produce important results, and ambitious experiments are planned with them in the next few years. However, several studies of the projected needs of nuclear physics carried out all around the world have made it quite clear that major breakthroughs towards the ultimate scientific goals will only be achieved by the next generation of RNB facilities.

The GSI laboratory in Germany has presented to the scientific community the conceptual design report of a powerful new accelerator facility generating intense high quality secondary beams, including radioactive nuclei and antiprotons. The new facility will provide beam energies of a factor 15 higher than presently available at GSI for all ions, for protons to uranium. Compared to the present GSI facility, the primary beam intensity will be a factor of 100 higher and a factor 10,000 in secondary radioactive beam intensities. The research program will be focused on the investigation of nuclei far from stability, dense hadronic matter and many-body nuclear physics.

2.4 Matter at Extreme Pressures and Temperatures

The quark gluon plasma is a state of matter that is predicted to have existed some 10 microseconds after the occurrence of the Big Bang. Seven different experiments at CERN, in which 33 TeV lead ions crashed into heavy element targets give strong hints of the existence of the quark gluon plasma. But confirmation of the existence of this new phase of matter awaits a new generation of results from the Relativistic-Heavy-Ion-Collider (RHIC) at Brookhaven and the future Large Hadron Collider at CERN. RHIC has now attained full energy with 100 GeV gold ions colliding with 100 GeV gold ions. The first experimental results were presented at INPC 2001 in Berkeley.

Beginning in 2007, part of the experimental program of the CERN Large Hadron Collider (LHC) will be devoted to the study of nuclear collisions in the special-purpose ALICE detector, at energies some thirty times higher than RHIC.

2.5 Nuclear Astrophysics

The origin of the universe is one of the most exciting question in physics. A new understanding of the history of our universe has emerged with the many discoveries that led to the theory of the Big Bang. The very light elements were created in the Big Bang and then the remaining elements arose from the processes involved in the birth, life and death of stars. Cosmology uses important inputs from nuclear physics. Many fundamental questions in the future will be addressed with nuclear astrophysics experiments.

Some of the most exciting results are the recent observations from the Sudbury Neutrino Observatory (SNO) in Canada. This laboratory, funded for a large part by the nuclear physics community, uses a considerable amount of heavy water to interact with cosmic neutrinos. The results have solved the mystery of the missing solar neutrinos, a puzzle for solar theory for more than 30 years. The results confirm that solar models are correct but give evidence that neutrinos decay and oscillate in their journey to the earth. Neutrinos transform from electron-neutrinos to muon- and/or tau-neutrinos. The flux of electron-neutrinos, measured in the charge current interaction $\nu_e + d \rightarrow p + p + e^-$, together with earlier results from Super-Kamiokande on the neutrino elastic scattering flux (encompassing all three neutrino types) $\nu_x + e^- \rightarrow \nu_x + e^-$ show that there is a non-electron type, active neutrino component in the solar flux. The total flux of active ^{8}B neutrinos that can be so deduced is in excellent agreement with the predictions of solar models.

3 Important Applications of Nuclear Physics

3.1 Cancer Therapy with Nuclear Beams

The goal of radiation therapy is to maximize the tumor dose without harming surrounding healthy tissues. The use of heavy particles in radiotherapy is motivated by a superior accuracy in the spatial dose distribution in the human body for deep seated tumors compared to photons and electrons, and an inverse dose profile depositing the highest dose at the end of the particle range in the tumor volume.

For proton therapy, the needed accelerators are, at present, industrial products, while optimized medical synchrotrons for light ion therapy have recently been designed by CERN and GSI. Active beam delivery systems using magnetic beam deflection and energy variation by the accelerator have recently been developed, and have been put into operation at PSI (Switzerland) for protons, and at GSI (Germany) for carbon beams.

Using positron emission tomography (PET), the small amounts of positron-emitting isotopes created by the carbon beam can be used to determine the exact beam location inside the patient's body. The new techniques of more accurate beam delivery and precise control permit the treatment of tumors in critical locations such as the brain, or the vicinity of the spinal cord.

3.2 *Nuclear Energy Research*

Fossil fuels, coal, oil and gas, currently meet more than 85% of world energy needs and will continue to dominate for some time. There is no longer any doubt that the increase in atmospheric CO_2 is due to our growing use of these fuels without any containment of the CO_2 waste. The latest report of the Intergovernmental Panel on Climate Change (IPCC) predicts that the effect of a continuation of this increase in CO_2 on the earth's climate will be significant and often damaging, with rising sea levels, more storms, floods and droughts, the destruction of precious habitats.

Large quantities of additional energy will be needed to fuel economic growth, especially in developing countries with large populations like China, India and Brazil. Currently, some two billion people have no access whatsoever to commercial energy; many more are quite poor by western standards and all will need more energy in the future. If recent trends in energy use continue, as most economic analysts expect, then worldwide demand will grow by about 50% by 2020 and will double by 2050. The growth will be even larger for electricity since, more than any other form of energy, electricity is an essential ingredient of economic development. Yet this growth with the present mix of fuels can only lead to more ecological problems.

Providing more energy economically while limiting the use of fossil fuels is difficult. There is no simple solution. All available options must be considered with an open mind. The world as a whole, therefore, needs to develop carbon-free energy sources.

The new 'renewable' sources of energy, solar power, wind power, biomass etc., are also carbon free and there is a widespread hope that they will supply higher and higher percentages of our energy mix, but it will not happen easily. This is not because of insufficient R&D. Solar photovoltaics, for instance, have benefited from large R&D investments because of their usefulness in space applications. Similarly, tens of thousands of wind generators have been built worldwide. The problem is the cost of the energy and resistance to deploying these low intensity sources, which inevitably impact significantly on the local environment. In 1980, 10% of the Swedish electricity in the year 2000 was foreseen to come from wind power. The correct number in 2000 was 3%.

Hydropower is cost effective, but potential sites are limited and often precious for other reasons so that its growth is also constrained. The OECD expects its contribution to primary energy to fall from 3% today to 2% by 2020. Renewable energies will not, at least for the foreseeable future, provide for the increased energy need

Nuclear fission is one of the few large-scale carbon-free energy sources and currently provides 7% of global primary energy (17% of electricity) without any CO_2 waste. Its costs are now well known and are unaffected by increases in oil and gas prices. It supplies 35% of the electricity generated in Europe, i.e., 75% of its CO_2free power.

Nuclear power does produce radioactive wastes. However, the shortlived wastes from operations are already disposed of safely in many countries. Comprehensive research for decades has led to a common view among international experts that the *"knowledge and technology exist"* for safe waste management, ready to be used by society. The final disposal of the longlived waste is not yet industrially implemented, but demonstrations are under way in several countries.

Today, overall, only 4% of the initial quantity of fuel is consumed in a reactor, i.e., less than 1% of the quantity of natural uranium needed for the production of enriched uranium. The spent fuels removed from the reactors contain 95% of uranium, 1% of plutonium and 4% of fission products. *Only fission products constitute waste. Uranium and plutonium can be re-used to produce energy.* With the dual aim of economizing natural resources and optimizing waste management, some countries, such as France, process the spent fuel to separate the energy-yielding materials from the waste. The recycled uranium is stored with the prospect of its use at a later date in fast breeder reactors, and the plutonium is recycled in today's reactors in the form of MOX fuel, a mix of uranium and plutonium. If the use of nuclear energy is to be greatly expanded to reduce man-made greenhouse gases, some such system will be needed.

To continue the development of nuclear energy, we must provide effective and acceptable technical solutions for the long-term management of the radioactive wastes produced by current reactors; solutions do exist and could be gradually implemented. Studies are underway on multiple recycling of plutonium in power reactors, thus destroying it and leaving the fission fragments and minor actinides for geological storage. Also under study are transmutation systems which convert the long-lived component of spent fuel to a form only requiring isolation for on the order of hundreds of years to a thousand years – a time span of already existing man-made structures.

Preparation for the future sustainable development of nuclear energy will involve a new generation of nuclear power generation systems, in an inclusive approach covering all the aspects of the reactor and fuel cycle. The "Generation IV" international initiative (Europe, United States, Japan, Russia, etc.), aims to develop, for deployment around 2030, new types of nuclear reactors which are simpler, completely free from coremeltdown, and competitive with the best fossilfired plants, as well as fuel cycles more resistant to proliferation. Comprehensive assessment studies have already demonstrated that these objectives are achievable.

Globally, the processing of spent fuels, the consumption of the plutonium in light water reactors, and the transmutation of long-life radiotoxic wastes (minor actinides) in the new generation reactors, could reduce the long-life radiotoxicity of the waste by a factor of 100, leaving a residual radioactivity that would then be comparable to that of the initial natural uranium after several hundred years.

Innovative nuclear energy research is of paramount importance to develop improved designs, maintain and renew expertise, whilst continuing to build competence in operation and decommissioning of the present generation of reactors.

4 Thermonuclear Fusion

Nuclear fusion has made significant progress in the last ten years. In 2002 Europe, Japan and Russia have decided to build ITER (International Thermonuclear Experimental Reactor), which will be the largest scientific facility built worldwide. ITER is a superconducting Tokamak that will be the next step, following the Joint European Torus (JET) facility in operation in England.

The ITER project has its origins in the world wide recognition that: fusion is a long-term energy source, with acceptable environmental characteristics. The goal of this new facility is to allow, in one device, full exploration of the physics issues as well as proof of principle, testing of key technological features of possible fusion power stations and demonstration of their safety and environmental characteristics. It is an international collaborative framework which allows participants to share costs and pool scientific and technological expertise towards a common goal.

Nine years of intensive joint work by the ITER teams under the auspices of the IAEA have yielded a mature design supported by a body of validating physics and technology R&D, safety and environmental analyses and industrial costing studies. After Conceptual Design Activities between 1988-1990, the Engineering Design Activities began in 1992 and are now completed with the ITER design. Following the choice of site and the commitment by the ITER parties of suitable funds, the construction phase (about 10 years) may start. This would be followed by an exploitation phase lasting roughly 20 years.

The United States and China have decided to join the ITER project together with Europe, Japan and Russia. Four sites have been proposed to host ITER, in Canada, France, Spain and Japan. The choice between these propositions is expected to take place in 2003. ITER will provide the integration step necessary to establish scientific and technical feasibility of fusion as an energy source.

5 Conclusions

The work of Peter Sauer and his group was a major step in our understanding of few-nucleon systems. It definitely established the presence of meson-exchange currents by the comparison between the electrodisintegration of the deuteron and the isovector $A = 3$ magnetic form factor. This was a fascinating period where experiments and theory interacted strongly. In a few years electron scattering data produced considerable advances in nuclear physics. I really enjoyed this period.

The frontiers of nuclear physics are continuously moving to new domains. Nuclear astrophysics, the study of hadrons and hadronic matter, new radioactive beam facilities will explore new limits. Society also expects more concern from scientists. Nuclear applications in particular concerning nuclear wastes, nuclear energy and cancer therapy are topics of increasing importance. Nuclear Physics has a bright future.

LIST OF PARTICIPANTS

Jiří Adam
Dept. of Theoretical Physics
Nuclear Physics Institute
250 68 Řež near Prague
Czech Republic
adam@ujf.cas.cz

Ana Arriaga
CFNUL/Universidade de Lisboa
Av. Prof. Gama Pinto, 2
1649-003 Lisboa
Portugal
arriaga@cii.fc.ul.pt

Zoltan Batiz
CFIF/Instituto Superior Técnico
Av. Rovisco Pais
1049-001 Lisboa
Portugal
zoltan@cfif.ist.utl.pt

Gordon Baym
Department of Physics
University of Illinois
at Urbana-Champaign
1110 West Green Street
Urbana, IL 61801-3080
USA
gbaym@uiuc.edu

Nicola Bianchi
INFN
Laboratori Nazionali di Frascati
Via E. Fermi 40
I-00044 Frascati
Italy
bianchi@hermes.desy.de

Jean-Christophe Caillon
Centre d'Etudes Nucléaires
Univ. Bordeaux I
IN2P3 Le Haut Vigneau
33175 Gradignan Cedex
France
caillon@cenbg.in2p3.fr

Juan Antonio Caballero
Dept. de Física Atómica,
Molecular y Nuclear
Universidad de Sevilla
E-41080 Sevilla
Spain
juan@nucle.us.es

Sidney A. Coon
D.O.E., Div. of Nuclear Physics
SC-23/Germantown Building
1000 Independence Avenue, SW
Washington, D.C. 20585-1290
USA
Sidney.A.Coon@science.doe.gov

Fritz Coester
Argonne National Laboratory
Physics Division
Argonne IL 60439
USA
coester@theory.phy.anl.gov

Arnoldas Deltuva
Institut für Theoretische Physik
Universität Hannover
Appelstrasse 2
30167 Hannover
Germany
deltuva@itp.uni-hannover.de

Jorge Dias de Deus
CENTRA/Instituto Superior Técnico
Av. Rovisco Pais
1049-001 Lisboa
Portugal
jdd@fisica.ist.utl.pt

Ana Eiró
CFNUL/Universidade de Lisboa
Av. Prof. Gama Pinto, 2
1649-003 Lisboa
Portugal
ana.eiro@fc.ul.pt

J.O. Fiase
Department of Physics
University of Botswana
P.B. 0022, Gaborone
Botswana
fiasejo@mopipi.ub.bw

Manuel Fiolhais
Departamento de Física
Universidade de Coimbra
3004-516 Coimbra
Portugal
tmanuel@teor.fis.uc.pt

Antonio C. Fonseca
CFNUL/Universidade de Lisboa
Av. Prof. Gama Pinto, 2
1649-003 Lisboa
Portugal
fonseca@cii.fc.ul.pt

Tobias Frederico
Departamento de Física
Instituto Tecnológico de Aeronáutica
12.228-900, São José dos Campos
São Paulo
Brazil
tobias@fis.ita.br

Bernard Frois
Directeur du DTA4
Ministère de la Recherche
1 rue Descartes
F-75231 Paris cedex 05
France
bernard.frois@technologie.gouv.fr

Humberto Garcilazo
Instituto Politécnico Nacional
Escuela Sup. de Física y Matemática
07738 México D.F.
Mexico
humberto@esfm.ipn.mx

Franz Gross
Jefferson Laboratory, Theory Group
12000 Jefferson Avenue
Newport News, VA 23606
USA
gross@jlab.gov

Johann Haidenbauer
Forschungszentrum Jülich GmbH
Leo-Brandt Strasse 1
52425 Jülich
Germany
j.haidenbauer@fz-juelich.de

Brigitte Hiller
Centro de Física Teórica
Departamento de Física
Universidade de Coimbra
3004-516 Coimbra
Portugal
brigitte@teor.fis.uc.pt

William Jacobs
Indiana University Cyclotron Facility
2401 Milo B. Sampson Lane
Bloomington, IN 47408-1368
USA
jacobs@iucf.indiana.edu

Frieder Kleefeld
CFIF/Instituto Superior Técnico
Av. Rovisco Pais
1049-001 Lisboa
Portugal
kleefeld@cfif.ist.utl.pt

Jean-Marc Laget
Service de Physique Nucléaire
Centre d'Etudes de Saclay
91191 Gif-sur-Yvette Cedex
France
jlaget@cea.fr

Andrea Lavagno
Politecnico di Torino
Dipartimento di Fisica
C.so Duca degli Abruzzi 24
I-10129, Torino
Italy
alavagno@polito.it

T.-S. H. Lee
Argonne National Laboratory
Physics Division
Argonne IL 60439
USA
lee@theory.phy.anl.gov

Verónica Malafaia
CFIF/Instituto Superior Técnico
Av. Rovisco Pais
1049-001 Lisboa
Portugal
malafaia@cfif.ist.utl.pt

Laura Elisa Marcucci
Dept. di Fisica
Università di Pisa
Via Buonarroti,2 56127 Pisa
Italy
marcucci@df.unipi.it

Gerald A. Miller
Department of Physics
University of Washington
Seattle, WA 98195-1560
USA
miller@phys.washington.edu

Alexei Nefediev
CFIF/Instituto Superior Técnico
Av. Rovisco Pais
1049-001 Lisboa
Portugal
nefediev@cfif.ist.utl.pt

Filomena Nunes
Universidade Fernando Pessoa
Praça 9 de Abril
P-4200 Porto
Portugal
filomena@wotan.ist.utl.pt

Uwe Oelfke
Deutsches Krebsforschungszentrum
Im Neuenheimer Feld 280
D-69120 Heidelberg
Germany
u.oelfke@dkfz.de

Shinsho Oryu
Department of Physics
Tokyo University of Science
Noda, Chiba 278-8510
Japan
oryu@ph.noda.tus.ac.jp

Teresa Peña
CFIF/Instituto Superior Técnico
Av. Rovisco Pais
1049-001 Lisboa
Portugal
teresa@cfif.ist.utl.pt

Sérgio Pinto
CFNUL/Universidade de Lisboa
Av. Prof. Gama Pinto, 2
1649-003 Lisboa
Portugal
sapinto@cii.fc.ul.pt

Willibald Plessas
Institut für Theoretische Physik
Karl-Franzens Universität Graz
A-8010 Graz
Austria
plessas@uni-graz.at

Constança Providência
Centro de Física Teórica
Departamento de Física
Universidade de Coimbra
3004-516 Coimbra
Portugal
constanca@teor.fis.uc.pt

Gilberto Ramalho
CFNUL/Universidade de Lisboa
Av. Prof. Gama Pinto, 2
1649-003 Lisboa
Portugal
Gilberto.Ramalho@cii.fc.ul.pt

Dan-Olof Riska
Helsinki Institute of Physics
University of Helsinki
P.O. Box 64, 00014
Finland
riska@pcu.helsinki.fi

Bidyut Roy
Forschungszentrum Jülich GmbH
Leo-Brandt Strasse 1
52425 Jülich
Germany
b.roy@fz-juelich.de

George Rupp
CFIF/Instituto Superior Técnico
Av. Rovisco Pais
1049-001 Lisboa
Portugal
george@ajax.ist.utl.pt

Arunava Saha
Jefferson Laboratory
12000 Jefferson Avenue
Newport News, VA 23606
USA
saha@jlab.org

Filipe Duarte Santos
CFNUL/Universidade de Lisboa
Av. Prof. Gama Pinto, 2
1649-003 Lisboa
Portugal
fdsantos@astro.oal.ul.pt

Peter U. Sauer
Institut für Theoretische Physik
Universität Hannover
Appelstrasse 2
30167 Hannover
Germany
sauer@itp.uni-hannover.de

Nobuyuki Sawado
Department of Physics
Tokyo University of Science
Noda, Chiba 278-8510
Japan
sawado@ph.noda.tus.ac.jp

Rocco Schiavilla
Jefferson Laboratory, Theory Group
12000 Jefferson Avenue
Newport News, VA 23606
USA
schiavil@jlab.org

Yu. M. Shabelski
Petersburg Nuclear Physics Institute
Gatchina
188350 St. Petersburg
Russia
shabelsk@thd.pnpi.spb.ru

Ingo Sick
Institut für Physik
Universität Basel
CH-4056 Basel
Switzerland
Ingo.Sick@unibas.ch

Alfred Stadler
CFNUL/Universidade de Lisboa
Av. Prof. Gama Pinto, 2
1649-003 Lisboa
Portugal
stadler@cii.fc.ul.pt

Rob Timmermans
Theory Group
Kernfysisch Versneller Instituut
Zernikelaan 25 9747 AA Groningen
The Netherlands
timmermans@kvi.nl

John Tjon
Jefferson Laboratory, Theory Group
12000 Jefferson Avenue
Newport News, VA 23606
USA
J.A.Tjon@phys.uu.nl

Alfredo Valcarce
Grupo de Física Nuclear
Universidad de Salamanca
Plaza de la Merced
E-37008 Salamanca
Spain
valcarce@mozart.usal.es

Luping Yuan
Institut für Theoretische Physik
Universität Hannover
Appelstrasse 2
30167 Hannover
Germany
Luping.Yuan@itp.uni-hannover.de

AUTHOR INDEX

Springer-Verlag
and the Environment

WE AT SPRINGER-VERLAG FIRMLY BELIEVE THAT AN international science publisher has a special obligation to the environment, and our corporate policies consistently reflect this conviction.

WE ALSO EXPECT OUR BUSINESS PARTNERS – PRINTERS, paper mills, packaging manufacturers, etc. – to commit themselves to using environmentally friendly materials and production processes.

THE PAPER IN THIS BOOK IS MADE FROM NO-CHLORINE pulp and is acid free, in conformance with international standards for paper permanency.

MIX
Papier aus verantwortungsvollen Quellen
Paper from responsible sources
FSC® C105338

If you have any concerns about our products,
you can contact us on
ProductSafety@springernature.com

In case Publisher is established outside the EU,
the EU authorized representative is:
Springer Nature Customer Service Center GmbH
Europaplatz 3, 69115 Heidelberg, Germany

Printed by Libri Plureos GmbH
in Hamburg, Germany